DATE DUE

JA 28 '94			

DEMCO 38-296

Dr Yockey has been one of the leading contributors to information theory in molecular biology for nearly three decades. The author now presents an introduction to the use of information theory in molecular biology. The book gives molecular biology a well developed mathematical foundation based on the digital aspects of the gene and the genetic apparatus. With an exciting blend of theory and controversial opinion, the author debates basic questions in molecular biology, tackling information, complexity, order, uncertainty, randomness and similarity. Dr Yockey communicates the excitement of the science through a clear and lively writing style. The book is essential reading for advanced undergraduates, graduates and research workers in the fields of molecular biology and molecular evolution.

Information theory and molecular biology

Information theory
and molecular biology

HUBERT P. YOCKEY

CAMBRIDGE
UNIVERSITY PRESS

Published by the Press Syndicate of the University of Cambridge
The Pitt Building, Trumpington Street, Cambridge CB2 1RP
40 West 20th Street, New York, NY 10011-4211, USA
10 Stamford Road, Oakleigh, Victoria 3166, Australia

First published 1992

Printed in Great Britain at the University Press, Cambridge

A catalogue record for this book is available from the British Library

Library of Congress cataloguing in publication data

Yockey, Hubert P.
Information theory and molecular biology/Hubert P. Yockey.
  p. cm.
Includes bibliographical references.
ISBN 0-521-35005-0.
1. Molecular biology. 2. Information theory in biology.
I. Title.
QH506.Y63   1992 574.8′8—dc20   91-38939

ISBN 0 521 35005 0 hardback

This book is dedicated to the memory of

Dr Henry Quastler M.D.
21 *May* 1908–4 *July* 1963
who was the first to see the importance of information
theory and coding theory in molecular biology

and

Professor Robert Leroy Platzman Ph.D.
23 *August* 1918–2 *July* 1973
my co-editors of Symposium on Information Theory in Biology

Contents

Contents   x

Contents xiii

Contents   xiv

Preface

This monograph is intended to introduce molecular biologists to the application of information theory and coding theory in molecular biology. A curious lack of communication between molecular biologists and the mathematicians and electrical engineers who developed information theory and coding theory has resulted in molecular biologists being unaware that great clarification and, in some cases, the solution of their problems has already been developed. By the same token, mathematicians are unaware of the very important opportunities for applied mathematics in molecular biology. This monograph is intended to be read by both groups with the expectation of significant mutual benefit.

Part I presents the mathematical ideas and background in sufficiently complete form that it is not necessary to refer to other books. Part II is devoted to the application of these mathematical ideas to problems in molecular biology. Those readers not familiar with a reasonably sophisticated presentation of probability theory are advised to study the material in Part I carefully. Mathematicians will miss σ-algebras, Borel sets and Lebesgue measure theory, but most readers will find quite enough mathematical sophistication in Part I. On the other hand, I suggest that mathematicians have a copy of a good text on molecular biology handy as they read this book. Mathematicians unfamiliar with molecular biology will find they need an explanation of the basic ideas and experimental results in molecular biology to understand the application of information theory and coding theory to problems in Part II.

I have tried to be reasonably consistent in the definition of mathematical symbols. Nevertheless, the notation may vary from chapter to chapter because I have adopted the notation found in the literature for the different subjects that are addressed. No confusion will result since the terms are defined in each chapter.

The theorems in Part I are referenced in Part II so that the reader may turn back to refresh his or her memory. In no case have I simplified or trivialized the sophistication of the mathematics. I have provided enough explanation of both the mathematics and the molecular biology so that an advanced senior student or graduate student in either biology or mathematics can understand the argument. This monograph may serve as a text for a graduate seminar attended by both molecular biologists and mathematicians. Established professionals also will find sufficient interest in information theory and coding theory and their application in molecular biology. They may wish to broaden their appreciation by referring to the reference material.

I have tried to make the references as complete as the subject demands. This is a monograph and not an encyclopedia so I have made no attempt to cite all references available. I have not regarded it necessary to call attention to every paper I believed did not make an important contribution or to those which are incorrect. I have included in the references only those I felt contributed to the point being made. Some readers may think that I have neglected an important paper here and there. I acknowledge that this may be the case but there are times when one must hew to the line and let the chips fall where they may. The reference material is up to date as of August 31, 1991.

This monograph follows my interest in the subject, which was first attracted by the work of Dr Henry Quastler. I am indebted to Professor Thomas H. Jukes, whose strong recommendations resulted in my original papers being published. Many of Professor Jukes' important contributions to molecular biology have shaped the ideas presented in this book, particularly those concerning the evolution of the genetic code. Professor Jukes also introduced me to Dr Adam S. Wilkins of Cambridge University Press, who saw the merit of my proposal to write this book. I also am grateful to Dr Gregory J. Chaitin, whose original and seminal work in algorithmic information theory is reflected throughout the book. I thank both Professor Jukes and Dr Chaitin for their careful review of a draft of this book. Each made important suggestions for its improvement.

I thank Dr Paul Woolley, who reviewed the penultimate draft in much detail and also contributed a number of suggestions that have removed errors and improved the clarity of the text. I would also like to thank Dr Robin C. Smith, who graciously completed the editing of this book after

Dr Wilkins left for other work. Dr Susan Parkinson copy-edited the book. She called my attention to a number of technical errors and contributed in an important measure to the clarity of the text. I am indebted to Dr Bryan MacKay of the Biology Department at the University of Maryland, Baltimore County campus, for use of the facilities at the Albin O. Kuhn Library and Gallery. My daughter, Cynthia Ann Yockey, edited this manuscript from proposal to final draft and contributed much to improve the clarity and organization of the material. Without the contribution of these people I would not have been able to write this book.

Hubert P. Yockey
Bel Air, Maryland USA

Prologue

It must be considered that there is nothing more difficult to carry out nor more doubtful of success, nor more dangerous to handle, than to initiate a new order of things. For the reformer has enemies in all those who profit by the old order, and only lukewarm defenders in all those who would profit by the new order, this lukewarmness arising partly from fear of their adversaries, who have the laws in their favor; and partly from the incredulity of men, who do not truly believe in anything new until they have had actual experience of it.

Niccolò Machiavelli, *The Prince*, Chapter 6.

The purpose of this book is to show that the sequence hypothesis and the genetic code, together with the principles of biochemistry, are essential to the mathematical foundations of molecular biology. The sequence hypothesis states that the sequences of nucleotides and amino acids carry the information that controls the folding and the activity of proteins (Watson & Crick, 1953a,b; Crick *et al.*, 1961; Anfinsen, 1973; Lesk, 1988). The essential difference between living and non-living matter is that the formation and function of important biological molecules is governed by genetic messages. No trace of a comparable specification by sequences or of a code between sequences exists in non-living matter (Mayr, 1988).

The issue of the sufficiency of physics and chemistry to deal with the complexities of biology was discussed long ago by Bohr (1933): 'The recognition of the essential importance of fundamentally atomistic features in the function of living organisms is by no means sufficient, however, for a comprehensive explanation of biological phenomena, before we can reach an understanding of life on the basis of physical experience'. Although molecular biology must be consistent with chemical and physical processes, biological principles are not derivable from physics and chemistry alone (Monod, 1972; Mayr, 1982, 1988; Küppers, 1986, 1990). Non-reducibility of this sort is not unusual in science. For instance, the second law of thermodynamics is consistent with Newton's laws of motion, except that it denies time reversal, but it is not derivable from them.

Bohr (1933) drew an analogy between the existence of life and the complementarity principle of quantum mechanics. The photon may act as a wave, for example in a diffraction experiment, or in the same diffraction experiment it may act as a particle. In experiments at very low light levels

1

the photon is detected at the positions in the diffraction pattern appropriate to its wave-like properties in a stochastic fashion as if it were a particle. The same wave–particle duality has been found in the case of electrons and neutrons. Bohr regarded these apparently contradictory properties as being complementary aspects of the same thing. The wave–particle duality imposes a limit to our ability to measure both the momentum and position of an atomic particle, such as an electron or a neutron. The experiment itself, that is, the measuring process, affects both the momentum, p, and the position, q, and introduces an experimental error in each. The relation between these errors is given by the Heisenberg uncertainty principle, $\Delta p \Delta q > h/2\pi$. Bohr extended the complementarity principle to the question of the ability of science to understand life. 'Thus, we should doubtless kill an animal if we tried to carry the investigation of its organs so far that we could describe the role played by single atoms in vital functions. In every experiment on living organisms, there must remain an uncertainty as regards the physical conditions to which they are subjected, and the idea suggests itself that the minimal freedom we must allow the organism in this respect is just large enough to permit it, so to say, to hide its ultimate secrets from us. On this view, the existence of life must be considered as an elementary fact that cannot be explained, but must be taken as a starting point in biology, in a similar way as the quantum of action, which appears as an irrational element from the point of view of classical mechanical physics, taken together with the existence of elementary particles, forms the foundations of atomic physics (Bohr, 1933).' In the nineteenth century, intuition led many to believe that the inheritance of characteristics, such as tall and short in Gregor Mendell's experiments with strains of pea, would yield a blending of these traits and produce plants of medium height. The facts, with which the reader is familiar, show that this is the case only when neither trait is dominant. The theory of blending inheritance predicts that experiments should show the variance of the offspring would be reduced to one-half of that of the parents in each generation (Fisher, 1930). This predicts either the disappearance of favored traits or that mutations must be several thousand times as frequent as they are known to be. Therefore, according to the blending theory of inheritance, Darwinian evolution would never occur. Küppers (1986, 1990) cites Fleeming Jenkin, a contemporary of Darwin, as having pointed this out. Mendel's laws of dominance, recessiveness and segregation are counter-intuitive and there-fore must be regarded as axioms or elementary facts that cannot be reduced to simpler ones. This may be one of the reasons that Mendel's work was ignored until nearly the end of the nineteenth century (Mayr, 1982).

To adopt this point of view may cause considerable *Angst* among molecular biologists, as was the case among classical physicists when Bohr proposed his idea (for which he was awarded the Nobel Prize in 1922) that negatively charged electrons, which are in orbital motion about the nucleus, do not collapse into the positively charged nucleus of the atom because they are excused by the quantum of action from radiating energy as required by Maxwell's laws. The sequence hypothesis may be taken as a starting point or axiom of molecular biology and play the role called for by Bohr (1933). Like the quantum of action, it is extremely well supported by experimental evidence. Stent (1988) has revisited Bohr's famous Light and Life lecture in which he drew attention to the implications for biology of the fundamental changes in natural law that quantum mechanics had brought to physics.

The sequence hypothesis, first suggested by Schrödinger (1944) and given a concrete formulation by Watson and Crick (Watson & Crick, 1953a, b; Crick, 1970), accounts for the diversity of living organisms in terms of the enormous number of genetic messages that may be recorded in sequences of nucleotides and amino acids. The phenomenon of reversible denaturation shows that there is enough information in a linear object, namely, a sequence of symbols selected from a finite alphabet, to determine a three-dimensional object. One is reminded of the linear sequence of signals from a television station which, because of the raster, defines a two-dimensional object, namely, the picture on the screen.

If the discussion can be carried on in terms of mathematics one may achieve for theoretical biology what has been achieved by theoretical physics. Reasoning with words alone often results in falling into a word-trap. Almost every word in the dictionary has multiple meanings. The proper meaning is evident only when the context is known. A writer may, and often does, use a word in different senses in the same paragraph or even in the same sentence. An example of reasoning with words alone that horrifies reasonable people today is the argument given by Aristotle (*The Politics*) in which he points out that people differ in ability and draws the conclusion that some people, especially barbarians (that is non-Greeks) are 'slaves by Nature.' Aristotle considered that nations also differ in ability and on this basis justified enslaving prisoners of war. He argued that the slave is actually benefited by being taken from his uncultured native wild heath and becoming part of a civilized Greek society, albeit at the bottom of the social scale. He took this position in spite of the fact that he was aware of objections by others, unnamed, that the rule of master over slave is based on force and is therefore wrong. Aristotle, the inventor of formal logic, based his arguments on the meanings of words (ancient Greek words, what else?) in order to show what is 'contrary to Nature' and to justify the social structure of his time.

The Greek mathematicians during the same period were discovering the axiomatic method of reasoning. They found that all geometry, both plane and solid, could be reduced to a set of statements or axioms from which all theorems could be proved (Euclid, *Elements of Geometry*). Thus Pythagorus found a proof that the many pleasant facts about the square on the hypotenuse that were well known to the Babylonians and the Egyptians were true in general. Mathematical symbols and the postulates on which one bases the argument do not change or have multiple meanings. It is not possible to draw unwarranted conclusions if the argument is carried out in the context of a mathematical discussion: Plato, in *Meno*, has Socrates draw from the slaveboy the correct way to double a square just by asking questions. (Quickly now, how does one double a square?) In this book I shall show a number of cases where reasoning with words alone has led many authors to wrong conclusions.

The idea that the DNA sequences of four letters determined the protein sequences of 20 letters by means of a code is so unconventional that had Gamow's paper been submitted by almost anyone else it would have most certainly been rejected. Peer review may take on the ecclesiastical function of protecting the readers of technical journals from heresy (Maddox, 1989a). But the world of science had learned either to take this big, genial Ukranian American seriously or to accept the consequences and, although his first code was soon shown to be incorrect, the general idea stuck. The search was on for the correct genetic code.

We must distinguish clearly between axioms and the theorems from which they are derived. For example, it is easy to show that the Central Dogma (Crick, 1970) is a theorem of coding theory and not a fundamental biological phenomenon. The Central Dogma states that information may be transferred from DNA to DNA, DNA to mRNA and mRNA to protein. Under special circumstances, information may be transferred from mRNA to mRNA, mRNA to DNA and DNA to protein. Three transfers that the Central Dogma states *never* occur are protein to protein, protein to DNA and protein to mRNA. The DNA and mRNA genetic codes have 64 code words made in triplets from an alphabet of four letters.

Information that has an alphabet of 20 letters, namely, the proteinous amino acids, is transferred from mRNA, as a source, to protein, as a destination. The question is: can information be transferred from a source with a 20-letter alphabet to a receiver with 64 letters or code words?

Let us use a dice game to illustrate this point. Suppose we have two dice that can be distinguished by color. The numbers on each die may be considered as members of an alphabet of six symbols. We can also make an alphabet of the 36 pairs of the numbers on two dice at the source and send messages to a receiver. Because of the colors of the dice, the source can distinguish between $(1+6)$ and $(6+1)$. The outcome of the game is determined only by the total of the dots on two dice after they have been cast. Therefore, let the receiver of the messages be sent only the total. The alphabet at the receiver has 11 members, that is, the numbers 2 through 12. Therefore, there is a code between sequences generated at the source and those at the receiver that is exemplified by the sum of the numbers on the two dice. The number 7, for example, is coded for by any of the following members $(1+6)$, $(2+5)$, $(3+4)$, $(4+3)$, $(5+2)$, $(6+1)$ of the source alphabet. Thus the code is redundant (degenerate) since there are six code words in the source alphabet for the one code word, 7, in the receiver's alphabet. The receiver cannot know which member of the source alphabet resulted from a toss of the two dice. This system is isomorphic with the genetic code in which there are 64 codons in the source alphabet in DNA and RNA and 20 amino acids in the receiver alphabet, namely, protein. The dice game, too, has a Central Dogma, namely, that information flows only in one direction. This is a property of all redundant codes, that is, those in which the alphabet at the source has more symbols than the one at the destination. Of course, when the number of symbols is equal information can flow in both directions. This illustrates the fact that the Central Dogma in molecular biology is a mathematical property that comes from the redundance of the genetic code and is not a fundamental property of nucleic acids and amino acids *per se* (Yockey, 1978).

Accepting Bohr's suggestion and the discussion by Monod (1972) and Mayr (1982, 1988) that physics and chemistry do not suffice to explain biology is to adopt the axiomatic method of reasoning and should not cause *Angst* among reductionists as might seem at first sight. One is simply using a method of reasoning that is the basis of all mathematics and mathematical physics. If it proves later that the sequence hypothesis and the genetic code can be shown to derive from physics and chemistry, then well and good. In that case, one would only need to rename them as theorems. No harm can come from this since they are so well supported by experiment.

The mathematical formalism that deals with the properties of sequences is called *information theory*. The formalism that deals with the relations between sequences is called *coding theory*. Terms such as 'information', 'complexity', 'uncertainty' and 'randomness' are often used with quite different meanings by different authors. Yet the precise meanings of these and other terms flow *from* the mathematical formalism not *to* it.

Great ideas and discoveries often have deep roots. So it is with information theory. One must have a means of measurement in order to put these concepts in a mathematical formalism. A logarithmic measure of 'information' was first suggested by R. V. L. Hartley, an engineer with the Bell Telephone Laboratories (Hartley, 1928). Important ideas were contributed by Norbert Wiener (1949b), who also used a logarithmic measure for information and noted that this measure had many of the properties of the Boltzmann expression for entropy in statistical mechanics.

According to Albert Szent-Györgyi, discovery is seeing what everyone has seen but thinking what no one has thought (cited by Mueckenheim, 1986). And so it fell to the lot of Claude E. Shannon (1948) to think what no one had thought and to supply the missing ideas for the mathematical formalisms we now know as information theory and coding theory. In these two papers he found a unique measure of information based on the mathematical principles of set theory and proved a number of theorems that provided an answer to the questions of how well the communication systems of the telephone and telegraph companies were being used and how these could be improved. Messages are written in a sequence of symbols chosen from a finite number of symbols called an alphabet. These symbols are selected at the source where p_i is the probability of the ith symbol. Shannon's measure of the *information per symbol*, H, is as follows:

$$H = - \sum_i p_i \log_2 p_i \qquad (0.1)$$

This expression is minus the expectation value of the logarithm of the probability p_i. The information per symbol in a sequence is given a quantitative definition by equation (0.1) and therefore information is measurable although it is not material. Since information is a function of probabilities, *information theory* is a branch of the mathematical theory of probability and so Chapter 1 is devoted to that subject.

Equation (0.1) shows that the word 'information' can be identified with a mathematical function that captures most, but not all, of what the word conveys. Shannon's *H* does not measure the *meaning* of a message as he pointed out at the beginning of his first paper (Shannon, 1948). Critics of information theory decry this as a defect and fail to notice the second paragraph of Shannon's paper, which explained why this is so. The communication system that the engineer designs must be capable of transmitting, decoding and recording any member of a large ensemble of messages. By the same token the biological information system must be able to accommodate the genetic messages of all organisms that have lived in the past, are now living or may live in the future. As we shall see in Chapter 2, both communication systems are essentially data recording and processing systems.

There is an anecdote (Tribus, 1979) that Shannon, desiring a name for the function given in equation (0.1), sought the advice of Professor John von Neumann, who replied, 'You should call it "entropy" for two reasons. First, the function is already in use in statistical mechanics under that name (Boltzmann, 1877a, b). Second, and more importantly, most people don't know what "entropy" really is. If you use "entropy" in an argument you will win every time!'. The Maxwell–Boltzmann–Gibbs entropy of thermodynamics is based on the probability of a selection of elements different from that of Shannon and the two have no relation. To reflect this fact, it is usual in the literature to call the entropy function described in equation (0.1) the Shannon entropy.

It is perhaps clear to the reader that the genetic system is, in principle, isomorphic with communication systems designed by communications engineers. As a matter of fact, genetical systems have historical priority since organisms have been using the principles of information theory and coding theory for at least 3.8×10^9 years! However, the interests peculiar to communication engineering colored the original papers by Shannon. These concepts have been found to be useful in the development of contiguous disciplines, namely, statistics, computer science, decision theory and of course molecular biology.

It will often be emphasized in this book that theorems lead to discoveries to which we would not be led by intuition. This is not unusual in science; quantum mechanics is full of it. X-rays and gamma rays have zero mass and are certainly electromagnetic waves. Yet, as was predicted, they collide with electrons like particles. Electrons, on the other hand, have mass and are certainly particles. Yet, as was predicted, in reflecting from a crystal they behave like waves. These very important results are contrary to intuition and take some getting used to. In information theory, Shannon's channel capacity theorem is an especially significant example of a counter-intuitive concept. This theorem proves that a code exists such that an original message may be recovered, by the proper use of redundancy, from the received message affected by noise with as small a probability of error as we please. It is on this theorem that the ability of the genetic message to preserve for so many million years the information needed to form a coelacanth is based. This result runs so much counter to naïve experience that it has been challenged publicly by communication engineers (McMillan, 1953). An *ad hoc* approach based on intuition would lead one to expect that the only way to overcome the effect of noise would be to send the message several times. There is no way of knowing how many times the message must be sent to achieve a given accuracy. Since most of the message is received correctly, this would in any event be very wasteful of the communication system. If simply repeating a message is the only way we have to reduce the effect of noise, 'channel capacity' is an ill-defined term. The channel capacity theorem shows that there is a well-defined information capacity for a communication channel.

The practical value of this, and other theorems in information theory and coding theory, to the world's economy is above measure. It permits the transfer of enormous funds between financial institutions and among nations at the speed of light. Obviously, errors could not be tolerated in such transactions. Spectacular color photographs have been received from space craft as far away as Neptune. These pictures were sent by use of the powerful Reed–Muller code. The brightness of each dot in the picture was encoded in a code word (or codon) of 32 bits having 26 redundant bits. These are only two examples of the use of Shannon's channel capacity theorem, which we now take for granted.

I shall replace the real amino acids and nucleotides by certain symbols in applying information theory and coding theory to molecular biology. These symbols represent what is important in the DNA, mRNA and protein molecules and are the elements that carry the genetic messages. The exact location of the various atoms that compose these informational molecules is unimportant and merely clutters our thinking. In molecular biology the molecules, each with its chemical and physical properties, are like chessmen, each with its inherent possible moves and rank. The essence of chess is in the rules and the strategy by which the game is played, not in the substance of the chessmen. The physical chessmen can be replaced by marks on paper, by electronic signals carrying messages between the players, or even the mental image of the board in the minds of the players, without changing the character of the game. The physics and chemistry play the role of the rule book, the genetic messages are the strategy. Thus the enormous diversity of living organisms is reflected in the enormous diversity of genetic messages that determine the life and characteristics of each organism.

The need for a fundamental theory in molecular biology, called for by Maddox (1988), can be satisfied by including information theory and coding theory as well as physics and chemistry. According to a strict reductionist paradigm, molecular biology is making all biology a branch of physical chemistry (Maddox, 1983). The strict materialist who insists that science must not be concerned with anything that is not matter or energy is already in serious trouble. In quantum theory one must use the *wave function* that is a solution to the Schrödinger equation. These wave functions often incorporate the square root of minus one, and are clearly not material. Wave functions are kind enough not to appear in the real world except in the form of the product with their complex conjugate, which must be a real number.

In considering the question of whether biology is totally contained in physics and chemistry let us ask: 'Is celestial mechanics totally contained in Newton's laws of motion and Newton's law of gravitation?'. An examination of these laws reveals no mention of the orbit of Mars. The problems that can be solved by Newton's mechanics include the pendulum, the gyroscope and the various problems in celestial mechanics. In each case one must specify the problem to be solved and then apply these laws to the solution. By the same token, in biology we must first define the problem and then apply the appropriate laws. As we shall see later in this book there must be considerable *information* in the message that defines the problem. For example, in celestial mechanics the Sun and the planets are first regarded as point masses. This is sufficient to derive Kepler's laws of planetary motion but it does not explain the precession of the equinoxes. To address that problem one must define the Earth as a rigid body with certain characteristics. A number of bodies in the solar system are close enough that the gravitational field varies appreciably over the dimension of the attracting bodies. The Earth–Moon system is a familiar example. The attraction of the Moon causes tides. Some flexibility in the Earth itself and some motion of the oceans must be allowed, in order to describe the tides. One must use more information about the two bodies in order to define the problem in such cases. We may find the consequences of the information in the definition of the problem but we cannot get more information out of the problem than we put in. By the same token, biology is not contained in physics and chemistry (Monod, 1972; Mayr, 1982, 1988). That which the sequence hypothesis and the genetic code can tell us must be included in the description of the problem. We shall find, in this book, that biology is not part of physics and chemistry but rather physics and chemistry are the hand-maidens of biology.

Jacques Loeb was the icon of the reductionist paradigm during the controversies with the vitalist philosophers Henri Bergson and Hans Driesch in the late nineteenth and early twentieth centuries (Pauly, 1987). The title of his book, *The Mechanistic Conception of Life: Biological Essays* (1912), was a deliberate pun. Loeb's experiments with artificial parthenogenesis and Muller's experiments with the generation of mutations with X-rays were prominent in supporting the reductionist paradigm. These experiments together with progress in all fields of biochemistry gradually diminished the influence of the vitalist philosophers. Bohr (1933) made it clear that he was not reviving vitalism in his *Light and Life* lecture. The information content of the genome is non-material but nevertheless measurable. Thus we can avoid the deficiencies of strict reductionism without adopting any of the romanticism, as Loeb called it, of vitalism.

The strict reductionist paradigm has been so widely accepted that the full role of the sequence hypothesis has remained largely undeveloped. Most of the reason is that information theory has been greatly misunderstood. In part this is due to a mistake, similar to that made by Aristotle, in attempting to base the argument on the meaning of words, namely, the anthropomorphism of the word 'information'. In order to force Shannon entropy H to measure *meaning*, critics have attempted to divide *ad hoc* the information content of a sequence into purposeful, purposeless, redundant and derivative information (Johnson, 1970). The notion of negative entropy has also haunted the literature. The anthropomorphism associated with the word 'information' is avoided if one regards the genetic system as a data recording and processing system.

It is appropriate to define, in the beginning, what is meant by 'theory' in this book. A *theory* is a set of elements and a set of axioms (or postulates) together with all the theorems that can be proved from these axioms. Familiar examples of theories are Euclidean geometry, Mendel's laws, thermodynamics, Newton's laws of mechanics and Maxwell's equations of electrodynamics. In developing and applying a theory we must discipline ourselves to stick to the original axioms and theorems. We may introduce knowledge about the problem to be solved at the beginning of the argument but we may not introduce *ad hoc* principles or anecdotal material at will as we go along. A theory, by this definition, is to be distinguished from scenarios, speculations, models, hypotheses and prescriptions.

Many molecular biologists may feel that mere scribbling with abstract symbols can't have anything to do with the practicalities of their science. 'It is always hard to realize that these numbers and equations we play with at our desks have something to do with the real world (Weinberg, 1977).' Such people are in good company. There are famous experimental physicists who had little use for mathematical theories, of whom two will be mentioned, namely, Sir Ernest Rutherford and Professor Ernest Lawrence. However, when theories are based on fundamental principles and not on *ad hoc* scenarios, the error is in not taking them seriously enough. Jaynes (1957a) points out that it is when theories fail to predict the results of experiment that they are most useful. Such discrepancies alert us to new knowledge. When a theory predicts correctly we are simply puzzle solving and confirming what we already know (Kuhn, 1970).

Those scientists who work in the laboratory need the guidance of theory to focus on important problems and to avoid collecting large amounts of data about the number of pickets in picket fences or attempting to build perpetual motion machines (Huszagh & Infante, 1989; Othmer, 1989). Reductionists must play the game by the same rules all the time. Theoretical biology is far behind theoretical physics in basing the explanation of data on axiomatic theories rather than on *post hoc* discussions of an essentially anecdotal or contrived character. The continual interaction between experiment and theory has often opened vast new fields of research that would otherwise have been overlooked. Otherwise, why worry about a mere 43 seconds of arc per century advance of the perihelion of Mercury?

In Part I, I develop the mathematical theory of probability and the fundamentals of information theory and coding theory. In Part II, I use this material to address and perhaps to solve some important problems in molecular biology. By employing information theory and coding theory as well as the principles of chemistry and physics, failed notions based on word traps can be disposed of quickly and effort directed to more promising ones. On the other hand, certain ideas receive substantial support and it can even be demonstrated that some scenarios are different aspects of the same thing. It is by producing results that are useful to the experimentalist that theoretical biology will take its place with theoretical physics. This monograph is limited to the mathematical foundations of molecular biology and therefore cannot present a solution for all the problems that need to be addressed. Nevertheless, the direction in which research has begun to go and should go in the future is considered.

There is much in this book that will be new to the reader. We are usually shocked by new ideas, especially when they do not appear to be extensions of our present philosophy. This is the reason for the warning from Niccolò Machiavelli in the epigraph at the beginning of this prologue. Nevertheless, it is the essence of progress in science that such shocks be experienced.

This is not a book on the philosophy of biology. I leave that to wiser heads than mine (Mayr, 1982, 1988). But philosophy is the helmsman of life† and perhaps also of research. The issues raised by Bohr (1933), Maddox (1983, 1988), Huszagh & Infante (1989) and Othmer (1989) must be addressed in molecular biology. The existence of alternate genetic codes in mitochondria shows that a unique genetic code is not derived directly from the first principles of physics and chemistry. Accordingly, let us take the sequence hypothesis as a starting point or axiom. If molecular biologists can persuade themselves to include information theory and coding theory, along with physics and chemistry, among those things twixt heaven and Earth known in their philosophy, one can then get on with the work of building a theoretical biology based on those mathematical foundations. That is what this book is all about.

† The motto of ΦBK, a scholastic honor society in the U.S.A.

I

The basic mathematical ideas

1

Basic ideas in probability theory

Then Samuel brought all the tribes of Israel near, and the tribe of Benjamin was taken by lot. He brought the tribe of Benjamin near by its families, and the family of Matrites was taken by lot; finally he brought the family of the Matrites near man by man, and Saul the son of Kish was taken by lot.

1 Samuel 10:20

And they cast lots for them and the lot fell on Mathias; and he was enrolled with the eleven apostles.

Acts 1:26

1.1 The origins and interpretation of probability

1.1.1 The origins of probability

The origins of the primitive ideas of chance and uncertainty seem to be lost in history. One of the earliest stories of a game of chance using dice known to me is in *The Iliad* (Homer, *c.* 850 BC a). Homer also relates in *The Iliad* (*c.* 850 BC b) several instances of the casting of lots to make an important decision. For example, in the tenth year of the Trojan War, the Greeks and Trojans, weary with endless fighting, agreed to let two protagonists, Menelaus and Paris, settle in single combat the possession of Helen, the return of the treasure stolen from Menelaus by Paris and the outcome of the war. Homer has Hektor shake two lots in a bronze helmet to determine whether Paris or Menelaus should cast the first spear. Later in *The Iliad*, Homer (*c.* 850 BC c) has Nestor shake lots in a bronze helmet with the outcome that the Greek hero, Ajax the Greater, was chosen as a suitable opponent in single combat with the Trojan hero, Hektor (no doubt greatly to the relief of the rest of the soldiers in the Greek army). Thus, the gods spoke through the casting of lots, and in each case an important decision was made without anyone taking the responsibility.

Chance and uncertainty were poetically personified in the Greek pantheon by the goddess Tyche. The Romans borrowed her and renamed her Fortuna. The Fates, three goddesses who controlled human life, were no doubt more ancient. They were Clotho, who spins the thread of life (now known to be DNA); Lachesis, who casts lots to determine how long

15

life shall be; and Atropos, who cuts the thread of life. These goddesses are sometimes referred to collectively and pejoratively as 'blind chance' or 'mere chance'. They may be called vindictive, callous, cruel and fickle, but not blind (see 'gambler's ruin', Feller, 1968, 1971). In our sophisticated modern world we still cast lots to determine events, especially those in which we wish to avoid responsibility. One hears of 'calculated risk' from politicians, military officers and business executives when the strategy has gone awry and of shrewd judgement when things have gone well; thus they take the credit for success while putting the responsibility for failure on the fickleness of Fortuna, Tyche or the three Fates, Clotho, Lachesis and Atropos, as did the ancient Greeks and Romans.

What we now call the theory of probability did not become a mathematical discipline until the mid sixteenth century with the publication of *Liber de Ludo Aleae* (The Book on Games of Chance) by Girolamo Cardano (1501–1576), an Italian mathematician, physician and gambler. Certain problems were solved by the famous mathematicians James Bernoulli (1654–1705), Daniel Bernoulli (1700–82), Pierre de Fermat (1601–65) and Blaise Pascal (1623–62). The publication of *Théorie Analytique des Probabilities* by Laplace (1796) put the subject on its first firm foundation. The hand of Pierre Simon Laplace (1749–1827) is still seen in the theory of probability.

1.1.2 The interpretation of probability

Before plunging into the theory it is as well to consider three important classes of interpretation of the formalism. These interpretations are not necessarily mutually exclusive and all three will be needed to address problems in this book.

(a) The classical or empirical frequency interpretation

This interpretation identifies probability directly with the empirical frequency of events. In the classical 'frequentist' or statistical interpretation, probability has meaning only in relation to an *ensemble* of trials. Suppose one has a repeatable event, A, such as obtaining a head in the tossing of a coin, drawing the ace of spades from a well shuffled deck, finding six dots on the top of a tossed die or finding a particular value, e.g. seven, for the sum of the numbers showing on two tossed dice. Let a be the number of times that event A is produced in the ensemble and n be the total number of tosses or draws. Then it is asserted that as n becomes very large:

$$\lim_{n \to \infty} \frac{a}{n} = p \tag{1.1}$$

where p is called the *probability* of the event. This is known as the weak law of large numbers (Feller, 1968). The *lim* is put in italics since it is not a limit in the usual mathematical meaning of the term (Lindley, 1965). Furthermore, no law of physics prevents large deviations from equation (1.1). Empirical tests show merely that such deviations are so very infrequent that for practical purposes they may be neglected. Should the theory of probability stay at this intuitive interpretation many of the problems in this book could not be addressed. In particular, the several scenarios for the origin of life could not be considered since that is not for us a repeatable event and therefore we cannot generate an ensemble of events.

The frequency definition of probability shown in equation (1.1) is circular since the idea of equally likely events is the same as equally probable events. Except in games of chance, which in their ideal form are specifically set up in this way, very few real problems have equally likely events. This difficulty of logical circularity is resolved by basing the theory of probability on a set of axioms.

Hilbert (1902) called for the development of axioms for the theory of probability in a paper that delineated his suggestions of the most important mathematical problems yet to be solved. His sixth problem suggests: 'To treat in the same manner (as geometry) by means of axioms, those physical sciences, in which mathematics plays an important part; in the first rank are the Theory of Probabilities and Mechanics'. This stimulated a great deal of creative activity and some of Hilbert's problems have been solved. In particular, Kolmogorov (1933) succeeded in setting up a set of axioms that solved Hilbert's sixth problem.

(b) The degree of belief or subjective interpretation

Other than in games of chance in their ideal form, most of the events with which one is confronted and for which one must make some sort of decision in advance are not repeatable, at least not exactly repeatable. Predictions of the weather, the course of market prices, one's employability and one's longevity are examples of events the probability of which one must perforce have some knowledge in order to make the best of certain possible decisions. A further example is the formation of probability risk assessments for large-scale accidents in complicated engineering systems such as passenger and military aircraft, chemical process facilities, nuclear power plants and their associated radioactive waste repositories. The object of probability risk assessments is to find the best administrative and engineering procedures to reduce the probability of catastrophic events below an acceptable figure (Apostolakis, 1990). The adherents of this interpretation regard probabilities as logical weights or degrees of belief.

(c) The probable inference or inductive logic interpretation

In this interpretation we are interested in the logical support the evidence has for selected hypotheses. The hypothesis might be the guilt or innocence of a person on trial in a court of law. In such cases it is reasonably certain that an event has occurred, i.e. that a crime has been committed. What must be established is the probability of the guilt or innocence of the person accused. This interpretation directs one to reject hypotheses that are believed to have a low probability in favor of ones that have an acceptably high probability.

It is important to distinguish between the mathematical theory of probability and the application to a model of the system under consideration. The model of the system is set up to reflect our best knowledge of the real world system. If the calculations prove to be incorrect then, assuming that one has made no blunder in the mathematics, it is the model or its parameters that are incorrect and not the mathematical theory.

One may illustrate the interpretation of probability and the model of the real world situation to which it has been applied by the following fable. Suppose a practical man, a true believer (Hoffer, 1951) and an independent observer watch the same sequence of ten heads for a coin tossed by the true believer's guru. Before the first toss each trusts the guru and believes the model of the system is that the coin is fair, although it hasn't been examined. If the coin is fair, the probability of heads or tails is 1/2 and after ten tosses the number of heads and tails should be close to five each. After the appearance of the tenth head the practical man continues to believe in the mathematics of probability and this leads him to become suspicious that the coin is two-headed. In another sense, the *meaning* of the sequence of ten heads to the practical man is that it is likely that the model of the real world system must be changed from that of a fair coin to that of a coin that is two-headed. If the sequence of heads continues as *n* increases without limit, his estimated probability that the coin is two-headed approaches unity (see equation (1.6) below). In more formal language, as *n* increases without limit, the practical man's degree of belief becomes a certainty in probability that the correct model of the system is a two-headed coin. On the other hand, the true believer will continue to believe that the coin is fair but that, in his model of the system, the continuing sequence of heads is evidence that the guru is capable of miracles. That is, the *meaning* of the sequence of heads to the true believer is that the guru is capable of controlling the fall of a fair coin by some mental or supernatural means available only to him. As the sequence continues to be all heads, these two beliefs solidify. Each ascribes an element of causality or meaning to the model of the system in which he believes and to the possible outcomes of an experiment or trial. The practical man becomes virtually certain that the coin is two-headed. The true believer, already certain that the guru is honest and capable of miracles, is further confirmed in his faith. The third observer will point out that a sequence of all heads has exactly the same probability as any other sequence and therefore should not be subjected to discrimination even from the point of view of the practical man. He may point out that long sequences of heads are nearly certain to happen in very long sequences of Bernoulli trials (section 1.2.9). The goddess Fortuna, according to her whim, provides very rare occurrences from time to time. The string of heads is simply one of these occurrences. No one feels a need to examine the coin.

Many people (Wald, 1954), who should know better, believe that a success is 'due' after a series of failures in independent trials. A baseball player who has struck out for a number of times is now believed to be 'due'. The manager of the team and the audience feel that he is virtually certain to get a hit. However, the batter can count on no help from Fortuna. Unless there has been considerable retraining, with performance demonstrated in practice, the probability for a hit in the next at-bat is not changed, *per se*, by a long sequence of strike-outs. The probability of the outcomes of experiments or trials that are independent continues to have the same value regardless of the outcomes of preceding experiments or trials.

It is worthwhile to notice at this point that the term probability as used in this book has taken on a mathematical meaning that is quantitative and not necessarily the same as that used in general conversation. This is common in science and we shall take this into account in later work.

1.2 The application of set theory to the axiomatic treatment of probability theory

1.2.1 A brief introduction to set theory; the algebra of sets

The modern theory of probability is based on the theory of sets, that is, the theory of collections or lists of objects, events, numbers or other conceptual entities. The Cretans seem to have had a bad reputation for veracity in the ancient world, even among themselves, according to the apostle Paul (Titus 1:12). Epimenides the Cretan said, 'All Cretans are liars'. Is that statement true or false? If the elements of the set of all Cretans are liars, including Mr. Epimenides, then his statement contradicts itself. This is one of a number of semantic paradoxes that have been constructed by logicians (Suppes, 1972). Mathematicians avoid these paradoxes by constructing a formalized language and basing set theory on axioms. Semantic paradoxes are the first encounter in this book with the fact that one must be careful to avoid anecdotal and contrived 'explanations' that are not based on mathematical logic. The theory of sets, when it is properly based on axioms, is of a generality such that it is fair to say that it is the foundation of all mathematical logic (Suppes, 1972).

We are concerned in this book with probabilities relating to objects, events or numbers of interest in molecular biology. For example, we are interested in the appearance of the amino acids that are functionally equivalent at a given site in a protein (section 6.1.1). In this case one may adopt the frequentist approach: the model of the system may be viewed mathematically as the *outcome* of a series of trials. All possible outcomes, or *elementary events*, are *the elements of a set* of amino acids. In general a particular set may be specified by the *roster method*; that is, by listing its elements. It may also be specified by the *property method*; that is, by stating the properties of the elements:

$$A = \{1, 3, 5, \ldots\} = \{a \,|\, \text{all odd positive real numbers}\}$$

This is mathematical shorthand for 'the element a is a member of the set of all odd positive real numbers'. By the same token,

$$A = \{\text{Ala}, \text{Arg}, \ldots, \text{Val}\} = \{a \,|\, \text{all proteinous amino acids}\}$$

is mathematical shorthand for 'the element a is a member of the set A of all proteinous amino acids'. Furthermore,

$$B = \{\text{Leu}, \text{Arg}, \text{Ser}\} = \{b \,|\, \text{all proteinous amino acids } \textit{that have}$$
$$\text{six codons}\}$$

is mathematical shorthand for 'the element b is a member of the subset of A containing amino acids that have six codons'. When the outcome of a trial is a member of the subset B we say that *event B has occurred*.

We shall call sets by the italic capitals A, B, C etc. The members or elements of set A will be called a and the following mathematical shorthand will be used:

$$a \in A$$

This is read, 'The element a is a member of set A'. If a is not a member of A, this is written as follows:

$$a \notin A$$

Mathematicians like to construct an algebra of the concept of sets and entertain an analogy to geometry. These algebras use as many of the ideas of the algebra of numbers as are appropriate, namely, the identity law, the associative law, the commutative law, the complement law and the idea of equality. The algebra of sets is constructed in the following manner: if every element of A is also an element of B, then A is called a subset of B. This is written

$$A \subset B$$

Two sets are equal if and only if all elements of one are also contained in the other. This is written

$$A = B \text{ if and only if } A \subset B \text{ and } B \subset A$$

Two sets may be combined. This is known as the *union* of sets A and B. The union contains all elements belonging either to A or B and is written

$$A \cup B$$

For short, this expression may be read 'A cup B'.

The set of events common to two sets is called the *intersection* and is written

$$A \cap B$$

For short, this may be read 'A cap B'.

The difference of A and B is the set of elements that belong to A but not to B. This will be designated

$$A - B$$

Definition: All elements under discussion are members of subsets of a fixed or universal set Θ. It is useful, in analogy with geometry, to call Θ the *space of events* or a *set space*. The elements in a set space are often referred to as *points* in the set space, according to the analogy with geometry.

In molecular biology, for example, we may consider Θ to be all chemically possible amino acids of each chirality. Then A might be the set of all proteinous amino acids.

Definition: We define $A^c = \Theta - A$ as the complement of A; i.e. the set of all sets contained in Θ but not contained in A.

The set operations in the algebra of sets may be visualized in analogy to the geometrical case by the use of Venn diagrams (Venn, 1888), as shown in Figure 1.1.

Definition: The set that contains no elements is called the impossible or null set and is designated by $\varnothing$. If $A \cap B = \varnothing$ then A and B are said to be *disjoint* and the elements of A and B are said to be *mutually exclusive*.

Definition: We may write the combinatorial or Cartesian product of two sets A and B as $A \times B$, which is defined to be the set space of all ordered pairs of the elements (a, b) where $a \in A$ and $b \in B$ (Khinchin, 1957; Feller 1968; Suppes, 1972). (This is not the same as the intersection $A \cap B$.)

Similarly, we may consider the Cartesian power of a set space. For example, A^2 is defined to be the set of all *ordered* pairs of the elements in set A and A^n is the set of all *ordered* n-tuples of the elements in set A. Thus ordered n-tuples in the set A^n are the elementary events of that set.

A useful visualization of the Cartesian product of two or more set spaces or the power of a set space may be shown by an example in a geometry of two or more dimensions. The product set space $X \times Y$ is a plane on which the elements of the set are all pairs (x, y) of the coordinates of the points. The product set space $X \times Y \times Z$ is the space whose elements are the triplets (x, y, z). This may be further generalized to spaces of any positive integer dimension n. We may regard the ordered n-tuples of elements of a set as defining an abstract set space of n dimensions. Thus the ordered n-tuples of the set A^n are mapped in one-to-one correspondence with points in the abstract set space. One-to-one correspondence is sometimes called a *bijection* and the corresponding elements are said to be *bijective* (MacLane, 1986).

The genetic code provides a specific example in molecular biology of a set raised to a Cartesian power. One has the set of all mRNA nucleotides,

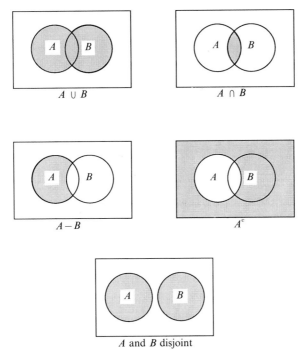

$A \cup B$

$A \cap B$

$A - B$

A^c

A and B disjoint

Figure 1.1 These Venn diagrams (Venn, 1888) illustrate the operations of set algebra discussed in section 1.2.1. The rectangle represents the universal or fixed set Θ. The areas labelled A and B represent subsets of Θ. The shaded regions are described by captions under the diagrams.

$G = \{U, A, G, C\}$. Then the set space $G \times G = G^2$ contains as elements the ordered pairs UU, UA, UC, UG, AA, AU, AC, AG, CC, CA, CG, CU, GG, GU, GA, GC. These are, of course, the doublet codons (7.2.3 *et seq.*). It follows that the set space G^3 contains all the ordered triplets of U, A, C, G, namely, the familiar triplet codons.

Definition: A set of events is called a *cylinder set* or, for short, a *cylinder* (Khinchin, 1957; Feller, 1968; Petersen, 1983) if pairs of events each satisfy restrictive conditions. For example, consider the set of points in a Euclidean space of three dimensions that lie within a square so that $0 < x < 1$ and $0 < y < 1$. If the values of z are unrestricted the points enclosed in the xyz-space with a square cross section define a *cylinder* in that space.

   The third position in eight of the familiar triplet codons in the set space G^3 may vary indefinitely among the four nucleotides in either the DNA or RNA alphabets without changing the read-off amino acid. The specificity of these codons is defined by the first two nucleotides. Such sequences meet the definition of a *cylinder set*. I propose to call these codons *cylinder codons* (sections 6.5.2 and 7.2.4; Epilogue).

1.2.2 The sample space in probability theory

As we have said, in molecular biology we are concerned with sets of elementary events such as nucleotides, codons and amino acids and with sequences or ordered n-tuples of these elements and the relations between them. We deal with them only as symbols. These elementary events are discrete and finite in number. They represent the 'graininess' with which we are concerned in applying information theory and coding theory to molecular biology. The exact position of every atom in the nucleotides, or amino acids, is irrelevant.

Definition: The set of all elementary events, a, in a set space A, *to each of which a probability has been assigned* is called a *probability sample space of elementary events* or simply a *sample space*.

According to the geometrical analogy, the elementary events in a sample space are often referred to as *points*. The sample space is a universal set designated by Ω. Some authors call Ω a *finite scheme*. The term *phase space* is used in mathematical physics. These terms are simply other ways of saying the same thing.

The Cartesian product of two finite probability sample spaces Ω_0 and Ω_1, with elements a and b respectively, is the finite probability space that contains all ordered pairs (a, b). This is written $\Omega_0 \times \Omega_1$. Probability sample spaces may be raised to a power. For example, suppose Ω contains the mRNA nucleotides U, A, C, G as elements or points. Then the sample space $\Omega \times \Omega = \Omega^2$ contains as elements the ordered pairs of U, A, C, G (listed earlier) with their associated probabilities. These are, of course, the mRNA doublet codons (section 7.2.3 *et seq.*). It follows that Ω^3 contains all the ordered triplets of U, A, C, G with their associated probabilities, namely, the familiar mRNA triplet codons.

1.2.3 The role of axioms in mathematics

Since before the time of Euclid, mathematicians have preferred to treat their subject in terms of axioms. This avoids an endless sequence of contrived and *ad hoc* assumptions being made in pursuit of the desired result. It prevents us from arguing a case, as it were, to achieve a foreordained result. The axiomatic treatment often reveals unsuspected theorems that are not intuitive. It discloses relations between problems that would otherwise not be realized. In particular, the axiomatic treatment of the theory of probability avoids logical circularities and allows the application of the theory to a broad range of problems.

Axioms are elementary facts that cannot be explained by reducing them to simpler ones; rather, they must be taken as a starting point. Axioms are not necessarily self-evident or representative of the real world. It is only in the application to real world problems that they are useful or useless, appropriate or inappropriate, interesting or dull, to the degree of the exactness that the set of axioms is a mathematical representation of the real world.

A set of axioms, or postulates if one prefers, must be *consistent*; that is, it must not be possible to prove a given statement or theorem both true and false. Axioms must be *independent*; that is, any axiom that can be proved from the others is a theorem and must be crossed off the list. The set of axioms must be *unique* and *finite* in number. The axioms must be *complete*; that is, one must not need to introduce *ad hoc* statements while proving theorems. I shall take up in sections 2.4.2 and 10.10.1 a theorem of Gödel (1931) that proves that for any set of axioms there are true statements that cannot be proved from the axioms. Furthermore, there are questions that are undecidable from the set of axioms. Reasoning from axioms is the highest form of human thought. Nevertheless, there are questions that are beyond human reasoning (Chaitin, 1988).

In axiomatic treatments throughout mathematics, one deals with undefined things. In geometry one does not define what a point or a line *really* is. The axioms state only the relations between these undefined things. 'Two points determine a line' and so forth. The undefined things about which the axioms deal receive their definition by application to a specific model of the system under consideration. A more extensive mathematical discussion of axioms and their role in the fundamentals of mathematics has been given by Hermes & Markwald (1974).

1.2.4 The axiomatic foundation of probability theory

So far everything that has been said derives from the frequentist interpretation of probability. Let us now write down the axioms that are an abstraction of that interpretation (Jaynes 1957a, 1988a; Lindley, 1965).

The additional interpretations of probability, discussed in section 1.1.2, are justified if they satisfy these axioms. We are concerned with events that either happen or do not happen. Therefore all probabilities are real non-negative numbers less than or equal to one. Negative probabilities have been defined by Bartlett (1945), and complex probability by Cox (1955). An interesting discussion is given by Mueckenheim (1986). These papers are concerned with the strange and wonderful world of quantum mechanics. They have no application in molecular biology.

Let a real non-negative number $p(A)$, less than or equal to one be established for each event A in the sample space Ω as a measure or description of the nature of the events in the sample space. Since probabilities are all non-negative numbers equal to or less than one, in order to establish a complete system their sum must be one. A probability of zero is associated with events that are impossible or, as I shall show later (section 1.2.6), are impossible except for elementary events of probability measure zero. On the other hand, a probability of one is associated with events that are certain except for elementary events of probability measure zero. This number will be called the probability of that event.

For specific pairs of events A and B a real number is defined according to the model of the system at hand and is called the *conditional probability* of event A given that event B has occurred. This probability will be written $p(A|B)$. For example, given that the ace of hearts has been played, event B, and that the ace of diamonds has not yet appeared, the probability that the ace of diamonds will be played next, event A, is $p(A|B)$. The probability that *both* the ace of hearts *and* the ace of diamonds have been played is the probability of the intersection $(A \cap B)$. This is written $p(A \cap B)$. A second example is that event B is the appearance of any odd number in the toss of a die and event A is the appearance on that toss of any number less than four. Event $(A \cap B)$ is the appearance of a number that is *both* odd *and* less than four. These probabilities must satisfy the following axioms:

Axiom 1: $0 \leqslant p(A) \leqslant 1$; if the events A_i are mutually exclusive then the sum of the probabilities of all n events in the sample space equals one: $\Sigma_n p(A_i) = 1$.

Axiom 2: If the events A_i are mutually exclusive then

$$p(A_1 \cup A_2 \cup \ldots \cup A_n) = p(A_1) + p(A_2) + \ldots + p(A_n)$$

Axiom 3: For two events A and B,

$$p(A \cap B) = p(B)\,p(A|B)$$

The first axiom is the *normalization* condition. The second axiom is called the *addition* law and the third axiom the *multiplication* law.

Theorem 1.1: The probability of the occurrence of two or more independent mutually exclusive events is the product of the probabilities of the individual events.

Proof: If event A is independent and mutually exclusive of event B, then $p(A|B)$ does not depend on the probability of event B. Accordingly, $p(A|B) = p(A)$. The proof is extended to more than two mutually exclusive events by mathematical induction.

Theorem 1.2: If A and B are any two events, then

$$p(A - B) = p(A) - p(A \cap B)$$

Proof: Event A can be decomposed into two mutually exclusive events $(A - B)$ and $(A \cap B)$. In set notation, $A = (A - B) \cup (A \cap B)$. Thus, by Theorem 1.1,

$$p(A) = p(A - B) + p(A \cap B)$$

which proves the theorem.

Theorem 1.3: If A and B are any two events, then

$$p(A \cup B) = p(A) + p(B) - p(A \cap B)$$

Proof: The event $(A \cup B)$ can be decomposed into mutually exclusive events $(A - B)$ and B, i.e. $(A \cup B) = (A - B) \cup B$. Therefore, by Axiom 2 and Theorem 1.2:

$$p(A \cup B) = p(A - B) + p(B)$$
$$p(A \cup B) = p(A) - p(A \cap B) + p(B)$$
$$p(A \cup B) = p(A) + p(B) - p(A \cap B)$$

The reader who feels the need for a more detailed discussion of the evolution of the basic ideas in probability theory from the simple frequentist notions derived from games of chance, in their ideal form, to the sophisticated axioms of the modern mathematical theory of probability may wish to refer to Jaynes (1957a, 1979, 1988a, b). He has shown that these axioms are consistent with reasonable rules of plausible reasoning and include deductive logic as a special case.

Often we learn more from exceptions to theories than from compliance. The exceptions lead us to new knowledge; compliance merely confirms knowledge we already have (Jaynes, 1957a, 1968; Kuhn, 1970).

1.2.5 The establishment of probabilities for events in a sample space

One first establishes a probability sample space Ω consisting of all the alternative events in the model of the system under discussion. Probabilities are then assigned to each event in such a way that the model is described. For example, the frequentist must establish that the problem requires the *a priori* probability of heads or tails in a fair coin be put at $1/2$. In fact, that is the definition of a fair coin. That is, probability is assigned as the number of ways the event can occur divided by the total number of elementary events in the sample space. In the case of the toss of a fair die the events may be *six* and *not six*. The *a priori* probabilities assigned to these events are $1/6$ and $5/6$, respectively. In the case of a die that is suspected of not being fair the frequentist defines probability experimentally by generating an ensemble of events according to equation (1.1). Thus a set of numbers that satisfies the axioms and is characteristic of the model represents the probabilities.

Let us now consider the application of the degree of belief or subjective interpretation to probability. Clearly, if the hypothesis is the guilt or innocence of a person on trial in a court of law, although there may be some similarity between cases, each must be considered on its own merits. One cannot even conceive of repeating the events under consideration. The same procedure of assigning probabilities applies to the probable inference or inductive logic interpretation. One constructs a sample space of all possible events pertinent to the case as reflected by the evidence. The event of guilt must be established, in legal language, 'beyond reasonable doubt'. In the language of probability a person may not be judged to be guilty unless his innocence is an event of probability measure nearly zero. If the numbers associated with the various possibilities in the case are in accord with the axioms, these numbers may be regarded as probabilities.

Apostolakis (1990) has given a discussion of the means by which probabilities are determined in the case of probability risk assessments associated with large complex engineering systems. The word *risk* in this context involves the concept of both the probability of a destructive event and the cost of the consequences. In this application probability is always a measure of the degree of belief that the destructive event will happen in the time assigned.

I shall give, in Chapter 3, further methods of determining the probabilities that are characteristic of the model under consideration to be associated with events in a sample space.

1.2.6 *The probabilities (or measures) zero and one*

Classical logic does not admit of exceptions. One exception, however rare, falsifies the argument. In probability theory it is always assumed and often stated that the theorem is true except for a set of probability (or measure) zero. Similarly, events of probability one may have exceptions of probability (or measure) zero. In that case one says 'the event is certain in probability'. It is often said that one cannot prove either a negative or a positive statement by statistics. But one can prove that the subject event has a probability measure of nearly zero or one, as the case may be. This is illustrated by the fable of the two-headed coin, the guru and the practical man in section 1.1.2. In another example, statistical mechanics tells us that the water in a pot put on a hot stove may freeze (Eddington, 1931a) but that the probability (or measure) of that event is effectively zero. This reflects the pragmatic attitude that we are not interested in exceptions so rare that they are not found in one's finite experience. However, there are events, which involve observing a violation of, say, the conservation of energy or momentum, that are not governed by probability. Such events may be said to be impossible or *strictly forbidden*.

Belief in hypotheses with very small or zero probability can be justified only by religious faith or faith in an infallible doctrine or ideology. This is not a part of science (Jukes, 1987a; Hoffer, 1951). I do not mean this in a pejorative sense. Science and religion or ideology have different belief systems. They are best kept apart. (Now faith is the substance of things hoped for, the evidence of things unseen. Hebrews 11:1.)

1.2.7 Bayes' theorem

Suppose we have two events A and B; then by the multiplication law

$$p(A \mid B) p(B) = p(B|A) p(A) \tag{1.2}$$

This is known as Bayes' theorem or Bayes' rule and has important applications in probability theory (Lindley, 1965; Feller, 1968; Hamming, 1986; Jaynes, 1988b). It will be used later, in section 5.3, to develop some important relations. The ideas of the prior probability $p(A)$ and the posterior probability $p(B|A)$ are so fundamental that the subjective interpretation is sometimes called the Bayesian theory of probability. A detailed discussion of the Bayesian view of probability is given by Lindley (1965) and by Jaynes (1968).

1.2.8 Random variables

Definition: A number x that is defined on the events A in a sample space Ω is called a *random variable*. That is, x, the random variable, is selected by the outcome of an experiment. Any function, $f(x)$, is also a random variable.

Suppose, for example, we consider the roll of a die. The number on the top face and all functions of that number are random variables. The probability of each event is established in the following way. Unless the dice belong to Mr. Epimenides the Cretan, we assume each die is fair, that is, that it is an accurate unweighted cube. There are six possible outcomes, each of which can happen in only one way. Therefore we assign the probability $1/6$ to the number that appears on each face. If we had prior knowledge of the behavior of the die showing that it is not an accurate cube, or has been weighted in some way, that would be reflected in the model of the system by assigning slightly different probabilities to each face.

Now consider the roll of two fair dice. The sample space is Ω^2 and there are 6^2 or 36 elementary events or outcomes. The sum of the numbers on the top face after a roll is the random variable x_i taking values from 2 to 12. Each value corresponds to an event. In this case some events can happen in more than one way. For example, the sum seven can happen in six ways. We establish the probability, p_i, of each value x_i of the random variable by dividing the number of ways in which each sum can occur by the total number of elementary events in the sample space. The set of probabilities

Table 1.1 *The Probability Distribution for Events in the Sample Space of two Fair Dice*

Value of random variable x_i	2	3	4	5	6	7	8	9	10	11	12
Number of ways x_i can occur	1	2	3	4	5	6	5	4	3	2	1
Probability p_i	1/36	2/36	3/36	4/36	5/36	6/36	5/36	4/36	3/36	2/36	1/36

for each event in the sample space will be called the probability distribution. This is exhibited in Table 1.1.

The probability distribution of random variables has a mean or *expectation value*. Notice that 7 is both the median and the mode since the probabilities are symmetric in this case. The expectation value $\langle x_i \rangle$ of a random variable x_i is defined as follows:

$$\langle x_i \rangle = \sum_i p_i x_i \qquad (1.3)$$

Thus the expectation value of any function $f(x_i)$ is

$$\langle f(x_i) \rangle = \sum_i p_i f(x_i) \qquad (1.4)$$

To explain this in a simple fashion, suppose a lottery has $1,000 to be awarded to the winner. A gambler knows, or believes, that 1000 tickets have been sold. What can he afford to pay for a ticket in order to break even in the long run? Since the probability of winning is 1/1000, he can afford to pay $1.00.

1.2.9 Some standard probability distributions

The ordered *n*-tuples of elementary events of *n* independent trials with two discrete and exhaustive outcomes form a product sample space of *n* dimensions. Call the outcome of one elementary event a *success* with probability *p* and the other outcome a *failure* with probability *q*. The probability of each outcome remains the same throughout the experiment. This process generates an ordered *n*-tuple of the outcomes.

The elementary events are called Bernoulli or binomial trials, after James Bernoulli (1654–1705). The ordered n-tuples of these outcomes are called Bernoulli sequences or *strings* by mathematicians but the term *sequence* is the one more familiar in molecular biology. The trials may have been going on in the indefinite past and may continue in the indefinite future. The sample space of n Bernoulli trials contains 2^n points. The toss of a die is not a Bernoulli trial since there are six outcomes. If the elementary events of the toss are limited to *odd* and *even* the experiment becomes a Bernoulli trial. Similarly, in the case of the toss of two dice, if the elementary events are *seven* and *not-seven* the sequence of trials is Bernoulli. Thus the probability of the outcomes *success* and *failure* are not necessarily equal; however

$$p + q = 1 \tag{1.5}$$

And, of course, in the case of an ordered n-tuple product sample space:

$$(p + q)^n = 1 \tag{1.6}$$

Expanding the left side by the binomial theorem:

$$(p + q)^n = p^n + np^{n-1}q$$

$$+ \frac{n(n-1)}{2} p^{n-2} q^2 + \frac{n!}{r!(n-r)!} p^r q^{n-r} + q^n = 1 \tag{1.7}$$

The first term in the binomial expansion is the probability of n successes, the second term is the probability of $(n-1)$ successes and one failure and so on for the remaining terms. The last term is the probability of n failures. Thus each term expresses the probability of a particular event, for example the probability of r successes in n Bernoulli trials. Here r is a random variable and an event is a combination of successes and failures without regard to their order of appearance. Of course the total probability of all the 2^n points in the sample space is unity.

We are arranging combinations of two subpopulations of n things, namely, successes and failures. It is therefore not surprising that the general form for the *binomial coefficient*, $C(n, r)$, is exactly the expression for the number of combinations of n things taken r at a time. (One must remember that $0! = 1$).

$$C(n, r) = \frac{n!}{r!(n-r)!} \tag{1.8}$$

If $p = 1/2$ the binomial probability distribution peaks where the number of successes and the number of failures are equal. If p is not equal to $1/2$, the peak of the distribution is near the maximum of the ratio of successes to the total number of trials. The binomial distribution has the following properties: the mean $\lambda = np$, the standard deviation $\sigma = (npq)^{\frac{1}{2}}$, the variance $\sigma^2 = npq$.

The binomial distribution may be generalized to any integer value k of outcomes in each of n trials. If there are k outcomes rather than just two, the probability distribution may be found by expanding $(p_1 + p_2 + p_k)$ $n = 1$. The sample space has k dimensions in this case. The coefficients of this expansion are known as the *multinomial coefficients*. They are as follows:

$$\frac{n!}{r_1! r_2! \ldots r_k} \tag{1.9}$$

Equation (1.9) gives the number of arrangements of n things in k subpopulations taken r_k at a time. This is easy to understand since we are dividing the n trial results between k outcomes without regard to the order of the appearance of the outcomes. Equation (1.9) will be used in section 7.1.4 to calculate the number of possible genetic codes with triplet codons.

If in the binomial distribution $C(n, r)$ the random variable r takes continuous values instead of discrete values, the binomial probability distribution becomes the Gaussian or normal probability distribution (replacing r by x to accord with the usual convention):

$$f(x) = \frac{1}{\sigma(2\pi)^{\frac{1}{2}}} e^{-(x-\lambda)^2/2\sigma^2} \tag{1.10}$$

Suppose p is small but n is large and $np = \lambda$. The binomial distribution may be approximated by the Poisson distribution (Siméon de Poisson 1781–1840) for k success events as follows:

$$P(k; \lambda) = \left(\frac{\lambda^k}{k!}\right) e^{-\lambda} \quad k = 0, 1, 2, \ldots \tag{1.11}$$

The Poisson distribution is often abused in the literature by applying it to compare cases where the probability, p is not constant (section 12.2.1). One must not use a formula plucked from a book without considering the conditions from which it was derived.

1.3 Probability matrices and probability vectors

1.3.1 Definitions of matrices and vectors

In order to deal with coding problems and the application of the theory of probability to molecular biology we shall need the powerful concepts of *matrices* and *vectors*. As I pointed out previously (Yockey, 1974), the genetic code is represented by a matrix and the codons and amino acids by vectors. This formalism will lead us to some very important theorems that will be useful in the discussion of phylogenetic trees and evolution in Chapter 12.

The theory of matrices and vectors comes from the properties of a set of m linear equations in n unknowns x_j where the a_{ij} and the y_i are specified by the problem:

$$\sum_{j=1}^{n} a_{ij} x_j = y_i \quad i = 1, \dots, m \tag{1.12}$$

Definition: A rectangular array of quantities a_{ij}, set out in m rows and n columns, is called an $m \times n$ *matrix*. Matrices will be indicated in this book in bold capital letters. The usual convention in the literature is to let the first subscript i refer to the rows of the matrix and the second subscript j refer to the columns. The quantities a_{ij} are called the *elements* of the matrix. Two matrices that have the same number of rows and columns are said to be the same size.

$$\begin{bmatrix} a_{11} & a_{12} & a_{13} & \cdots & \cdots & \cdots & a_{1n} \\ a_{21} & a_{22} & \cdots & \cdots & \cdots & \cdots & \cdots \\ a_{31} & \cdots & \cdots & \cdots & a_{ij} & \cdots & \cdots \\ \cdots & \cdots & \cdots & \cdots & \cdots & \cdots & \cdots \\ a_{m1} & \cdots & \cdots & \cdots & \cdots & \cdots & a_{mn} \end{bmatrix} = A$$

If $m = n$ the matrix is a *square matrix* of order n. If a matrix is square, the elements where $i = j$ are called the *diagonal elements*.

Definition: Given a matrix **A** for which $m = n$ then the *determinant* of **A**, written either $|\mathbf{A}|$ or $|a_{ij}|$, is, in the case where $n = 3$:

$$\begin{vmatrix} a_{11} & a_{12} & a_{13} \\ a_{21} & a_{22} & a_{23} \\ a_{31} & a_{32} & a_{33} \end{vmatrix} = \begin{aligned} & a_{11} a_{22} a_{33} + a_{12} a_{23} a_{31} + a_{13} a_{32} a_{21} \\ & - a_{31} a_{22} a_{13} - a_{32} a_{23} a_{11} - a_{33} a_{12} a_{21} \end{aligned}$$

In other cases the rule for forming these products is the same. One starts at a_{11} and forms the product of each of the n elements going diagonally down. One then begins with a_{12} multiplying together all elements, going diagonally down; a term starting with a_{13} then follows. Each term is the product of n elements. The last term begins with a_{1n}. Each of these terms carries a plus sign. One then begins at a_{m1} and forms the products of all elements going up along the diagonal. Then one begins again at a_{m2} multiplying together all the $n-1$ diagonal elements with a_{11} so that each term is the product of n elements. One continues in this manner until a_{mn} is reached. Each of these terms carries a minus sign. The sum of all terms so formed is the value of the determinant $|a_{ij}|$.

Definition: Cramér's rule

If $m = n$ and if the determinant $|a_{ij}| \neq 0$ the set of equations (1.12) can be solved by means of Cramér's rule (Gabriel Cramér, 1704–52). This rule states that the value of x_j is given by replacing the jth column in the determinant $|a_{ij}|$ by the column y_i and dividing that determinant by $|a_{ij}|$.

If there are more than three equations this becomes tedious. However, programs are available for personal computers and even for sophisticated pocket calculators that calculate determinants and accomplish this task quite easily.

Definition: Given an $m \times n$ matrix **A**, the $n \times m$ matrix resulting from the interchange of rows and columns is called the transpose of **A**. That is, the first row of **A** becomes the first column of the transpose, the second row of **A** becomes the second column of the transpose and so forth. The usual notation for the transpose of **A** is $\mathbf{A}^{\mathrm{T}}$.

Definition: A square matrix, **P**, is called a *stochastic matrix* if the elements of each of its rows is non-negative and their sum is equal to one. If in addition the sum of the elements of each of its columns is equal to one the matrix is called *doubly stochastic* (Feller, 1968; Lancaster & Tismenetsky, 1985; Hamming, 1986). Some authors, for example Moran (1986) define a stochastic matrix as one in which the columns, not the rows, are non-negative and sum to one.

Some authors (Feller, 1968; Hamming, 1986), define a stochastic matrix as the transpose so that each of the rows of $\mathbf{P}^{\mathrm{T}}$, is non-negative and their sum is equal to 1.

Definition: A square matrix of order *n* that has all elements on the diagonal equal to one and all others equal to zero is called an *identity matrix*, $\mathbf{I}_n$.

Definition: If all the elements of a matrix are zero it is called the *null matrix* and written **O**.

Definition: Any ordered sequence of *n* numbers is called an ordered *n*-tuple. An ordered *n*-tuple is also called an *n*-dimensional vector. Each of the rows and columns of a matrix is a *vector*. In this book vectors will be indicated by bold lower case letters. The numbers in the sequence are known as the components of the vector.

Definition: Matrices that have only one row are called *row* or $1 \times n$ matrices. A $1 \times n$ matrix is also referred to as a *row* vector of dimension or order *n*. Matrices that have only one column are called *column* or $m \times 1$ matrices. An $m \times 1$ matrix is also referred to as a *column* vector of dimension or order *m*. A 1×1 matrix is called a *scalar*.

Definition: A vector that has all non-negative real number components, the sum of which is 1, is called a *stochastic vector*.

Each of the rows of a stochastic matrix is a stochastic vector.

1.3.2 Algebraic properties of matrices and vectors

The mathematical power and usefulness of matrices and vectors comes largely from the fact that one may construct an algebra of these arrays that has many of the properties of the algebra of numbers. The algebra of matrices enables one to manipulate large collections of data displayed in these arrays without *ad hoc* assumptions and in a simple and convenient fashion. This avoids the usual necessity of resorting to averages. As I shall show later in this book, averaging destroys information. There is a large number of powerful and useful theorems in the algebra of matrices and vectors. We shall see that many of these theorems are not intuitive and indeed some are counter-intuitive. Therefore their existence would not be suspected without the development of this algebra.

Two matrices of the same size may be added or subtracted by adding or subtracting each element with the same indices *i,j*. A matrix may be multiplied by a scalar by multiplying each element by the scalar. These rules apply to vectors also.

We are led to a natural definition for the algebraic operation of *matrix multiplication* by the following argument: suppose we consider the set of two linear equations in three unknowns x_1, x_2 and x_3:

$$a_{11} x_1 + a_{12} x_2 + a_{13} x_3 = y_1 \qquad (1.13)$$

$$a_{21} x_1 + a_{22} x_2 + a_{23} x_3 = y_2 \qquad (1.14)$$

Let us make a change of variables to three equations in two unknowns y_1 and y_2:

$$b_{11} y_1 + b_{12} y_2 = z_1 \qquad (1.15)$$

$$b_{21} y_1 + b_{22} y_2 = z_2 \qquad (1.16)$$

$$b_{31} y_1 + b_{32} y_2 = z_3 \qquad (1.17)$$

Upon substituting in equations (1.15), (1.16) and (1.17) for y_1 and y_2 from equations (1.13) and (1.14) one has

$$(b_{11} a_{11} + b_{12} a_{21}) x_1 + (b_{11} a_{12} + b_{12} a_{22}) x_2 + (b_{11} a_{13} + b_{12} a_{23}) x_3 = z_1 \qquad (1.18)$$

$$(b_{21} a_{11} + b_{22} a_{21}) x_1 + (b_{21} a_{12} + b_{22} a_{22}) x_2 + (b_{21} a_{13} + b_{22} a_{23}) x_3 = z_2 \qquad (1.19)$$

$$(b_{31} a_{11} + b_{32} a_{21}) x_1 + (b_{31} a_{12} + b_{32} a_{22}) x_2 + (b_{31} a_{13} + b_{32} a_{23}) x_3 = z_3 \qquad (1.20)$$

Equations (1.13)–(1.17) define the algebraic operation of the product of a matrix and a column vector to yield another column vector:

$$\begin{bmatrix} a_{11} & a_{12} & a_{13} \\ a_{21} & a_{22} & a_{23} \end{bmatrix} \begin{bmatrix} x_1 \\ x_2 \\ x_3 \end{bmatrix} = \begin{bmatrix} y_1 \\ y_2 \end{bmatrix} \qquad (1.21)$$

$$\begin{bmatrix} b_{11} & b_{12} \\ b_{21} & b_{22} \\ b_{31} & b_{32} \end{bmatrix} \begin{bmatrix} y_1 \\ y_2 \end{bmatrix} = \begin{bmatrix} z_1 \\ z_2 \\ z_3 \end{bmatrix} \qquad (1.22)$$

In similar fashion equations (1.18), (1.19) and (1.20) may be written in matrix notation:

$$\begin{bmatrix} b_{11} a_{11} + b_{12} a_{21} & b_{11} a_{12} + b_{12} a_{22} & b_{11} a_{13} + b_{12} a_{23} \\ b_{21} a_{11} + b_{22} a_{21} & b_{21} a_{12} + b_{22} a_{22} & b_{21} a_{13} + b_{22} a_{23} \\ b_{31} a_{11} + b_{32} a_{21} & b_{31} a_{12} + b_{32} a_{22} & b_{31} a_{13} + b_{32} a_{23} \end{bmatrix} \begin{bmatrix} x_1 \\ x_2 \\ x_3 \end{bmatrix} = \begin{bmatrix} z_1 \\ z_2 \\ z_3 \end{bmatrix} \qquad (1.23)$$

Equation (1.21) may be written in the formal matrix style:

$$\mathbf{Ax} = \mathbf{y} \tag{1.24}$$

The matrix $\mathbf{A}$ may be thought of as *operating* on column vector $\mathbf{x}$ to change it to column vector $\mathbf{y}$. That is, the matrix $\mathbf{A}$ can be thought of as a mapping of the points in space X into the points of the space Y. By the same token the matrix $\mathbf{B}$ may be thought of as operating on the column vector $\mathbf{y}$ to change it to column vector $\mathbf{z}$ in equation (1.22). That is, the mapping of the points of space X on the points of space Y may be followed by another mapping of the points of space Y onto the points of space Z. Thus equation (1.22) can be written in the compact matrix form

$$\mathbf{By} = \mathbf{z} \tag{1.25}$$

It is natural to substitute for $\mathbf{y}$ in equation (1.25) from equation (1.24):

$$\mathbf{BAx} = \mathbf{Cx} = \mathbf{z} \tag{1.26}$$

where $\mathbf{BA} = \mathbf{C}$ and the elements of $\mathbf{C}$ are given in equation (1.23).

One notices in equation (1.23) that the elements of $\mathbf{C}$ are the sums of the arithmetical products of the elements in the rows of $\mathbf{B}$ and the elements in the columns of $\mathbf{A}$. This may be shown to be true in general and forms the basis of a definition of matrix multiplication. Write equation (1.25) as a set of m linear equations in p unknowns:

$$\sum_{k=1}^{p} b_{ik} y_k = z_i \quad i = 1, \ldots, m \tag{1.27}$$

Equation 1.12 may be rewritten using k rather than i as a suffix; we may substitute for the y_k in equation (1.27) from equation (1.12):

$$\sum_{k=1}^{p} b_{ik} \left(\sum_{j=1}^{n} a_{kj} x_j \right) = z_i \quad i = 1, \ldots, m \tag{1.28}$$

Equation 1.28 may be rewritten:

$$\sum_{j=1}^{n} \sum_{k=1}^{p} b_{ik} a_{kj} x_j = z_i \quad i = 1, \ldots, m \tag{1.29}$$

Therefore successive mappings $\mathbf{Ax} = \mathbf{y}$ and $\mathbf{By} = \mathbf{z}$ yield a mapping $\mathbf{Cx} = \mathbf{z}$ where $\mathbf{C}$ is an $m \times n$ matrix with the elements c_{ij}:

$$c_{ij} = \sum_{k=1}^{p} b_{ik} a_{kj} \quad i = 1, \ldots, m \quad \text{and} \quad j = 1, \ldots, n \tag{1.30}$$

Definition: The product matrix **C** is called the *Cayley product* (Arthur Cayley, 1821–95). Two matrices **B** and **A** may be multiplied if **B** is $m \times p$ and **A** is $p \times n$ where m, p and n are positive real numbers. It is easy to see, in the simple case of equation (1.23) and more generally in equation (1.30), that unless the number of columns in **B** is the same as the number of rows in **A** the product **BA** is not defined.

Definition: Two matrices that may be multiplied are said to be *conformable*. When two matrices do not meet these conditions the product is not defined and they are said to be *non-conformable*.

The multiplication of matrices is not commutative in general. Even if **AB** are conformable, **BA** may not be. For example if **A** is 2×3 and **B** is 5×2 then **BA** is conformable but **AB** is not. If **AB** = **O** it is not possible to assume in general that either **A** = **O** or that **B** = **O**. Multiplication is, however, associative if the matrices are conformable in sequence:

$$(\mathbf{AB})\,\mathbf{C} = \mathbf{A}(\mathbf{BC}) \tag{1.31}$$

Column vectors are $m \times 1$ matrices. Row vectors are $1 \times n$ matrices. They have two kinds of products. The first is called the *inner*, *scalar* or *dot product*. It is the product of a $1 \times n$ matrix by an $n \times 1$ matrix, which is a 1×1 matrix and therefore a scalar. It is defined as follows:

$$\mathbf{x} \cdot \mathbf{y} = \sum_{i=1}^{n} x_i y_i \tag{1.32}$$

If the inner product of two vectors vanishes they are said to be orthogonal. From equation (1.32) one may see that the elements c_{ij} of **C** in equation (1.30) are the inner products of the ith row vector of **B** and the jth column vector of **A**.

The second vector product is the product of an $n \times 1$ matrix by a $1 \times n$ matrix and is therefore an $n \times n$ matrix. It is called the *vector product*.

Theorem 1.4: The product of two stochastic matrices is also a stochastic matrix.

The proof of this theorem can be done in a few lines (Hamming, 1986).

Square matrices may be raised to positive integer powers. A stochastic matrix of any power is also a stochastic matrix. The usual laws of exponents hold.

$$\mathbf{A}^\alpha \mathbf{A}^\beta = \mathbf{A}^{\alpha+\beta}; (\mathbf{A}^\alpha)^\beta = \mathbf{A}^{\alpha\beta} \tag{1.33}$$

Some matrices have a square root. If $\mathbf{A}^2 = \mathbf{B}$ then $\mathbf{B}^{\frac{1}{2}} = \mathbf{A}$. In some cases other fractional roots and exponents exist. Some matrices have no square root, others have an infinite number (Eves, 1966).

$\mathbf{A}^0$ is taken to be the identity matrix, $\mathbf{I}_n$.

Definition: Given a sample space Ω with events A_i where A_i occurs with probability $p(A_i)$, then the ordered sequence $\{p(A_1), p(A_2), ..., p(A_n)\}$ is called a *probability vector* $\mathbf{p}$. The probability space may be denoted (Ω, A, p).

For an evolving system, such as one undergoing mutations, the different possible states of the system may be described as different events, and at a given time the probability vector for the various states or events may be specified. As the system evolves further, the probability vector will, in general, change, as will be seen in the example of section 1.3.3.

Definition: A matrix $\mathbf{A}$, which changes one vector to another by multiplication, is called a *transition matrix* by mathematicians and theoretical physicists.

In this book we are concerned only with *transition probability matrices* that change one probability vector to another by multiplication. The word *transition* has assumed a special meaning in molecular biology. It means a nucleotide change that does not change the purine–pyrimidine orientation. A mutation that changes purine–pyrimidine to pyrimidine–purine or pyrimidine–purine to purine–pyrimidine is called a *transversion*. In this book, to avoid the awkward terminology *transition/transversion*, a matrix that changes one probability vector to another will be called a *probability transition matrix*. The meaning will be clear depending on whether the context is mutations or matrices.

1.3.3 The Perron–Frobenius theorem and its application in molecular biology

Let us consider the Leu codon CUG to be component 1 in the probability vector for a simple molecular biological example. Suppose we consider only mutations between the first nucleotides and that there are no mutations to or between any of the second and third nucleotides. Then the mutation transition matrix is a mapping of the element CUG in the sample space onto all other elements, namely, AUG, UUG and GUG, to be called components 2, 3 and 4. The transitions are from state i to state j. From axioms 2 and 3 the probability, $p_j(t+1)$, of occupying the jth state after step $(t+1)$ is:

$$p_j(t+1) = \sum_i p_i(t)p_{ij} \tag{1.34}$$

The p_{ij} are the matrix elements of the transition probability matrix $\mathbf{P}$; p_{ij} is the probability of a transition from state i to state j. Here i refers to row, numbered down from the top; j refers to column, numbered from left to right. The vectors whose components are $p_j(t+1)$ and $p_i(t)$ must be row vectors so that the matrices may be conformable, that is, may be multiplied according to equation (1.30). We may write (1.34) as

$$\mathbf{p}(t+1) = \mathbf{p}(t)\mathbf{P} \tag{1.34a}$$

In general, when a vector $\mathbf{q}$ is multiplied by a matrix $\mathbf{P}$ the order of the multiplication $\mathbf{Pq}$ is that of a $p \times 1$ matrix, or column vector $\mathbf{q}$, multiplying an $m \times p$ matrix $\mathbf{P}$. If the order is reversed giving $\mathbf{qP}$, then $\mathbf{q}$ must be written as a $1 \times m$ matrix, or *row* vector.

We note also that a $1 \times m$ matrix (row vector) is the transpose of an $m \times 1$ matrix (column vector).

Returning to our example, the transitions are illustrated in the transition graph in Figure 1.2. Let $\mathbf{p}(0)$ be the row probability vector that reflects the initial situation. If $\mathbf{p}(1)$ is the row probability vector that expresses the probabilities of occupation of states 1, 2, 3, 4 after the first step then, from equation (1.34a),

$$\mathbf{p}(1) = \mathbf{p}(0)\mathbf{P} \tag{1.35}$$

Since we suppose that the state of the probability vector is CUG initially, then

$$\mathbf{p}(0) = [1 \quad 0 \quad 0 \quad 0] \tag{1.36}$$

Substitute the transition probability matrix, $\mathbf{P}$, and the probability vector $\mathbf{p}(0)$ into equation (1.35). In this case the diagonal elements of $\mathbf{P}$ are zero because a transition from the first state to a second state must have been made. The other matrix elements are all $1/3$ since we assume that the transition probabilities to any of the three other states are all equal. We can, of course, assign non-zero elements for the diagonals. This would reflect the possibility that no transition had occurred. I shall address that problem in section 1.3.5 after more mathematics has been developed. We find $\mathbf{p}(1)$ by performing the multiplication as indicated in equation (1.35):

$$\mathbf{p}(1) = \mathbf{p}(0)\mathbf{P} = [1 \quad 0 \quad 0 \quad 0]\begin{bmatrix} 0 & 1/3 & 1/3 & 1/3 \\ 1/3 & 0 & 1/3 & 1/3 \\ 1/3 & 1/3 & 0 & 1/3 \\ 1/3 & 1/3 & 1/3 & 0 \end{bmatrix}$$

$$= [0 \quad 1/3 \quad 1/3 \quad 1/3] \tag{1.37}$$

If the transition probability of going from one state to the next does not change as the number of steps increases, the process is known as a homogeneous (Lancaster & Tismenetsky, 1985) or stationary (Feller, 1968) Markov chain after the mathematician A. A. Markov (1856–1922). In that case the t-step transition probability matrix is $\mathbf{P}^t$. We may multiply equation (1.37) through by $\mathbf{P}$ to obtain the probability vector representing the second mutation:

$$\mathbf{p}(2) = \mathbf{p}(1)\mathbf{P} = \mathbf{p}(0)\mathbf{P}^2 \tag{1.38}$$

The probability transition matrix for the second transition is $\mathbf{P}^2$.

$$\mathbf{p}(2) = \mathbf{p}(0)\mathbf{P} = [1 \quad 0 \quad 0 \quad 0]\begin{bmatrix} 3/9 & 2/9 & 2/9 & 2/9 \\ 2/9 & 3/9 & 2/9 & 2/9 \\ 2/9 & 2/9 & 3/9 & 2/9 \\ 2/9 & 2/9 & 2/9 & 3/9 \end{bmatrix}$$

$$= [3/9 \quad 2/9 \quad 2/9 \quad 2/9] \tag{1.39}$$

Notice that the transition probability matrix elements are more nearly equal to each other in $\mathbf{P}^2$ than in $\mathbf{P}$. If $\mathbf{P}$ is raised to the fourth power, reflecting a fourth mutation step, the matrix elements of this matrix are all nearly equal to 1/4.

$$\mathbf{p}(0)\mathbf{P}^4 = [1 \quad 0 \quad 0 \quad 0]\begin{bmatrix} 21/81 & 20/81 & 20/81 & 20/81 \\ 20/81 & 21/81 & 20/81 & 20/81 \\ 20/81 & 20/81 & 21/81 & 20/81 \\ 20/81 & 20/81 & 20/81 & 21/81 \end{bmatrix}$$

$$= [21/81 \quad 20/81 \quad 20/81 \quad 20/81] \tag{1.40}$$

It is apparent that the matrix elements and consequently the probability vector components are approaching a limit of 1/4 at higher powers. This observation leads us to the Perron–Frobenius theorem, which will be of great value in considerations later in this book.

Definition: A stochastic matrix is said to be *regular* if all elements at some power $\mathbf{A}^p$ are non-zero.

Theorem 1.5 (The Perron–Frobenius theorem) (Perron, 1907; Frobenius, 1908, 1912; Seneta, 1973; Petersen, 1983; Moran, 1986):

Let $\mathbf{A}$ be a regular stochastic matrix. Then:

(a) A fixed probability vector $\mathbf{t}$, none of whose components is zero, is associated with $\mathbf{A}$.

(b) The sequences of powers $\mathbf{A}, \mathbf{A}^2, \mathbf{A}^3, \ldots$ approaches the matrix $\mathbf{T}$ whose rows are each the vector $\mathbf{t}$.

(c) If $\mathbf{p}$ is any probability vector, then the sequence of vectors, $\mathbf{pA}$, $\mathbf{pA}^2, \mathbf{pA}^3, \ldots$ approaches the fixed probability vector $\mathbf{t}$.

Proof: Because of the normalization condition $\Sigma_i p_{ij} = 1$, we have

$$\max_{ij} p_{ij} \leqslant 1 - (n-1)\delta \tag{1.41}$$

where

$$\delta = \min_{ij} p_{ij}$$

Consider the equation:

$$p_{ij}^{(t+1)} = \sum_k p_{ik}^{(t)} p_{kj} \tag{1.42}$$

Let

$$m_i^{(t)} = \min_j p_{ij}^{(t)}$$
$$M_i^{(t)} = \max_j p_{ij}^{(t)}$$

Then

$$m_i^{(t+1)} = \min_j \sum_k p_{ik}^{(t)} p_{kj}$$

$$m_i^{(t+1)} \geqslant m_i^{(t)} \min_j \sum_k p_{kj}$$

$$m_i^{(t+1)} \geqslant m_i^{(t)}$$

Thus the sequence $m_i^{(1)}, m_i^{(2)}, \ldots$ is non-decreasing, and by the same token, $M_i^{(1)}, M_i^{(2)}, \ldots$ is non-decreasing. Therefore, both sequences tend to a limit. We now prove these limits to be equal. Some terms in the equation $\Sigma_i(p_{ij}-p_{ik}) = 0$ for a given j and k will be such that $p_{ij} \geqslant p_{ik}$ and some may be such that $p_{ij} < p_{ik}$. Let Σ^+ indicate summation of the first set and Σ^- summation over the second set. Then the two sets may be arranged in this manner:

$$\sum_i{}^+(p_{ij}-p_{ik}) = \sum_i{}^+ p_{ij} - \sum_i{}^+ p_{ik} \tag{1.43}$$

$$\sum_i{}^+(p_{ij}-p_{ik}) = 1 - \sum_i{}^- p_{ij} - \sum_i{}^+ p_{ik} \tag{1.44}$$

Let s be the number of values of i in the first set,

$$\sum_i{}^+(p_{ij}-p_{ik}) \leqslant 1-(n-s)\delta - s\delta \tag{1.45}$$

$$\sum_i{}^+(p_{ij}-p_{ik}) \leqslant 1-n\delta$$

This leads to

$$M^{(t+1)} - m_l^{(t+1)} = \max_{jk} \sum_i p_{li}^{(t)}(p_{ij}-p_{ik}) \tag{1.46}$$

$$M^{(t+1)} - m_l^{(t+1)} \leqslant \max_{jk}\left[\sum_i{}^+ p_{li}^{(t)}(p_{ij}-p_{ik}) + \sum_i{}^- p_{li}^{(t)}(p_{ij}-p_{ik})\right] \tag{1.47}$$

$$M^{(t+1)} - m_l^{(t+1)} \leqslant \max_{jk}\left[\sum_i{}^+ M_l^{(t)}(p_{ij}-p_{ik}) + \sum_i{}^- m_l^{(t)}(p_{ij}-p_{ik})\right] \tag{1.48}$$

$$M^{(t+1)} - m_l^{(t+1)} \leqslant \max_{jk} \sum_i{}^+ (M_l^{(t)} - m_l^{(t)})(p_{ij}-p_{ik}) \tag{1.49}$$

$$M^{(t+1)} - m_l^{(t+1)} \leqslant (1-n\delta)(M_l^{(t)} - m_l^{(t)}) \tag{1.50}$$

We can now conclude that

$$M_l^{(t)} - m_l^{(t)} \leqslant (1-n\delta)^t \tag{1.51}$$

Therefore $M_l^{(t)}$ and $m_l^{(t)}$ approach the same non-zero constant as t increases. If the probability matrix is doubly stochastic and the vector $\mathbf{p}$ satisfies the equation $\mathbf{pP} = \mathbf{p}$ then each component of $\mathbf{p}$ is $1/n$.

This proof and further developments can be found in Moran (1986).

1.3.4 The eigenvalues and eigenvectors associated with a matrix

I pointed out in section 1.3.2 that the matrix $\mathbf{A}$ may be thought of as operating on vector $\mathbf{x}$ to change it to vector $\mathbf{y}$. That is, the matrix $\mathbf{A}$ can be thought of as a mapping of the points in space X of the n-component vector

x onto the points in the space Y of the *p*-component vector **y**. The situation
is very complicated if one considers all vectors in these spaces. However,
there are specific vectors that behave in a very simple way when operated
on by the matrix **A** that maps space X into itself. Such a vector, **x**, when
multiplied by a matrix becomes a multiple of itself, $\lambda \mathbf{x}$. It is called a
characteristic vector, latent vector or *proper vector* of the matrix. The more
frequent designation is the Deutsch–English hybrid *eigenvector*. The
German word *eigen* is translated as *peculiar, proper, special* or *specific*. This
describes the relationship of the eigenvector **x** to the matrix **A**. Eigenvectors
have a number of important properties among which is that they are
mutually orthogonal if **A** is symmetric.

Definition: An $m \times n$ matrix $\mathbf{A}(\lambda)$ whose elements are polynomials in λ is
called a λ matrix. $\mathbf{A}(\lambda)$ is said to be *singular* or *non-singular* according to
whether the determinant $|\mathbf{A}(\lambda)|$ is zero or not.

Definition: If **A** is a square matrix of order *n* and **x** is a non-trivial column
vector that satisfies the equation

$$\mathbf{A}\mathbf{x} = \lambda \mathbf{x}$$

then **x** is called a ***right*** *characteristic vector, latent vector* or *proper vector*
but more frequently the ***right*** *eigenvector* of **A** corresponding to λ. Non-
trivial means that at least one component of **x** is not zero.

If **A** is a square matrix of the order *n* and **x** is a non-trivial row vector that
satisfies the equation

$$\mathbf{x}\mathbf{A} = \lambda \mathbf{x}$$

then **x** is called a ***left*** *characteristic vector, latent vector* or *proper vector* but
more frequently the ***left*** *eigenvector* corresponding to λ. Since matrix
multiplication is not commutative, left and right eigenvectors are not
necessarily equal.

Definition: If **A** is a square matrix of order *n* then the λ matrix

$$[\mathbf{A} - \lambda \mathbf{I}_n]$$

is called the *characteristic matrix* of **A**. The determinant $|\mathbf{A} - \lambda \mathbf{I}_n|$ is
called the *characteristic determinant* of **A**. The expansion of this de-
terminant is a polynomial of degree *n* in λ that is called the *characteristic
function* of **A**. When set equal to zero this polynomial is called the

characteristic equation of **A**. The roots of this polynomial are called the *characteristic roots* of **A**. Other terms that appear in the literature are: *latent roots, proper roots,* and *secular values.* The most popular term in the current literature, however, is again the Deutsch–English hybrid *eigenvalues.* The set of all eigenvalues is called the *spectrum* of **A**. If no two eigenvalues are equal the spectrum is said to be *distinct.* If two or more eigenvalues are equal the spectrum is said to be *degenerate.*

Theorem 1.6: There is a non-trivial solution to the equation $\mathbf{A}\mathbf{x} = \lambda\mathbf{x}$ if and only if the determinant $|\mathbf{A} - \lambda\mathbf{I}_n|$ vanishes.

Proof: Setting the characteristic determinant equal to zero gives the characteristic equation $\mathbf{A}\mathbf{x} = \lambda\mathbf{x}$. This equation may be solved for each eigenvalue λ_l. Since **A** is a square matrix $\mathbf{A}\mathbf{x} = \lambda\mathbf{x}$ may be written as a set of simultaneous equations and solved by Cramérs rule.

Corollary: **A** has the eigenvector **x** if and only if λ is an eigenvalue of **A**.

Theorem 1.7: If **A** is symmetric (i.e. $a_{ij} = a_{ji}$) and if its eigenvalues are distinct then its eigenvectors are orthogonal.

Proof: Given the n distinct eigenvalues λ_l of **A**, then $[\mathbf{A} - \lambda\mathbf{I}_n]\mathbf{x} = \mathbf{O}$ may be written as a system of n simultaneous equations in n unknowns for the components of **x**. Let the components of **x**, which corresponds to the eigenvalue λ_l of **A**, be designated as x_{il}. Then, calling the elements of **A** a_{ij} we may write

$$\sum_j a_{ij} x_{jl} = \lambda_l x_{il} \tag{1.52}$$

Multiply both sides of equation (1.52) by x_{ik} and sum over i:

$$\sum_{i,j} x_{ik} a_{ij} x_{jl} = \lambda_l \sum_i x_{ik} x_{il} \tag{1.53}$$

These indices are only labels; therefore, we may write the same equation interchanging k and l:

$$\sum_{i,j} x_{il} a_{ij} x_{jk} = \lambda_k \sum_i x_{il} x_{ik} \tag{1.54}$$

Again, interchange i and j on the left of this equation:

$$\sum_{i,j} x_{jl} a_{ji} x_{ik} = \lambda_k \sum_i x_{ik} x_{il} \tag{1.55}$$

Since **A** is symmetric ($a_{ij} = a_{ji}$) the left-hand sides of equation (1.53) and equation (1.55) are equal, so that

$$(\lambda_l - \lambda_k) \sum_i x_{il} x_{ik} = 0 \tag{1.56}$$

Since the eigenvalues λ_l of **A** are distinct, then

$$\sum_i x_{il} x_{ik} = 0 \tag{1.57}$$

The expression $\Sigma_i x_{ik} x_{il}$ is the inner product of the eigenvectors $\mathbf{x}_k$ and $\mathbf{x}_l$, which vanishes when two vectors are orthogonal.

Since the eigenvectors of a symmetric matrix are orthogonal if the eigenvalues are distinct, they form the basis of an abstract Euclidean space with which all other vectors in that space can be expressed by linear combination. The axioms of Euclidean geometry are obeyed in this abstract space as they are in the space in which we live. This is another illustration of the power and generality of the axiomatic method of reasoning. If one takes a literal interpretation of Euclid's axioms this point would be missed. Generalizations of this kind are frequently made in mathematics and theoretical physics.

Borstnik & Hofacker (1985) have applied the eigenvectors of the mutation probability matrix of Dayhoff, Eck & Park (1972), which is approximately symmetric, to the construction of an abstract Euclidean property space of amino acids. The position of the amino acids and the distance between them is a measure of their functional equivalence as replacements in mutation (section 6.3.2).

The reader will be delighted (or appalled) to know that not only is there a vector and matrix algebra but also a differential and integral calculus of vectors and matrices (Bellman, 1970). Furthermore, mathematicians have more complicated arrays of numbers, called tensors, which also have an algebra and a calculus.

The algebra and calculus of matrices and tensors will find many applications in theoretical molecular biology because they enable one to make computations with large amounts of data that conserve all the knowledge in that data. Most quantities in molecular biology cannot be expressed by a single number without destroying knowledge through averaging.

1.3.5 *Markov chains and the random walk*

In Chapter 12 I shall apply matrix algebra to the construction of phylogenetic chains. Phylogenetic chains, which relate one protein to another by means of mutational steps, are examples of the general mathematical theory of the transitions between successive states of a system. The sequence of *Markov states* (or events) relating the transitions between the initial and final states of a system is known as a *Markov process* or Markov chain. In molecular biology we are concerned only with discrete and finite Markov chains (Kemeny, Snell & Knapp, 1976). In general, the probability of the final state is governed by the probabilities of a sequence of states preceding the final state. If there are k such states the Markov process is known as a kth-order Markov process. In information theory, where one is interested in the generation of a sequence of symbols, known as an alphabet, that generate a genetic message, a kth-order Markov process is known as a k-memory source. We write the conditional transition probability that state E_j will be occupied after the sequence E_{j1}, $E_{j2}, \ldots E_{jk}$ has occurred as

$$p_j(E_j \mid E_{j1}, E_{j2}, \ldots E_{jk})$$

Note the reverse order of the listing. This is standard notation in information theory.

The transition graphs for two kinds of Markov states are shown in Figures 1.2 and 1.3. As another example, consider the sequence of mutations $\text{Lys}^{\text{AAG}} \rightarrow \text{Asn}^{\text{AAC}} \rightarrow \text{Tyr}^{\text{UAC}} \rightarrow \text{Phe}^{\text{UUC}}$. The conditional transition probability for this sequence of mutations is written:

$$p(\text{UUC} \mid \text{AAG}, \text{AAC}, \text{UAC})$$

Since all three nucleotides have been changed this is called a three-memory source.

Markov processes are an example of the fact that the probability of an event is often strongly influenced by events occurring previously. In general, several previous events may affect the conditional transition probability. Two states are said to *communicate* if one can be reached from the other. If a state is such that once entered it cannot be exited, it is called an *absorbing state*. A state through which a system may move is called a *transient state*. A *reflecting* state is one that sends the system into another state without being occupied. In molecular biology the nucleotides in DNA or mRNA sequences and the codons in the genetic code may be regarded as Markov states. All codons communicate, even the stop codons, by means of a sequence of single base interchanges, that is, by a Markov chain. The stop codons are usually absorbing states, but they may be transient states (section 12.3).

The elements of the transition probability matrix of Markov states may change as the process proceeds from one step to the next. If the elements of the transition probability matrix do not change, the Markov process is called homogeneous or stationary. Bernoulli processes or Bernoulli trials in which succeeding events are independent are simple examples of stationary Markov processes. The sequences so produced are called Bernoulli sequences (Solomonoff, 1964). The sequence of outcomes of the toss of a coin is a Bernoulli sequence because there are only two outcomes, the trials are independent and the probability distribution is stationary. The sequence of the numbers generated by a sequence of throws of a die is not a Bernoulli sequence since there are six outcomes.

Among the examples of the Markov state process, for which we shall have use, is one known as a *random walk*. Suppose a man is standing on the midfield line of a football field. He flips a coin to determine which way to go: heads one way, tails the other. The total distance moved in one direction or the other is a random variable. A number of questions can be asked. For example, what is the probability of crossing a goal line? This is an example of a one-dimensional random walk. The scenario may be elaborated to more than one dimension. The man might be on the corner of two city streets. He has forgotten where his hotel is. It is very late at night and there is no one to ask. Here the Markov state system is the pair of coordinates giving the man's position on the city map. He goes one block at a time and at each corner he flips a coin twice, to determine which of the four ways to go. He starts out with the following code: HT = north, HH = south, TH = west, TT = east. All streets going in the north direction end at a river, which is an absorbing state. If he goes that far, he will fall in. There is a wall on the south end of the north–south streets, which is a reflecting state. Of course, if he does find his hotel he goes in and so the hotel is also an absorbing state. All other states are transient states. Suppose his wife is at the hotel and she goes out to look for him. She would need the probability distribution so she could go to the more probable street intersections first. He might forget his code between corners. This change in code is a mutation after which, for example, HH = north and HT = south. There are many other questions that could be asked and such problems may get quite complicated.

There are many applications of the random walk in science. Perhaps the most famous is the solution to the problem of Brownian motion. This was solved by Einstein (1905, 1906) and proved to be the final nail in the coffin of the continuous matter theory. After millennia of philosophical and scientific arguments, atoms did indeed exist!

To illustrate a Markov chain in molecular biology, let us begin with an initial probability matrix vector $\mathbf{p}(0)$. As we learned in section 1.3.3, if the transition probability matrix, $\mathbf{P}$, does not change between steps the process is stationary and the subsequent probability vector after t steps, $\mathbf{p}_t$, is found by equation (1.58):

$$\mathbf{p}(t) = \mathbf{p}(0)\mathbf{P}^t \tag{1.58}$$

For illustration I consider again the simplest case, where the single base interchanges occur only on one nucleotide. Let us choose transitions of the third nucleotide of the codons UGA, UGC, UGU and UGG, listed as 1, 2, 3, 4. Since there is only one transition/transversion between these codons this is a one-memory source or first-order Markov system. UGA is an absorbing state because it is a termination codon, therefore any transition to that state cannot lead to another. The transition graph of the Markov process is illustrated in Figure 1.2; we shall take the transition probability α to be $1/3$. Suppose the system is at first in the Cys codon UGC, then

$$\mathbf{p}(0) = [0 \quad 1 \quad 0 \quad 0] \tag{1.59}$$

The state i is absorbing if and only if the ith **row** of the transition probability matrix $\mathbf{P}$ has a one on the main diagonal and zero for all other elements. Using equation 1.35 $\mathbf{p}(1)$ is calculated as follows:

$$\mathbf{p}(1) = \mathbf{p}(0)\mathbf{P} = [0 \quad 1 \quad 0 \quad 0] \begin{bmatrix} 1 & 0 & 0 & 0 \\ 1/3 & 0 & 1/3 & 1/3 \\ 1/3 & 1/3 & 0 & 1/3 \\ 1/3 & 1/3 & 1/3 & 0 \end{bmatrix}$$

$$= [1/3 \quad 0 \quad 1/3 \quad 1/3] \tag{1.60}$$

The first row of the matrix says that the transition probability of UGA to any other codon is zero. A transition must take place and the transition probabilities between the other codons and to UGA are all equal to $1/3$.

All powers of $\mathbf{P}$ have zero elements. It is therefore not a regular stochastic matrix and the Perron–Frobenius theorem does not apply.

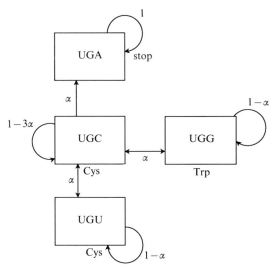

Figure 1.2 Transition graph of codons in a Markov process. The transitions shown are limited to (A|C), (U|C), (G|C). The flow to the state UGA is unidirectional since this is an absorbing state.

There is no fixed vector **t** and the matrix elements do not approach equality. Nevertheless, we may calculate the progress of the system as far as we wish by this procedure. Therefore at mutation step two we find:

$$\mathbf{p}(2) = \mathbf{p}(0)\mathbf{P}^2 = [0 \quad 1 \quad 0 \quad 0] \begin{bmatrix} 1 & 0 & 0 & 0 \\ 5/9 & 2/9 & 1/9 & 1/9 \\ 5/9 & 1/9 & 2/9 & 1/9 \\ 5/9 & 1/9 & 1/9 & 2/9 \end{bmatrix}$$

$$= [5/9 \quad 2/9 \quad 1/9 \quad 1/9] \tag{1.61}$$

At mutation step three we have:

$$\mathbf{p}(3) = \mathbf{p}(0)\mathbf{P}^3 = [0 \quad 1 \quad 0 \quad 0] \begin{bmatrix} 1 & 0 & 0 & 0 \\ 19/27 & 2/27 & 3/27 & 3/27 \\ 19/27 & 3/27 & 2/27 & 3/27 \\ 19/27 & 3/27 & 3/27 & 2/27 \end{bmatrix}$$

$$= [19/27 \quad 2/27 \quad 3/27 \quad 3/27] \tag{1.62}$$

If we square $\mathbf{P}^3$ we find the following equation at step six:

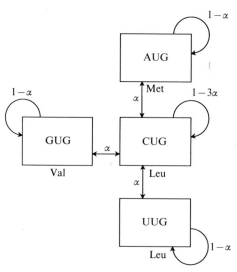

Figure 1.3 Transition graph of codons in a Markov process. The transitions shown are limited to $(A|C)$, $(G|C)$, $(U|C)$.

$$\mathbf{p}(6) = \mathbf{p}(0)\mathbf{P}^6 = \begin{bmatrix} 0 & 1 & 0 & 0 \end{bmatrix} \begin{bmatrix} 1 & 0 & 0 & 0 \\ 665/729 & 22/729 & 21/729 & 21/729 \\ 665/729 & 21/729 & 22/729 & 21/729 \\ 665/729 & 21/729 & 21/729 & 22/729 \end{bmatrix}$$

$$= \begin{bmatrix} 665/729 & 22/729 & 21/729 & 21/729 \end{bmatrix} \tag{1.63}$$

Since the matrix element p_{i1} represents transitions to an absorbing state the transition probabilities in the first row become larger at each increase in power of the transition probability matrix. This absorbing state (UGA) will eventually become the only occupied state, independently of the initial probability vector $\mathbf{p}(0)$. This is shown by the progression of the components of the probability vectors $\mathbf{p}(0)$, $\mathbf{p}(1)$, $\mathbf{p}(2)$, $\mathbf{p}(3)$ and $\mathbf{p}(6)$. It is noteworthy that at the third mutation step 0.296 of the codons have still not mutated to the absorbing state UGA. Even at the sixth mutation step 0.088 of the codons have still not mutated to UGA. Intuition and the principle of maximum parsimony would predict that, in this rather artificial example, progress to the absorbing state would be much faster. It is always important to follow the mathematics rather than intuition or *ad hoc* procedures.

What is the probability *vector* of the codons as time passes if there are no stop codons? Let us illustrate this by considering another case in which only single base interchanges are allowed. This is illustrated in Figure 1.3 for the codons CUG, AUG, UUG and GUG. Since there is a change in only one nucleotide, this is also a one-memory source or a first-order Markov system. Let us now suppose that mutations may occur with probability α. The transition probability λ matrix is $\mathbf{P}(\alpha)$, where α is the probability of a mutation:

$$\mathbf{P}(\alpha) = \begin{bmatrix} 1-3\alpha & \alpha & \alpha & \alpha \\ \alpha & 1-3\alpha & \alpha & \alpha \\ \alpha & \alpha & 1-3\alpha & \alpha \\ \alpha & \alpha & \alpha & 1-3\alpha \end{bmatrix} \tag{1.64}$$

Calculate $\mathbf{p}(1) = \mathbf{p}(0)\mathbf{P}$. Then square $\mathbf{P}$ and calculate $\mathbf{p}(2)$ to illustrate the effect of mutations at the second step.

$$\mathbf{P}^2 = \begin{bmatrix} 1-6\alpha+12\alpha^2 & 2\alpha(1-2\alpha) & 2\alpha(1-2\alpha) & 2\alpha(1-2\alpha) \\ 2\alpha(1-2\alpha) & 1-6\alpha+12\alpha^2 & 2\alpha(1-2\alpha) & 2\alpha(1-2\alpha) \\ 2\alpha(1-2\alpha) & 2\alpha(1-2\alpha) & 1-6\alpha+12\alpha^2 & 2\alpha(1-2\alpha) \\ 2\alpha(1-2\alpha) & 2\alpha(1-2\alpha) & 2\alpha(1-2\alpha) & 1-6\alpha+12\alpha^2 \end{bmatrix}$$
$$\tag{1.65}$$

This produces the sequence of probability vectors:

$$\mathbf{p}(0) = [1 \quad 0 \quad 0 \quad 0] \tag{1.66}$$

$$\mathbf{p}(1) = [1-3\alpha \quad \alpha \quad \alpha \quad \alpha] \tag{1.67}$$

$$\mathbf{p}(2) = [1-6\alpha+12\alpha^2 \quad 2\alpha-4\alpha^2 \quad 2\alpha-4\alpha^2 \quad 2\alpha-4\alpha^2] \tag{1.68}$$

In this case, if $\alpha \neq 0$, there are no elements of $\mathbf{P}$ that are zero. $\mathbf{P}$ is a regular matrix and the Perron–Frobenius theorem applies. The Perron–Frobenius theorem tells us that the components of $\mathbf{p}(t)$ approach $1/n$ as t increases without limit. This means that all influence of the original condition has been lost. We now calculate the fixed probability vector. Let the components of $\mathbf{t}$ be $\{x, y, z, (1-x-y-z)\}$. We form the product $\mathbf{tP} = \mathbf{t}$. This will give us three equations in three unknowns. The first equation is representative of the others.

$$x = (1-3\alpha)x + \alpha y + \alpha z + \alpha(1-x-y-z) \tag{1.69}$$

$$0 = -4\alpha x + \alpha \tag{1.70}$$

$$x = 1/4 \tag{1.71}$$

If $\alpha \neq 0$ it may be cancelled out. Otherwise, since division by zero is not permitted, x, y, and z would be indeterminate. It is easy to see that

$$x = 1/4, y = 1/4, z = 1/4 \tag{1.72}$$

Thus the matrix **T** (see Theorem 1.5) has all elements equal to $1/4$, independently of the value of α if and only if $\alpha \neq 0$. The Markov process is stationary and as the process progresses eventually all influence of the original state is lost.

The Perron–Frobenius theorem is not intuitive, so that if we attempt to find an equilibrium fixed probability vector by raising **P** to higher powers, we are in for a formidable amount of computation. However, the mathematics leads us to the correct result with unseemly ease and, in particular in the last step, reminds us that α must not be equal to zero.

Notice that as a result of the second step, in equation (1.68) both positive and negative terms in α and α^2 appear in the components of the probability vectors. This illustrates the fact that, in proceeding from the original condition to any subsequent step, the mathematics takes into consideration the probabilities of all possible routes to the final step in evaluating the components of the probability vector of the final state (Eves, 1966). The terms in α and α^2 and, in higher steps, the higher powers of α represent the flow into and out of each state by multiple back mutations in a very natural manner without the need to introduce *ad hoc* considerations. This is applied to the discussion of phylogenetic trees in section 12.2.6.

The formalism of the Markov process can be incorporated, by straightforward procedures, into computer programs handling a large number of states with as many transition probabilities between mutation steps as desired. The best estimate of the progress of evolution will be made by constructing a transition probability matrix that reflects the realities of molecular biology. Remember that the real system will be in a definite state. The components of the probability vector are only probabilities that have the interpretation discussed in this chapter.

1.3.6 Ergodic and regular sequences

A sequence generated by a regular matrix is called a regular sequence. A Markov system is the source of a Markov chain or sequence of symbols. A Markov chain or sequence is called *ergodic* if the transition probability matrix is such that the system can progress from any state to any other by a finite number of steps. That is, there can be no absorbing states in this

Markov system. Regular sequences are always ergodic but ergodic chains are not always regular. A Markov system that is the source of an ergodic Markov chain or sequence is called an ergodic source. Ergodic sequences are of great significance in communication theory and by the same token also in molecular biology. Every ergodic sequence, except for a set of probability zero, produced by an ergodic source has the same statistical properties. In the simplest terms the ergodic property means probabilistic homogeneity. There is an enormous mathematical literature on ergodic theory (Petersen, 1983). In generating a protein sequence, most of the time one does not have one of the three termination codons that are absorbing states. Protein sequences are often very long and for most practical problems we can consider protein sequences to be ergodic. After all, every message has an end.

This discussion of the theory of probability is the minimum one should know to follow the arguments in the rest of this book. It is not intended to be complete. The reader is encouraged to consult the various excellent texts (Lindley, 1965; Feller, 1968, 1971; Moran, 1986; Lancaster & Tismenetsky, 1985) and papers in the field. I have found the work of E. T. Jaynes (1957a, 1968, 1979) very helpful.

Exercises

1. Make a Venn diagram (Venn, 1888) showing the sequence of sets and the conditional probabilities used by the prophet Samuel, quoted in the epigram, in the selection process by which Saul was selected as king of Israel.
2. Show that Axiom 2 is reasonable by breaking down the events, A, in equation (1.1) into several mutually exclusive subevents, each with a number of occurrences, a_i, which sum to a.
3. Show that Axiom 3 is reasonable by showing that in an experiment similar to the experiment that leads to equation (1.1):

    $$p(A|B) \approx N_{A \cap B}/N_B = (N_{A \cap B}/N)/(N_B/N)$$

    where N_A, N_B indicate the numbers of occurrences of events A, B and N is as large as one chooses.
4. Use equation (1.7) to calculate the probability of all 'heads' in 10, 11, 12, 13, 14, 15, 16, 17, 18, 19 and 20 tosses of the coin in the fable in section 1.1.2.
5. The slingers in ancient armies were a formidable force. A modern professional baseball pitcher can throw a baseball accurately at a speed of 144 km/hour. The slingers, with the advantage of the longer swing like that of a modern hammer thrower, could certainly propel a stone somewhat faster. A stone, being more dense than a baseball, would have a devastating effect even on an armored soldier. Before David went out to

meet Goliath, the occasion not being one for peaceful coexistence, he selected five smooth stones from the brook. Presuming that one hit would incapacitate Goliath, what probability per throw did David need to have at least a 95 % probability of one hit in five tries? Did David know what he was doing in picking only five stones? David succeeded on the first trial, but what was the most probable number of stones thrown?

Note: the same problem confronted Doc Holliday and Wyatt, Virgil and Morgan Earp in a gun fight with Billy Clairborne, Tom and Frank McLaury, and Billy and Ike Clanton. The dispute took place at the OK Coral in Tombstone, Arizona Territory, USA on October 26, 1881. The famous encounter lasted less than 30 seconds, allowing little time to reload a Colt .45. Are six bullets enough for a 95 % probability of a hit?

6. Random walk problems are related to the problem of the gambler's ruin. A crime was committed in Minsk. The Czar's police exonerated all suspects but two, Iranov and Bouljanov. Let the probability that each was guilty be 1/2. This posed several problems. If the Czar's police continued to work on the case they would waste time that could be better spent otherwise. If the charges against both were dropped, a guilty man would escape punishment. Police Chief Kopalski and Czar Nicholas would appear impotent. If both were sent to Siberia, then it would be certain that an innocent man was being punished. This would make the Czar appear to be a heartless tyrant. The lawyers for the two suspects quoted both scripture (1 Samuel 10:20–21; Psalms 22:18; Jonah 1:7; Matthew 27:35; Mark 15:24; Luke 23:34; John 19:23; Acts 1:26) and Homer's *Iliad* and *Odyssey* for authority. They proposed that Fortuna be the one to decide as in the precedents established in these other important cases. The lawyers proposed that each man be given 1000 rubles. They would bet one ruble on each flip of a coin 100 times a day. The first to lose all his money would be the guilty party. This seemed reasonable to Police Chief Kopalski, who supposed it would take only a few days. The approval of Czar Nicholas, who was glad to be relieved of further responsibility, was obtained. So it was done. Find the expected number of trials and the expected time for Fortuna to decide which man was to go to Siberia. Did the guilty man escape punishment? Try some unequal probabilities, for example, let the probability that Iranov was guilty be 1/10. How long does it take Fortuna to decide the guilty man? See Feller (1968), Volume I, 3rd edition, Chapter XIV, section 3, pp. 344–9. This is one of many examples of the fact that intuition is often a poor guide and one should stick to the mathematics.

7. Given the matrices

$$A = \begin{bmatrix} 1 & 2 & 3 \\ 0 & -1 & 1 \\ 2 & 3 & 0 \end{bmatrix} \text{ and } B = \begin{bmatrix} 3 & 1 & 0 \\ 1 & -1 & 2 \\ 0 & 2 & 1 \end{bmatrix}$$

show that

$$AB = \begin{bmatrix} 5 & 5 & 7 \\ -1 & 3 & -1 \\ 9 & -1 & 6 \end{bmatrix} \text{ and that } BA = \begin{bmatrix} 3 & 5 & 10 \\ 5 & 9 & 2 \\ 2 & 1 & 2 \end{bmatrix}$$

8. Perform the following matrix multiplications:

$$\begin{bmatrix} 0 & 1 & 2 & 3 \\ 3 & 2 & 1 & 0 \end{bmatrix} \begin{bmatrix} 0 & 3 \\ 1 & 2 \\ 2 & 1 \\ 3 & 0 \end{bmatrix} = \begin{bmatrix} 14 & 4 \\ 4 & 14 \end{bmatrix}$$

$$\begin{bmatrix} 0 & 3 \\ 1 & 2 \\ 2 & 1 \\ 3 & 0 \end{bmatrix} \begin{bmatrix} 0 & 1 & 2 & 3 \\ 3 & 2 & 1 & 0 \end{bmatrix} = \begin{bmatrix} 9 & 6 & 3 & 0 \\ 6 & 5 & 4 & 3 \\ 3 & 4 & 5 & 6 \\ 0 & 3 & 6 & 9 \end{bmatrix}$$

9. Extend the Markov state diagram in Figure 1.2 to include all the transitions you can get on the paper. Notice how quickly this diagram gets busy. One gets lost very soon without the mathematical methods developed in this chapter.

10. Consider the fable in section 1.3.5 in which the man is lost in a city and tries to find his hotel. Lay out the grid of the city and locate the man, his hotel, a river on the north side and a wall on the south side of town. Choose whatever transition probability matrix you wish for his choice of direction at each intersection. The elements are not necessarily all equal. Will the man choose the path of maximum parsimony on his way to the hotel?

11. Matrices may be added and subtracted by adding or subtracting their corresponding elements. Similarly an α matrix may be expanded in a series of powers of α. Expand the α matrix in equation 1.65 in powers of α.

12. Mark five points distributed about on a paper and label them as Markov states (Eves, 1966). Draw lines between nearest neighbors to serve as paths between the Markov states. Construct the transition matrix that reflects the one-step transitions in this diagram. Record 1 for the matrix element if the transition can be made in one step and 0 if more than one step is necessary. Raise this matrix to the second power. This reflects the transitions between states in two steps. Count the number of steps between each pair of states and compare this number with the elements of the transition matrix in the second power. One could find the matrix elements of, say the eighth step, by counting the number of routes but this would be tedious and subject to blunders. Raising the transition matrix to the eighth power in a sequence of square, fourth power and eighth power accomplishes this in a straightforward manner. Computer programs exist for personal computers that can accomplish the task of raising transition matrices to any integral power.

13. Apply the methods in Exercise 12 to finding all the routes in the sequence of mutations $Lys^{AAG} \rightarrow Asn^{AAC} \rightarrow Tyr^{UAC} \rightarrow Phe^{UUC}$.

2

The role of entropy: a quantitative measure of information, uncertainty and complexity

Ah, but my Computations, People say,
Reduced the year to better reckoning? – Nay,
T'was only striking from the Calendar
Unborn tomorrow and dead Yesterday

Omar Kháyyám, The Rubáiyát

2.1 The concepts of information and uncertainty

2.1.1 Information, meaning, knowledge and uncertainty

The messages conveyed by sequences of symbols sent through a communication system have meaning (otherwise why are we sending them?). By the same token, the sequences of nucleotides or amino acids that carry a genetic message have specificity (otherwise how does the organism live?). Natural languages are written with a set of elementary events or symbols called an alphabet. Sequences of the symbols of the alphabet are called 'words', and are listed in a dictionary in correspondence with other words. These correspondences are called definitions. We make choices from this dictionary to form sequences of words in order to convey to a receiver knowledge, meaning or information. The reception of the sequence at the receiver removes uncertainty because of the information that it conveys. So, then, one sees that information and uncertainty are two aspects of the same thing.

One must know the language in which a word is being used: 'O singe fort' may be read in French or German with entirely different meanings. The reader may find it amusing to list all the words in languages that he knows that are spelled the same but have different meanings. For example, a German-speaking visitor to the United States might have all his suspicions about America confirmed when he finds that there is a *Gift* shop in every airport, hotel and shopping center. The message 'mayday, mayday' may mean a distress signal, a Bolshevik holiday or a party for children in the spring, all depending on the context. Entirely different sequences of words may have exactly the same meaning. For example: 'I am his mother's brother'; 'He is my sister's son'. Usually the meaning is dependent on word order, but not always. The reader may wish to find how many ways there are to rearrange the words in a line from Thomas Gray's 'Elegy Written in a Country Churchyard' that preserve the meaning: 'The ploughman homeward plods his weary way'. I have written down 20 ways, some of which fit the rhythm of the poem – but I have no doubt that the poet picked the best of the ensemble.

The examples cited above show that the meaning of a sequence of symbols in natural languages is subject to the arbitrary agreement between source and receiver. This question of meaning is best left to philosophers, linguists and semanticists (Nauta, 1972). The communications engineer, in designing his equipment, need not concern himself with the meaning of the sequence of symbols. Indeed, no humans may be involved. The message may be, for example, a computer communicating data to another machine or a spacecraft sending data from which pictures of other planets will be made.

A great deal of arbitrariness is also found in the sequences that carry specificity in a protein, as we shall see in Chapter 6. In fact, as I shall show in Chapter 9, there are 9.737×10^{93} iso-1-cytochrome c sequences that differ in at least one amino acid, each carrying the same specificity. Like a good communications system, the genetic information storage and transmission apparatus is independent of the specificity of the genetic messages. It deals with specificity of the genetic messages only through the information those messages carry. Nature uses the same system, with a few codon changes in the case of mitochondria, for all the enormous variety of organisms that now exist, have existed or will exist in the future.

We shall find, however, that meaning is not entirely unrelated to information, uncertainty and specificity. Many messages that carry a great deal of meaning also have a large information content. For that and other reasons, information theory has many applications in molecular biology. In each of the numerous kinds of communication system, something we call information is being sent by the source and uncertainty is being removed at the receiver. If we can't deal directly with the meaning of such messages perhaps we can find a measure, which will be useful in all cases, of the amount of information in the message. Not all sequences are messages that convey meaning. The meaning of some messages is more important than that of others. This has prompted some authors to break up the information in the genetic message into bound information, meaningful information, structural information, functional information, valued information etc. (Johnson, 1970; Eigen, 1971; Ryan, 1972, 1980; Wicken, 1984). This is simply *ad hoc* and should be avoided.

There is a great deal of misunderstanding about the relation of the semantic aspects of a message and the problems of the design and operation of communication systems, especially by authors who attempt to use information theory and coding theory or that portion which suits them. I can't do better to explain this point than to quote the second paragraph of Shannon's seminal paper (Shannon, 1948; Shannon & Weaver, 1949): 'The fundamental problem of communication is that of reproducing at one point either exactly or approximately a message selected at another point. Frequently the messages have *meaning*; that is they refer to or are correlated according to some system with certain physical or conceptual entities. These semantic aspects of communication are irrelevant to the engineering problem. The significant aspect is that the actual message is one *selected from a set* of possible messages. The system must be designed to operate for each possible selection, not just the one which will actually be chosen since this is unknown at the time of design'.

One need only change the one word *engineering* to *biological* to make this paragraph apply to the application of information theory and coding theory in molecular biology. The messages of interest in molecular biology are those in the genome. There is a clear isomorphism between electronic communication systems and the genetic information system because in molecular biology the genetic information system must be capable of handling *all* genetic messages of *all* organisms extinct and living as well as those not yet evolved. It is therefore independent of any particular message. In Chapter 8 I shall show that the evidence indicates that life originated on Earth between 4.0×10^8 and 3.8×10^8 years ago. The genetic information system, which must have existed at that time, has well served *all* organisms that have evolved since. Although we cannot deal with the semantic – or in molecular biology, the specificity – aspects of the genetic message, we can establish a measure of the amount of information in an ensemble of messages. I shall show how Shannon (1948) accomplished this in the next section.

2.1.2 *Establishment of a quantitative measure of information*

Signals are no longer sent through communications systems in the continuous analogue form. The continuous signal is sampled and converted to a digitized code in binary numbers by a process known as pulse code modulation. By applying the theorems of information theory and coding theory, it is straightforward to recover the original signal from a digitized and coded signal in the presence of noise but difficult or impossible to do so from an analogue signal. The genetic information processing system is digitized and has been so for at least 3.8×10^9 years (Schidlowski, 1988)!

In the first approach to the question of establishing a measure of information, I shall follow Shannon's original paper (Shannon, 1948). A finite stationary Markov process may be chosen, without loss of generality, as the model of the source generating an ergodic chain or sequence of events associated with a finite alphabet of n elements or letters. With every step of the Markov chain there is associated uncertainty. The 'uncertainty' of an event is small when its probability of occurrence is large. For example our uncertainty of rolling a 7 is less than of rolling a 2 or 12 with a pair of dice. Our uncertainty of the outcome at the receiver is greatest when the probabilities of all events are equal and least when all but one is very small or zero. The uncertainty is also greatest when there are many outcomes in the probability space and least when there are few outcomes. By the same token, there is more choice at the source if there are many states than if there are few. Consequently, if there is a unique measure $H(p_i)$ of uncertainty and information, it is a function of the components of the probability vector $\mathbf{p}$ in the probability space Ω.

The first condition we wish to require is that when all $p_i = 1/n$ the measure has its maximum value. By the same token, the measure must vanish if all but one of the p_i are zero. This means, of course, no choice for the source and no uncertainty at the receiver. A second requirement is that of continuity. Small changes in the components of $\mathbf{p}$ must result in small changes in the measure $H(p_i)$. The third condition to be expected of a measure of uncertainty or information is that if the events in the chain are composed of two or more contingent events the measure of uncertainty is the weighted sum of the expectation values of the uncertainty of the events of which it is composed. This is a condition of conservation. As the message moves through the communication system and is coded and decoded in alphabets of different numbers of letters, the measure of uncertainty or information must be conserved. For example, suppose a message is first written in a 26-letter alphabet and then coded in, say, the four-letter (dot, dash, letter space, word space) Morse code, and again written in a 26-letter alphabet at the receiver. There must be just as much information in the message expressed in the Morse code as in the alphabetical form.

It is always essential in mathematics to prove an existence theorem for the concept under consideration. We must first prove that the measure of which we are speaking exists at all. If no such measure exists then we are fooling ourselves if we imagine that one does exist. To summarize the requirements in the paragraph above, if a measure, $H(p_1, p_2, \dots p_n)$, of the information generated by the stationary Markov process at the source or received by the receiver exists, then it is reasonable to require these properties

    (1) H must be continuous in the p_i

    (2) If all the n probabilities are equal, $p_i = 1/n$, then H must be a monotonic increasing function of n.

    (3) If an event is composed of two or more successive conditional events, the overall value of H must be the sum of the expectation values of the individual values, H_i.

One who takes the skeptic's role, and assumes that the remarks in the paragraphs above will result only in arm-waving platitudes, greatly underestimates the cleverness of mathematicians. When put in mathematical language, the three requirements have real teeth, as we shall soon see.

Theorem 2.1: The only function satisfying the conditions (1), (2) and (3) has the form

$$H = -K \sum_i p_i \log p_i \qquad (2.1)$$

where K is a positive constant.

There are several proofs of this theorem in the literature in addition to the one given by Shannon (1948), namely, those by Fadiev (1956) and Khinchin (1957). The proof by Fadiev may be found in English translation in Feinstein (1958). The proof given below is adapted from that given by Khinchin (1957). Additional steps and explanation are provided in order to make the argument more easily understood.

Proof: First let us establish $H = K \log n$ for the case where all $p_i = 1/n$. Let

$H(1/n, 1/n, ..., 1/n) = L(n)$. Since by condition (2) adding a Markov state with probability zero does not change the value of H, therefore

$$L(n) = H\left(\frac{1}{n}, \frac{1}{n}, ..., \frac{1}{n}, 0\right) \leqslant H\left[\frac{1}{n+1}, \frac{1}{n+1}, ..., \frac{1}{n+1}\right] = L(n+1)$$

(2.2)

$L(n)$ is a nondecreasing function of n. Let us consider m mutually independent probability spaces $\Omega_1, \Omega_2, ..., \Omega_k, ..., \Omega_m$ each of which contains r equally probable events. Then we may write

$$H(\Omega_k) = H\left(\frac{1}{r}, \frac{1}{r}, ..., \frac{1}{r}\right) = L(r)$$

(2.3)

where $1 \leqslant k \leqslant m$. By condition (3) and the independence of the probability spaces Ω_k we may write the following equation:

$$H(\Omega_1, \Omega_2, ..., \Omega_k, ..., \Omega_m) = \sum_{k=1}^{m} H(\Omega_k) = mL(r)$$

(2.4)

since each probability space contains r equally probable events. The product space (section 1.2.2) of all the m mutually independent probability spaces $\Omega_1 \times \Omega_2 \times ... \times \Omega_k \times ... \times \Omega_m$ consists of r^m equally probable events so that its entropy is $L(r^m)$. Therefore we have

$$L(r^m) = mL(r)$$

(2.5)

By the same token for any other pair of positive integers we may write

$$L(s^n) = nL(s)$$

(2.6)

We may choose the numbers r, s and n arbitrarily, but let us choose m so that it satisfies the following inequalities:

$$r^m \leqslant s^n \leqslant r^{m+1}$$

(2.7)

We may take the logarithm of each term and the inequalities remain as follows:

$$m \log r \leqslant n \log s \leqslant (m+1) \log r$$

(2.8)

We may divide through each term by $n \log r$:

$$\frac{m}{n} \leqslant \frac{\log s}{\log r} \leqslant \frac{m}{n} + \frac{1}{n}$$

(2.9)

From the monotonicity of entropy and from inequalities (2.7):

$$L(r^m) \leqslant L(s^n) \leqslant 2L(r^{m+1}) \tag{2.10}$$

Consequently, remembering equations (2.5) and (2.6):

$$\frac{m}{n} \leqslant \frac{L(s)}{L(r)} \leqslant \frac{m}{n} + \frac{1}{n} \tag{2.11}$$

Subtracting (2.9) from (2.11) we may conclude that

$$\left| \frac{L(s)}{L(r)} - \frac{\log s}{\log r} \right| \leqslant \frac{1}{n} \tag{2.12}$$

Because the left-hand side of this inequality is independent of m and because n can be chosen arbitrarily large on the right-hand side, the functions between the absolute value bars may be squeezed as small as one wishes. We may conclude that

$$\frac{L(s)}{L(r)} = \frac{\log s}{\log r} \tag{2.13}$$

This means that these functions are proportional and so we have proved the first part of the theorem where, because of the monotonicity condition (2), $K \geqslant 1$:

$$L(n) = K \log n \tag{2.14}$$

Let us now consider the more general case where the probabilities p_i are rational numbers. Then the p_i are formed from the positive integers g_i chosen such that

$$p_i = \frac{g_i}{g} \tag{2.15}$$

where

$$g = \sum_{i=1}^{n} g_i \tag{2.16}$$

Consider a finite probability space Ω_A consisting of n events A_i with probabilities p_i. We must now define the entropy of this probability space. Consider a second probability space Ω_B, which is dependent on Ω_A. Let Ω_B be composed of g events that we shall divide into n groups so that group i

contains g_i events. Because of the dependency of Ω_B on Ω_A, if the event A_i in Ω_A occurred, then all the g_i events in the ith group of Ω_B have the same probability $1/g_i$ and all the other groups have zero probability. For that reason, given any event A_i in Ω_A, Ω_B becomes a system of g_i equally probable events. Therefore the conditional entropy $H_i(\Omega_B)$ may be written, according to equation (2.14), as

$$H_i(\Omega_B) = H(1/g_i, 1/g_i, \ldots, 1/g_i) = K \log g_i \tag{2.17}$$

We may take the expectation value of $H_i(\Omega_B)$ as follows:

$$H_A(\Omega_B) = \sum_{i=1}^{n} p_i H_i(\Omega_B) = K \sum_{i=1}^{n} p_i \log g_i \tag{2.18}$$

We can substitute the g_i from equation (2.15):

$$H_A(\Omega_B) = K \sum_{i=1}^{n} p_i \log p_i + K \log g \tag{2.19}$$

The function we are looking for appears on the right-hand side of equation (2.19). Now consider the product probability space $\Omega_A \times \Omega_B$, which consists of the pairs of events (A_i, B_l), where $1 \leqslant i \leqslant n$ and $1 \leqslant l \leqslant g$. These events have a non-vanishing probability only if B_l belongs to the ith group. Thus the number of events for a given i is g_i, and the total number of events with a non-vanishing probability is g, from equation (2.16). Therefore the non-vanishing probability of each event in $\Omega_A \times \Omega_B$, from equation (2.15), is

$$\frac{p_i}{g_i} = \frac{1}{g} \tag{2.20}$$

Thus $\Omega_A \times \Omega_B$ consists of g equally probable events and by equation (2.14) the entropy is

$$H(\Omega_A \times \Omega_B) = K \log g \tag{2.21}$$

Because of the composition property (3) we may write

$$H(\Omega_A \times \Omega_B) = H(\Omega_A) + H_A(\Omega_B) \tag{2.22}$$

Substituting from equation (2.21) and (2.19) we find that $K \log g$ drops out and we have

$$H(\Omega_A) = -K \sum_{i=1}^{n} p_i \log p_i \tag{2.23}$$

Finally we can extend the validity of equation (2.23) to all real numbers for p_i because the expression on the right-hand side of the equation is continuous and because all real numbers can be approximated as closely as we please by rational numbers. This completes the proof of Theorem 2.1.

In order to measure the entropy in terms of bits, that is, the answer to a yes or no question where each alternative is equally probable, we take the logarithm in equation (2.23) to the base 2 and set K equal to 1. The expression in equation (2.23) now takes the form

$$H(p_i) = -\sum_i p_i \log_2 p_i \tag{2.24}$$

Thus $-\log_2 p_i$, by the definition given in Chapter 1, is a random variable and $H(p_i)$ is the expectation value $\langle -\log_2 p_i \rangle$ of the random variable. As p_i approaches zero, $\log_2 p_i$ approaches minus infinity. The limit of the product is zero; however, it is customary to take $0 \log 0 = 0$. (See Exercise 4.) In addition, $\log 1 = 0$. Logarithms to the base 2 can be calculated by the use of many pocket calculators using the following formula:

$$\log_2 y = \frac{\log_e y}{\log_e 2} \tag{2.25}$$

2.1.3 The entropy of probability distributions

The expression in equation (2.24) is of the same form as the function (S) for entropy in statistical mechanics due to Ludwig Boltzmann (1872). It is convenient and customary, in the theory of probability, to call $H(p_i)$ the entropy of the probability distribution. Often in the literature equation (2.24) is called the Shannon entropy to distinguish it from the Maxwell–Boltzmann–Gibbs entropy of statistical mechanics.

The Maxwell–Boltzmann–Gibbs entropy is derived by the following argument. Suppose we have N atoms in N_i infinitesimal cells in the six-dimensional space defined by position and momentum vectors. This is known in theoretical physics as a *phase space* which is simply a name that theoretical physicists use for this probability space Ω. Let W be the number of ways in which the N atoms can be arranged in the N_i cells. S is defined as follows:

$$S = \log_e W \tag{2.26}$$

From equation (1.7) we have

$$W = \frac{N!}{\prod_i N_i!} \tag{2.27}$$

$$S = \log_e W = \log_e N! - \prod_i \log_e N_i! \tag{2.28}$$

Stirling's approximation for the factorial is

$$N! \cong N^N e^{-n} (2\pi N)^{\frac{1}{2}} \tag{2.29}$$

Using the law of large numbers (Feller, 1968, equation 1.1), let $p_i = N_i/N$, then

$$S = \log_e W = - \sum_i p_i \log_e p_i \tag{2.30}$$

This expression, first given by Planck in this form (Planck, 1906) applies only in the case of classical mechanics. In a few lines of algebra the Maxwell–Boltzmann probability distribution for the kinetic energy of atoms in a perfect gas is derived from (2.30). This derivation of the distribution is far simpler and more plausible than the derivation that was found by Maxwell. In quantum mechanics there are different expressions for entropy and the reader is referred to publications by Jaynes (1963), Wehrl (1978) and others.

Brillouin (1962) finds the expression for entropy in information theory with an argument by Boltzmann parallel to that just given. Let N be the total number of sites in a long genetic message in DNA. Let N_i be the number of codons of the ith kind. Let one and only one codon occupy a given site. (This is known in physics as Fermi–Dirac statistics, which is followed by particles such as electrons that have $1/2$ integral spin. The other kind of statistics allows multiple occupation of each site and is known as Bose–Einstein statistics. Photons and α-particles that have integral spin follow Bose–Einstein statistics). Let W be the number of ways the codons can be arranged in the sequence. We may then follow the same argument for the entropy of the ensemble of sequences of codons in DNA that led to equation (2.30) for atoms. The expression must be unique and positive because we are dealing with events that either happen or do not happen. Having taken this route, Brillouin (1962) is obliged to prove that equation (2.30) satisfies the conditions (1), (2) and (3) that were stipulated for a measure of information. This approach is not quite so short as may appear in the paragraph above.

Jaynes (1957a) adopted a third point of view, namely, to introduce entropy as a primitive concept and to regard it as an axiom according to equation (2.24). This method is satisfactory if we prove that it is unique and satisfies the conditions established as a measure of information.

The rather innocent-looking expression in equation (2.24) is rich with theorems. It has found many important roles and applications in the theory of probability and in ergodic theory. Petersen (1983) goes into more deep mathematical questions than I have found necessary in this book. Professional mathematicians will find his two chapters on entropy a worthwhile supplement to the material in his book. Petersen (1983) shows that entropy has many interesting and useful properties that are outside the context that led us to it. This is in contrast to *ad hoc* models and scenarios that often contain less than we hoped they would.

Entropy will play a central role in the application of information theory to molecular biology in this book. Many of these roles are surprising in that they are contrary to intuition. One should think first of entropy and then of what entropy is measuring or the role it is playing in the particular circumstances. Sometimes entropy measures information and sometimes it measures uncertainty, but it has other properties as well.

One must further remember that entropy is a name of a mathematical function, nothing more. One must not ascribe meaning to the function that is not in the mathematics. For example, the word '*information*' in this book is never used to connote knowledge or any other dictionary meaning of the word not specifically stated here. It will be helpful to remember that a communication system, including the genetic information system, is essentially a data recording and processing system. Thus one avoids much of the anthropomorphism that words convey.

2.1.4 Conditional entropy

Suppose we have two probability spaces X and Y. We may form the product probability space $X \times Y$, which consists of the pairs of elements (x, y), and we may then form the entropy of this product probability space in accordance with equation (2.24):

$$H(X, Y) = - \sum_{i,j} p(x_i, y_j) \log_2 p(x_i, y_j) \qquad (2.31)$$

We may also define the *conditional entropy* of X given an element in Y.

Let $p(x_i, y_j)$ be the probability of the pair (x_i, y_j) and let $p(x_i|y_j)$ be the probability of x_i given y_j. Then

$$H(X \mid Y) = - \sum_{i,j} p(x_i, y_j) \log_2 p(x_i|y_j) \tag{2.32}$$

Since one has the following relation between $p(x_i, y_j)$ and $p(x_i|y_j)$

$$p(x_i, y_j) = p(y_j) p(x_i|y_j) \tag{2.33}$$

equation (2.32) may be expressed in this form:

$$H(X|Y) = - \sum_{i,j} p(y_j) p(x_i|y_j) \log_2 p(x_i|y_j) \tag{2.34}$$

2.2 The criterion for isomorphism between probability spaces: when are they the same?

2.2.1 The concept of isomorphism in set theory and probability theory

In set theory two systems are *isomorphic* if and only if there is a one-to-one correspondence or mapping, sometimes called a *bijection*, from one system to the other, where the elements that are mapped or correspond have the same probability (Pickert & Görke, 1974; Shields, 1973; Petersen, 1983; MacLane, 1986; Ornstein, 1989), ignoring, of course, elements of probability measure zero. Systems that appear very different but which are isomorphic may be thought of, in an abstract sense, as being the same system. Let us illustrate what is meant by isomorphism by an example from the study of crystal structure. There are 14, and only 14, different possible Bravais space lattices defined such that two lattices are different when the environment of their points is different. All crystals can be divided into 32 and only 32 crystal classes if the point group is taken as the criterion. In the sodium chloride crystal each positive ion is surrounded by six negative ions and vice versa. This crystal structure is not *isomorphic* with, say, the hexagonal close-packed structure. Because of the one-to-one mapping or bijection property, isomorphous substances form mixed crystals, while non-isomorphous substances do not.

What is the criterion that probability spaces are isomorphic? Two probability spaces Ω_0 and Ω_1 are isomorphic if and only if they have the same entropy (Kolmogorov, 1958, 1965; Ornstein, 1970, 1974; Shields, 1973). This matter is discussed in detail by Petersen (1983). In particular, the probability distributions of codons and amino acids are not isomorphic with the probability distributions in statistical mechanics. The sequences generated by tossing a coin and a die or cards drawn from a deck with replacement are also not isomorphic for the same reason. However, if the numbers on the die are reported only as odd or even and if the cards are reported only as black or red, then the systems are isomorphic. These three systems are elements of a set whose members have the same entropy. Each could be replaced with another member of the set without changing the entropy or any other stochastic property. They are simply examples of generators of random binary numbers.

2.2.2 Is there a relation between Shannon entropy and Maxwell–Boltzmann–Gibbs entropy?

There are a number of papers in the literature in which an attempt is made to establish a relation between Shannon entropy and Maxwell–Boltzmann–Gibbs entropy. Eigen (1971) follows Brillouin (1962) in confusing Maxwell–Boltzmann–Gibbs entropy with Shannon entropy. This point is discussed in section 12.1. There is such a relationship if and only if the probability spaces are isomorphic; in that case they have the same entropy (section 2.2.1). For example, the entropy of the probability space characteristic of the tossing of two dice is not equal to the entropy of the probability space of the tossing of a fair coin because these probability spaces are not isomorphic. Furthermore, there is a Shannon entropy for each alphabet depending on the arbitrary choice of the number of letters. There is a relationship between the Shannon entropies of different alphabets if and only if they are related by a code. It is clear from section 2.2.1 that Shannon entropy and Maxwell–Boltzmann–Gibbs entropy do not pertain to probability spaces that are isomorphic. In addition, these probability spaces are not related by a code. For that reason Maxwell–Boltzmann–Gibbs entropy is not applicable to the probability spaces used in communication theory or in the genetic information system of molecular biology. Furthermore, in thermodynamics and statistical mechanics there is an integral of the motion of the system, namely, the conservation of energy, which has no counterpart in information theory. For these reasons there is, therefore, no relation between Maxwell–Boltzmann–Gibbs entropy of statistical mechanics and the Shannon entropy of communication systems (Yockey, 1974, 1977c).

2.3 The entropy of sequences

2.3.1 The entropy of Markov chains

Suppose we have the events E_i from a Markov source with a finite number of states $E_1, E_2, \ldots, E_n$ generating a Markov chain. Let $E_{i1}, E_{i2}, \ldots, E_{ij}$ be the previous j states. The transition probability matrix elements are

$$p_i(E_i|E_{i1}, E_{i2}, \ldots, E_{ik}, \ldots, E_{ij})$$

Then it is clear that we may form the entropy $H_i(E_i|E_{i1}, E_{i2}, \ldots, E_{ik}, \ldots, E_{ij})$ by using equation (2.24):

$$H_i = -\sum_j p_i(E_i|E_{i1}, E_{i2}, \ldots, E_{ik}, \ldots, E_{ij}) \log_2 p_i(E_i|E_{i1}, E_{i2}, \ldots,$$
$$E_{ik}, \ldots, E_{ij}) \quad (2.35)$$

H_i is the entropy or the information obtained when the Markov process moves one step or shift ahead starting from E_i. The expectation value of the random variable H_i is

$$H^{(1)} = -\sum_i \sum_j p_i(E_i \,|\, E_{i1}, E_{i2}, \ldots, E_{ik}, \ldots, E_{ij}) \log_2 p_i(E_i \,|\, E_{i1}, E_{i2}, \ldots,$$
$$E_{ik}, \ldots, E_{ij}) \quad (2.36)$$

Let us now consider a first-order Markov process and generalize to the case of moving r steps ahead. Let the system be in state E_i; then the probability that in the next r steps it will be in states $E_{k_1}, E_{k_2}, \ldots, E_{k_r}$ where the numbers $k_1, k_2, \ldots, k_r$ are arbitrary numbers from 1 to n is calculated from the conditional transition probability. The quantity

$$H^{(r)} = \sum_{i=1}^{n} p_i H_i^{(r)} \quad (2.37)$$

is the expectation value of the random variable $H_i^{(r)}$, that is, the entropy or information generated by the Markov process moving ahead r steps. It will be called the *r-step entropy* of the chain. In view of the additive property of entropy one expects that the r-step entropy can be added to, say, the s-step entropy:

$$H^{(r+s)} = H^{(r)} + H^{(s)} \quad (2.38)$$

or for any positive integer r, $H^{(r)} = rH^{(1)}$.

In order to prove equation (2.38) let the system be in the state E_i. The system may be considered to be the product of two dependent probability spaces, $\Omega_0 \times \Omega_i$, the probability space corresponding to the first trial with entropy $H_i^{(1)}$ and Ω_r is the probability space describing the next r trials. Because of the composition property given in equation (2.22) we have

$$H_i^{(r+1)} = H_i^{(1)} + \sum_{k=1}^{n} p(i|k) H_k^{(r)} \tag{2.39}$$

The expectation value of $H_i^{(r+1)}$ is given by

$$H^{(r+1)} = \sum_i p_i H_i^{(r+1)} = \sum_i p_i H_i^{(1)} + \sum_{k=1}^{n} H_{k-1}^{(r)} \sum_i p_i p(i|k) \tag{2.40}$$

Therefore

$$H^{(r+1)} = H^{(1)} + \sum_{k=1}^{n} p_k H_k^{(r)} = H^{(1)} + H^{(r)} \tag{2.41}$$

Since equation (2.41) is true for any r, then it is true for $r = 1, 2, \ldots$ and therefore by mathematical induction we have the final result (Khinchin, 1957):

$$H^{(r+1)} = (r+1)H^{(1)} \tag{2.42}$$

2.3.2 Is the number of messages in a sequence the same as the total number possible?

Let C be a sequence generated by a stationary ergodic Markov process with states $E_{k_1}, E_{k_2}, \ldots, E_{k_N}$ where the subscripts refer to the state E_{k_s} at the sth step in the sequence and the numbers k_s run from 1 to n. Let $p(C)$ be the probability of such a long message of N symbols where the symbols are mutually exclusive and occur independently at each site as they do in amino acid sequences in proteins (but not in natural languages!). These sequences may be regarded as compound events or *cylinders* in the probability space. Therefore, the probability of each sequence is the product of the probability of each letter i in the alphabet. Since there are N symbols, the expected number of occurrences of letter i is $p_i N$. Then $p(C)$ of *this particular message*, a cylinder in probability space, will be given with close approximation by the following equation (Shannon, 1948; Shannon & Weaver, 1949):

$$p(C) = \prod_i p_i^{p_i N} \tag{2.43}$$

Thus

$$\log_2 p(C) = N \sum_i p_i \log_2 p_i = -NH \qquad (2.44)$$

and therefore

$$p(C) = 2^{-NH} \qquad (2.45)$$

The number of sequences is approximately $1/p(C)$ or 2^{NH}. One might expect this result to be the total number of *possible* sequences, that is, for an alphabet of n symbols, n^N. It is easy to show that $n^N \geqslant 2^{NH}$, the equality pertaining only if all $p_i = 1/n$. What, then, is the significance of the number of sequences calculated from $1/p(C)$ and how did entropy get into the equation? We shall learn from the following theorem, the Shannon–McMillan theorem (Shannon, 1948; McMillan, 1953; Khinchin, 1957), that 2^{NH} is approximately the number of sequences in the high probability group.

Theorem 2.2: Let H be the one-step entropy of the stationary ergodic Markov chain that generates sequences of length N. Given $\varepsilon > 0$ and $\eta > 0$, no matter how small, then for the length of the sequence, N, being sufficiently large, all sequences chosen from an alphabet of n symbols can be divided into two groups such that the following hold.

   (i) The probability $p(C)$ of any sequence in the first group satisfies

$$|[\log 1/p(C)]/N - \mathrm{H}| < \eta \qquad (2.46)$$

   (ii) The *sum of the probabilities* of *all* sequences in the second group is less than ε.

Proof: The Markov process is ergodic, therefore the statistical properties of the sequence are stationary. The probability of any sequence does not depend on the starting point. Suppose we start at E_{k_1}, which has probability P_{k_1}. Let i and l be two arbitrary numbers each ranging from 1 to n and let m_{il} be the numbers of pairs of the form $k_r k_{r+1}$, where $1 \leqslant r \leqslant N-1$, in which $k_r = i$ and $k_{r+1} = l$. Then $p(l|i)$ is the transition probability from state i to state l. The values of the k's run from 1 to n at each step. The subscripts indicate the position in the sequence and run from 1 to N. Then we have, for m_{il} pairs,

$$p(C) = P_{k_1} \prod_{i=1}^{n} \prod_{l=1}^{n} p(l|i)^{m_{il}} \qquad (2.47)$$

We now assign the sequence to the first group if it has the following two properties.

(1) It is a possible outcome, that is, all $p(l|i) > 0$. For a DNA sequence that means that there are no terminator codons.

(2) For any i, l the following inequality holds:

$$|m_{il} - NP_i p(l|i)| < N\delta \qquad (2.48)$$

The second term in the absolute value bars is the estimate of m_{il} from the law of large numbers. In order to make relation (2.48) an equality let us find a number h_{il} the absolute value of which is less than unity for any pair m_{il} and write as follows:

$$m_{il} = NP_i + N\delta h_{il} \qquad (2.49)$$

Substitute the expression for m_{il} in equation (2.47), then take the logarithm to the base 2:

$$-\log_2 P(C) = -\log_2 P_{k_1} - N \sum_i \sum_l P_i p(l|i) \log_2 p(l|i)$$

$$-N\delta \sum_i \sum_l h_{il} p(l|i) \quad (2.50)$$

The second term on the right is simply N times the one-step entropy in the chain, as we have seen above. Equation (2.50) may be rewritten in the following form, returning to the inequality by replacing all h_{il} by 1:

$$|(-\log_2 p(C)/N) - H| < -(1/N) \log_2 P_{k_1} - \delta \sum_i \sum_l \log_2 p(l|i) \quad (2.51)$$

The left-hand side of the inequality is positive but can be made as small as we please by choosing N sufficiently large. By property (2) we have chosen δ sufficiently small. This proves the first part of the theorem and we see that the procedure leading to equations (2.44) and (2.45) is justified.

To prove the second part of the theorem we must find the sum of the probabilities of those sequences that do not satisfy inequality (2.48). That is:

$$\sum_{i=1}^{n} \sum_{l=1}^{n} P\{|m_{il} - NP_i p(l|i)| \geq N\delta\} \qquad (2.52)$$

Let us start by selecting any pair i, l. By the law of large numbers we can say that the probability that $|m_i - NP_i|$ is less than $N\delta$ is less than $1 - \varepsilon$. In mathematical symbolism this is written as follows:

$$P\{|m_i - NP_i| < N\delta\} > 1 - \varepsilon \qquad (2.53)$$

where m_i is the frequency of events in state i.

By the same token, if N is sufficiently large we can write

$$P\{|(m_{il}/m_i) - p(l|i)| < \delta\} > 1 - \varepsilon \qquad (2.54)$$

Therefore the probability of satisfying both of the inequalities in the brackets is the product of the separate probabilities and is greater than $(1 - \varepsilon)^2 > 1 - 2\varepsilon$. If we multiply the inequality of (2.53) through by $p(l|i)$ it follows that, since $1 > p(l|i) > 0$

$$|p(l|i)m_i - NP_i p(l|i)| < p(l|i)N\delta < N\delta \qquad (2.55)$$

If we add this inequality to the inequality that comes from expression (2.53), thus

$$|m_{il} - m_i p(l|i)| < \delta m_i \leqslant N\delta \qquad (2.56)$$

we have

$$|m_{il} - NP_i p(l|i)| < 2N\delta \qquad (2.57)$$

Thus for any i and l and for N sufficiently large we have that the probability

$$P\{|m_{il} - NP_i p(l|i)| < 2N\delta\} > 1 - 2\varepsilon \qquad (2.58)$$

which is the same statement as

$$P\{|m_{il} - NP_i p(l|i)| > 2N\delta\} < 2\varepsilon \qquad (2.59)$$

We can now carry out a summation of expression (2.59) and find

$$\sum_{i=1}^{n} \sum_{l=1}^{n} P\{|m_{il} - NP_i p(l|i)| > 2N\delta\} < 2n^2\varepsilon \qquad (2.60)$$

Since the right-hand side of this inequality can be made as small as we please by choosing ε sufficiently small, the sum of the probabilities of all the sequences in the second group can be made as small as we please by choosing N sufficiently large. This meets the requirement of the second part of the theorem and completes the proof.

Theorem 2.2 proves the statement made above, after the derivation of equation (2.45), that the number of sequences in the high probability group, (i), is very close to 2^{NH}. The entropy H controls the fraction of the total number of possible sequences that are in the high probability group. This is proved in Khinchin (1957) by an argument that closely parallels that given above. Let us now look at the significance of this result. Since $2^H \leqslant N$, the ratio of the number of sequences in the high probability group to the total number possible approaches zero as N becomes very large. To illustrate this, let us calculate both functions and the ratio for a simple, easily understood probability space. In Chapter 1, Exercise 7 considers a random variable on the sample space of all pairs of numbers of readings of two dice, namely, the sum of the pairs of numbers. In this chapter, Exercise 2 suggests that the reader should calculate the entropy of that probability distribution and evaluate both n^N and 2^{NH}. The result, if $N = 100$, is that the high probability group is only 2.69×10^{-6} of the total number of possible sequences.

Probability theory contains many theorems that are contrary to most people's intuition. Authors in molecular biology, almost without exception, are unaware of the Shannon–McMillan theorem, and have been led to false conclusions. They do not realize that, by the same token as the sequences in the dice throws, all DNA, mRNA and protein sequences are in the high probability group and are a very tiny fraction of the total possible number of such sequences.

2.3.3 The entropy of a sequence and the compression coefficient

Suppose we have an ensemble of ergodic sequences with entropy H generated by a stationary Markov source as discussed above. If this ensemble of sequences is a text in a language, English for example, we can shorten the sequence and preserve the meaning by establishing abbreviations, or a code, for frequently used words or short sequences of words. In molecular biology we do not write *deoxyribonucleic acid* each time we need the word. We use a code word, DNA, and put *deoxyribonucleic acid* in a glossary.

The question then arises, what is the limit to which we can compress an ensemble of ergodic sequences by choosing an optimum compression code (Blahut, 1987) that has the same number of symbols as the input code? Each sequence of s symbols of the input text has a probability $p(C)$. The sequence into which it is coded has a length $N(C)$. Each sequence can be regarded as a compound event or *cylinder* in the probability space (section 1.2.3) (Khinchin, 1957). The ratio $N(C)/N$ is the compression ratio for a given sequence of length N. The expectation value of this ratio is

$$\lambda_N = \sum_C p(C)[N(C)/N] \tag{2.61}$$

The summation is over all sequences of length N. As N becomes very large λ_N approaches λ, which we shall call the *compression coefficient* of a given text for a given code. The answer to the question asked above is in the following remarkable and useful theorem.

Theorem 2.3: If the entropy of a given ergodic sequence of symbols is H, then the greatest lower bound of the compression coefficient λ for all possible codes is $H/\log_2 n$, where n is the number of different symbols both in the input sequence and in the compressed sequence.

A sequence that results in a compression in this fashion is indistinguishable from a random sequence since all the patterns and regularities have been removed. The proof is not especially instructive and can be found in Khinchin (1957). But what is significant is that we need to consider only the entropy and the number of symbols. The theorem instructs us to find a code such that the ensemble of compressed messages has an entropy as close as we can find to $\log_2 n$. Shannon (1948) has called this ratio the *relative entropy*. This theorem will be applied to the genetic code in section 7.4.2.

2.4 Algorithmic information theory

2.4.1 The definition of algorithmic information

The preceding discussion of probability theory is based on measure theory and the notion of the ensemble. There is another conceptual foundation that will prove useful in understanding the concepts in this book, namely, *algorithmic information theory*. This theory is based on the theory of algorithms.

Definition: An algorithm is a mathematical procedure by which a desired result can be achieved in a finite number of computable operations.

Definition: Numbers or functions that can be calculated by an algorithm are described as *recursively enumerable* or as belonging to a recursively enumerable set.

Mathematicians were led to algorithmic information theory by the search for a fundamental definition and measure of randomness. Up to this point I have used the term '*random*' without specifying exactly what I mean. In the past a random choice of events has usually meant that they are chosen with equal probability and there is no discernible pattern or orderliness (Feller, 1968). However events, even in some games of chance, are often not of equal probability, for example in the tossing of two dice. The probabilities in problems of practical interest are almost always not all equal, yet one feels that the sequences generated by Markov processes with such probability distributions have a hierarchy of randomness that requires a measure. For this reason, the definition of randomness must be replaced by a more sophisticated definition in order to come abreast of developments in mathematics and to provide a foundation for the discussion of a number of practical problems in this book. Consider the following two sequences in the alphabet (0, 1)

010101010101010101010101

011011001101111100001000

The first sequence clearly shows a pattern of orderliness and one is inclined to expect that the same pattern will persist indefinitely. A pattern is not so clear in the second sequence. However, most gamblers have a system that they believe helps them beat the odds in spite of the fact that it can be shown that no system will do so (Feller, 1968). A gambler, looking for a system in the second sequence, will notice a clustering of 11 separated by one or two 0's. If he applies this system after the thirteenth event he will begin to lose. To save his system he will complicate and embellish it, but eventually he will lose. Fortuna will punish him for his impertinence by playing the Lorelei and will lead him to the gambler's ruin (Feller, 1968). On the other hand, the first sequence, as well as the second, has an equal number of 0's and 1's and therefore they each have exactly the same *a priori* probability. Since the first sequence is an honest member of the ensemble of sequences and could be generated by a 'random' process, one comes to the rather unsatisfactory conclusion that, although it is orderly, it, as well as the second sequence, must be regarded as 'random'.

Let me give several more sophisticated examples to show why we need a more careful means of distinguishing between orderliness and randomness. Rational fractions are represented by an infinite sequence of decimal digits. The computer program for calculating the decimal sequences of rational fractions is very simple and short, for example, divide 17 by 39. The answer is the repeating sequence: 0.435897435897435897435897...which, although it is an infinite sequence, clearly has an orderly pattern from which the rest of the sequence can be predicted. We can use the orderliness of this sequence to compose a message specifying an algorithm of finite length that generates the infinite sequence: 'Type 0.435897, and repeat 435897 indefinitely'. Thus a finite sequence can contain all the information in an infinite one.

In the course of preparing tables of random numbers the question arose as to whether the sequence of decimal digits representing transcendental numbers such as π and e could pass the standard tests for randomness. It was found that the decimal sequences representing some transcendental numbers did, indeed, pass such tests, π and e included (Parthia, 1962; Stoneham, 1965). These sequences therefore exhibit no orderly pattern from which the rest of the sequence can be predicted. Yet clearly the numerical sequences representing π and e are not determined by a stochastic process. Each digit is computable by an algorithm of finite information content. It is a pity that a responsible scientist would imply, even in a science fiction novel, that there is a message of salvation hidden somewhere in the numerical sequence that represents π. We may conclude, in general, that a finite message, specified by an algorithm, may exist that carries all the information in an infinite sequence even though there is no discernible pattern.

Every method of computation must be capable of being described by a finite number of symbols, that is, by an algorithm. Since a description or algorithm consists of a finite number of symbols there are only countably many algorithms. The total number of functions, say transcendental numbers is, however, transfinite, that is, uncountably infinite. There are, therefore, an infinite number of functions or transcendental numbers that cannot be calculated by an algorithm. Numbers or functions that can be calculated by an algorithm are described as *recursively enumerable* or as belonging to a recursively enumerable set (Hermes & Markwald, 1974).

Returning to the system or algorithm the gambler is trying to develop, we realize that, although any short sequence generated by a truly stochastic process will show pseudo-regularities, eventually so many exceptions to these regularities accumulate that the algorithm representing the sequence approaches or exceeds the length of the sequence itself. At that point one may just as well forget about trying to develop the algorithm and send the sequence.

How can these considerations lead to a definition of randomness that does more than classify a sequence as random or not random and provides a measure of the amount of randomness in the sequence? An answer to this question emerged from the initially independent work of Solomonoff (1964), Kolmogorov (1965), Chaitin (1966, 1969) and Martin-Löf (1966) on the foundations of information theory and the foundations of probability theory. The current state of the theory is found in Chaitin (1987a, 1988) and is all collected in Chaitin (1987b). The germ of the idea is that the length of the sequence, in bits, describing the shortest algorithm by which the sequence is specified is, in analogy only with Theorem 2.3, the entropy of the sequence. This idea is based on the theory of algorithms, not on the theory of stochastic ensembles. Consequently one can establish a hierarchy of randomness according to the entropy of the compressed sequence. Therefore a very long sequence that may be compressed to a much shorter one is an orderly sequence and not a random one. The size in bits of the shortest algorithm describing the sequence or bit string is a measure of its entropy or randomness. The amount of randomness in π is just that in the shortest algorithm by which it is calculated. By the same token, we now have a definition and a measure of 'order', namely, that an *orderly* sequence is one that can be specified by a short algorithm.

Let us estimate the number of sequences that have a large amount of randomness. Suppose we have expressed a sequence in a binary alphabet $(0, 1)$ suitable for the input of a computer. Let the length of the sequence be *l*. A very orderly sequence can be generated by a program of one symbol, either 0 or 1. There are two such programs. There are 2^2 programs that can be specified by two symbols, namely $[(0,0), (0,1), (1,0), (1,1)]$. In general there are 2^i programs specified by i symbols. The total number, S, of such programs is the sum of a geometric series:

$$S = 2^1 + 2^2 + \dots + 2^l \tag{2.62}$$

$$S = \frac{2(2^l - 1)}{(2-1)} = 2^{l+1} - 2 \tag{2.63}$$

The sum of the number of programs 10 less than *l* is $(2^{l-9} - 2)$. Thus for long sequences there is a *factor* of 1024 fewer programs in sequences just 10 shorter than *l*. This can easily be generalized for alphabets of any finite discrete number of letters. It is easy to see that in general, almost all long sequences have an entropy nearly equal to that of random sequences.

2.4.2 Can a given sequence be proven to be random?

The considerations in section 2.4.1 might lead us to believe that, having a definition and a measure of randomness, one could prove a given sequence to be random. In fact it is impossible to do so. It can easily be shown that a given sequence is not random. We need only find a program that compresses the sequence to one substantially shorter than the sequence itself. We need not prove that the program is the shortest one, just that it is shorter than the original sequence. On the other hand to prove that a given sequence is indeed random one must prove that no shorter program exists. This leads immediately to the paradoxes of sets that have themselves as members. Epimenides, the Cretan, alleged, 'All Cretans are liars'. Is that statement true or false? Consider the Berry paradox: 'Find the smallest positive integer that to be specified requires more characters than there are in this sentence'. The sentence supposedly specifies a positive integer of more characters than there are in the sentence. We are now in the midst of a subtle and fundamental anomaly in the foundations of mathematics, discovered by Chaitin (1975a, b) that is related to a famous theorem due to Kurt Gödel called Gödel's incompleteness theorem. Gödel showed that for the axioms defining number theory there are always theorems that can neither be proven or disproven from the set of axioms. This theorem caused considerable *Angst* among mathematicians comparable to that among physicists caused by quantum theory and relativity theory.

The question of incompleteness or undecidability in arithmetic has been considered by Chaitin (1987a, 1988) by means of diophantine equations. A simple example is equation (2.64). The famous conjecture known as 'Fermat's last theorem' is that there are no positive integer solutions

$$(x+1)^{n+3} + (y+1)^{n+3} = (z+1)^{n+3} \tag{2.64}$$

Only addition, multiplication and exponentiation (which is an extension of multiplication) are allowed. The extensive search for non-negative solutions of equation (2.64) has found none, although no proof is known that there are none. Chaitin (1988) wrote a universal exponential diophantine equation in 17 000 variables. The number of solutions of this equation when a parameter is varied jumps from finite to infinite in a manner that is indistinguishable from flipping a fair coin. Since the outcome of flipping a fair coin is *undecidable* one has a problem in pure mathematics, the solution of which is random. Thus it is fundamentally undecidable whether a given sequence is random or not.

2.4.3 *Entropy as a measure of complexity, order, orderliness and information content*

We are now in a position to place a quantitative mathematical meaning on several words that are often used in discussions in molecular biology and in evolution. Organisms are often characterized as being 'highly ordered' and in the same paragraph as being 'highly organized'. Clearly these terms have opposite meanings in the context of this chapter. The first message discussed in section 2.4.1 is highly ordered and has a low entropy. Being 'highly organized' means that a long algorithm is needed to describe the sequence and therefore highly organized systems have a large entropy. Therefore highly ordered systems and highly organized ones occupy opposite ends of the entropy scale and must not be confused. Since highly organized systems have a high entropy, they are found embedded among the random sequences that occupy the high end of the entropy scale.

Kolmogorov (1965, 1968) and Chaitin (1966, 1969) have called the entropy of the shortest algorithm needed to compute a sequence its *complexity*. Chaitin (1975b) proposed a definition of complexity that has the formal properties of the entropy concept in information theory. Chaitin (1970, 1979) and Yockey (1974, 1977c) pointed out the applicability of this concept in establishing a measure of the complexity or information content of the genome. This definition is not based on probability ensemble concepts but rather pertains to the theory of algorithms. Some equivalences between the entropy based on the theory of stochastic ensembles and Kolmogorov–Chaitin complexity are discussed by Chaitin (1974) and by Sik, Leung-Yan-Cheong & Cover (1978).

Thus both random sequences and highly organized sequences are *complex* because a long algorithm is needed to describe each one. Information theory shows that it is *fundamentally undecidable* whether a given sequence has been generated by a stochastic process or by a highly organized process. This is in contrast with the classical law of the excluded middle (*tertium non datur*), that is, the doctrine that a statement or theorem must be either true or false. Algorithmic information theory shows that truth or validity may also be indeterminate or fundamentally undecidable.

We can interpret entropy as a measure of the intuitive concepts randomness, complexity, order, orderliness and information content in much the same way as we did for information and uncertainty in section 2.1.1. We can regard these words as merely names for the entropy of the sequence. The word entropy, in turn, is merely the name of a mathematical function.

In Chapter 6, I calculate an estimate of the complexity of the cytochrome c homologous family. To do this I use all the regularities I can find to shorten the algorithm that defines cytochrome c. If one finds other regularities later they may be used to shorten the algorithm further. I need not prove that I have found the shortest algorithm. This is an example of how to find and estimate the complexity of any homologous protein family. In doing so I follow the philosophy discussed by Solomonoff (1964) and Chaitin (1966). We always attempt to find a theory that explains our data. This is another way of saying that we try to find an algorithm that is shorter and of more general application than the data of the experiment. The value of a scientific theory is that it provides a compact description of a phenomenon or several phenomena which may, at first sight, appear to be distinct. Thus if we wish to convey our knowledge of the general theory of relativity to our little green friends on some 'suitable planet' near Arcturus, we simply send Einstein's equations. Since they are more advanced than we, they would recognize these equations immediately and draw the conclusion that we know about the expanding universe and the Big Bang. The scientist is performing in the same manner as the gambler who tries to find a system to beat the odds of some game of chance. The scientist, however, will find a theory that explains his data unless the data is extremely complex or even random. Some randomness will always remain because of the requirement to describe the theory.

The best possible theories will always give only probabilities, not exact predictions, although some predictions may have exceptions with probability zero. It is essential to be aware of the fact that some problems in physics and indeed in the mathematics of number theory are fundamentally intractable. For example, the problem of proving whether particular diophantine equations have finitely or infinitely many solutions is absolutely intractable (Chaitin, 1987a). Mathematical reasoning is the most powerful available to man yet because such questions have no discernable structure or pattern they evade the capability of mathematics. We are left, as were the ancient Greeks, Romans and Hebrews, with statistical limits to our knowledge and to the goddess Fortuna to decide important questions.

2.4.4 *The question of negentropy: can entropy be negative?*

The idea of negative entropy was introduced by Schrödinger (1944) to explain the appearance of what he thought was *order* during evolution. Brillouin (1953, 1962) used the concept also to address certain problems in physics. Simpson (1964) had the following remark: 'A fully living system must be capable of energy conversion in such a way as to accumulate negentropy, that is, it must produce a less probable, less random organization of matter and must cause the increase of available energy in the local system rather than the decrease demanded in closed systems by the second law of thermodynamics'. On the contrary, we have learned in section 2.4.3 that highly organized systems have a high entropy and they are found embedded among the random sequences that occupy the high end of the entropy scale. Furthermore, one is concerned with the Shannon information theory entropy not the Maxwell–Boltzmann–Gibbs entropy of statistical mechanics.

There are, I believe, serious mathematical objections to the notion of negentropy. The constant K in equation (2.24) must be positive in order to satisfy condition (b). Taking K positive, some of the probabilities p_i would have to be negative to make $H(p_i)$ negative. Negative probabilities are not defined in the probability space we are using. Refer to the comments made in section 1.2.4 on the question of probabilities that are not non-negative real numbers.

There is a further mathematical difficulty. The logarithm is defined according to the following equation:

$$y = \int_1^x \mathrm{d}f/f = \log_e x \tag{2.65}$$

The logarithm of any negative number cannot be found from equation (2.65) because the integrand $1/f$ approaches infinity as f approaches zero. The logarithm is undefined for negative real numbers, therefore equation (2.24) is undefined for negative values of p_i.

For the reasons given above, I am of the opinion that negentropy is an *ad hoc* notion that need not be introduced. I have eschewed the introduction in this book of any ideas or definitions that cannot be justified from first principles, are not mathematically sound and do not lead to useful results. We are not at liberty to change cavalierly a sign or otherwise modify a mathematical expression without proving theorems. It will be seen that the need for negentropy that was felt by Schrödinger (1944) and Brillouin (1953, 1962) need not be met by *ad hoc* conjectures but can be met by straightforward analysis. Introducing negentropy is an attempt to adapt the mathematics to a Procrustean bed of words. I prefer to make the words fit the mathematics. A clear understanding, which is developed from algorithmic information theory, of what is meant by such words as *highly ordered, highly organized, random* and *complex* makes the introduction of '*negentropy*' unnecessary.

2.4.5 *Reflections on complexity in molecular biology*

Chaitin (1970, 1979) noted that a precise concept of complexity would contribute to a definition of 'living organism'. Saunders & Ho (1976) proposed that 'complexity' should replace 'fitness' or 'organization' as the quantity that increases during evolution without giving a quantitative definition. Although Saunders & Ho (1981) were aware of the work of Kolmogorov and Chaitin they stated that: 'There almost certainly does not exist a unique definition of complexity suitable for all systems and all problems'. I was the first to relate the Kolmogorov–Chaitin definition of complexity to protein sequences in molecular biology (Yockey, 1977c). Ebling & Mahnke (1978) followed the ideas of Kolmogorov and Chaitin and calculated the 'complexity of a sequence' as a measure of the quantity of information stored in the sequence' (not by the method used in Chapter 6). Ebling & Jimenez-Montano (1980) considered the 'grammar complexity' of protein sequences in terms of the Kolmogorov–Chaitin complexity. Later Ebling, Feistel & Herzel (1987) adopted the point of view that there is no universal and exhaustive calculable measure of complexity. Kampis & Csányi (1987) complained that complexity is a vague and ill-defined concept and questioned whether it can be properly defined, and thereby *measured* or not. This paper confuses *order* and *orderliness* with *complexity* although the authors are aware of the work of Kolmogorov and Chaitin. The question of whether complexity increases during evolution is also discussed by these authors. This will be taken up in Chapter 12, which is on evolution and information theory.

It is fair to say that the definition given by Kolmogorov, Chaitin and Martin-Löf has been subjected to some criticism in the computer community as well. The field of information-based complexity is an expanding area of research concerned with the intrinsic difficulty of solving problems with a satisfactory dègree of approximation when the data is partial, contaminated or priced. Computational complexity is illustrated by the deceptively simple traveling salesman problem. Given a set of n cities in which one visit must be made and the distance between each pair of cities, find the shortest tour that visits each city once and returns to the starting point. The number of computational steps in the algorithm that accomplishes this computation defines the complexity of the problem. In spite of intensive study the best algorithms for finding the shortest tour require a number of steps that is at least exponential in n (Johnson & Papadimitriou, 1985; Lawler *et al.*, 1985; Packel & Traub, 1987).

Computational complexity that is exponential in *n* where *n* is the size of the problem, or greater than this, is called *intractable*. These problems are solvable in principle but not in practice because the time for solution and the required computer memory escalates so rapidly.

Measures of computational complexity have been proposed by a number of authors (Traub, Wasilkowski & Wazniakowski, 1983). Lempel & Ziv (1976) believe that no absolute measure of complexity exists. Other important contributions have been made by Levin (1984) and by Rissanen (1986). In spite of the opinion that an absolute measure of complexity does not exist, other measures proposed owe much to the Kolmogorov–Chaitin definition.

I pointed out above in the discussion about finding a definition and a measure of '*information*' that the mathematical definition does not capture all the meanings that are associated intuitively with that word. By the same token, the Kolmogorov–Chaitin complexity may likewise not capture all that one thinks of in the word '*complexity*'; nevertheless, it will prove very useful in molecular biology. Not the least of this usefulness is to put into focus the role of entropy in defining the meaning of such words as complexity, order, orderliness, randomness, highly ordered, highly organized and so forth. Critics of Kolmogorov–Chaitin complexity attempt to squeeze a mathematical meaning from words. We see again that words find their meaning from the mathematics.

2.4.6 Algorithmic information theory and the isomorphism between communication systems, computers, the mathematical logic system and the genetic logic system.

I have pointed out the relation or correspondence between the genetic logic system and communication systems such as the telephone and telegraph in section 2.1.1 (see also section 5.1.1 and Figure 5.1). In section 2.2.1, I discussed the criteria for *isomorphism* between systems (Ornstein, 1989). The idea of isomorphism is ubiquitous in biology. The science of taxonomy is based on the isomorphism of organs. The wings of a bat, the wings of a bird, the paw of a gorilla and the hands of Vladimir Horowitz are isomorphic although their function is quite disparate. Let us now use this more general and exact terminology for relation or correspondence and point out in addition the isomorphism between the genetic logic system, the logic systems of communication systems, the logic systems of computers and the logic system in mathematics by which theorems are proved from a list of axioms. These systems may be regarded as being the same abstract system (Ornstein, 1989).

The basic principle on which computers operate is that of the Turing machine (Turing, 1937). Turing conceived this abstract model of a computing machine to solve problems in the foundations of mathematics. In particular he addressed the question of computable numbers and the application to David Hilbert's tenth problem, known as the *Entscheidungsproblem* (decision problem) that Hilbert proposed mathematicians would solve in the twentieth century. The problem is to find some general mechanical procedure by which, *in principle*, all problems of mathematics formulated in axiomatic systems could be solved. In addressing the problem, Turing imagined an abstract machine in which a message or sequence is recorded on an input tape that could be weightless and of infinite length. In computer terminology these messages or sequences are called *bit strings* because they are expressed in a string of the alphabet (0, 1). The algorithmic information content of these bit strings is, of course, measured in bits, just as we have discussed in section 2.1.2. There is a reading head that may move in either direction along the tape to read the input, which interacts with a finite number of internal states. These states are called a program in modern computer terminology. The program carries out its instructions on the message read from the input tape. The results are read out on an output tape and the machine stops when the program is executed. The problem of deciding whether an algorithm will halt on a universal Turing calculating machine is related to Gödel's incompleteness theorem. That is, the question of whether the machine will eventually solve the problem or will go on indefinitely was proved by Turing (1937) to be unsolvable in general. If the Turing machine stops, the number is computable, otherwise the number is random and un-computable. By the same token, no computable number is random. A more detailed description of Turing machines and their use in the study of the computability of numbers would be diversion. However, a rather detailed but accurate account of Turing machines is given in the popular book by Penrose (1989).

The logic of Turing machines has an isomorphism with the logic of the genetic information system. The input tape is DNA and the bit string recorded is the genetic message. The internal states are the tRNA, mRNA, synthetases and other factors that implement the genetic code and constitute the genetic logic system. The output tape is the family of proteins specified by the genetic message recorded in DNA. There is also an isomorphism between the information in the instructions on the Turing machine tape and the information in the list of axioms from which theorems are proved. Without noticing these isomorphisms, corresponding properties would appear to be unrelated. But in each of these four cases one has an information source, a transmission of information, a set of instructions or tasks to be completed and an output. By the same token we shall find examples of properties in each system to have corresponding properties in the others. I shall take advantage of these isomorphisms to borrow concepts that have been developed in communication systems and computers and apply them to problems in molecular biology. I shall take up the isomorphism between the genetic logic system and the standard communicating system again with more detail in sections 5.1.1–5.1.2.

2.5 Do the gods cast lots?

Einstein (Pais, 1982) could never give up strict Aristotelian causality or accept the stochastic character of quantum mechanics. His famous complaint was 'Der liebe Gott würfelt nicht mit der Welt'. On the contrary, Homer (Homer, *The Iliad*, c. 850 BC) tells us that, as a matter of fact, the gods did indeed cast lots for the world. Zeus sent his messenger Iris to order Poseidon to cease helping the Greeks during a battle in the Trojan War. Poseidon's reply was that he, Zeus and Hades, three brothers born to Kronus by Rhea, were all equal in station and they had divided the world by lot between them. Zeus drew the sky, Poseidon the sea and Hades the underworld of the dead. Thus, if even the gods cast lots to decide important matters then mortals cannot do better. Not only does God cast lots in quantum mechanics but also in Newtonian mechanics, molecular biology and even in arithmetic as well (Chaitin, 1987a).

Exercises

1. Carry out all the steps needed to arrive at equation (2.30).
2. In Exercise 7 of Chapter 1 the reader calculated the probability of each event in the probability space that is the sum of the numbers on the dice when two dice are thrown. Calculate the entropy of that probability space and from that then calculate 2^{NH} for $N = 100$. Calculate 11^N to get the total number of *possible* sequences of 100 rolls of the dice. Calculate the ratio of 2^{NH} and 11^N. The result illustrates that, since amino acids are not equally probable, the number of such sequences with which we need to be concerned as calculated from the entropy of the sequence is far smaller than the total possible.

3. Are the probability spaces of a coin toss and the toss of two dice isomorphic? What does the Kolmogorov–Sinai theorem say about this question (Petersen, 1983, pp. 244–6)? Can you establish a conversion factor between these two spaces such as Brillouin (1953, 1962) and others claim must exist between Shannon entropy and Maxwell–Boltzmann–Gibbs entropy? Can you supply the connection between Shannon entropy and Maxwell–Boltzmann–Gibbs entropy?

4. Prove that $\lim_{x \to 0}(x \log_e x) = 0$. Hint: write this expression as follows: $\lim_{x \to 0}(\log_e x/(1/x))$; apply l'Hôpital's rule, differentiate both numerator and denominator separately and go to the limit.

5. Apply entropy as a measure of a student's uncertainty of the correct answers in a true–false test. What is a perfect score, that is, no uncertainty? What is the score of a student who deliberately wishes to fail and so answers all questions incorrectly? Does he succeed in failing? Should this method of scoring replace the present one? Is the entropy of the tests a random variable?

6. You have 32 coins, one of which is counterfeit, that is, over weight. If you weigh each coin separately against a standard, what is the average number of weighings required to find the counterfeit? Find a weighing strategy (algorithm) that finds the answer using the minimum number of weighings. In other words, how much can the procedure of weighing each coin separately be compressed? What is the information content in bits of the weighings of the 32 coins? What is the information content of the weighings of 64 coins? Suppose you are allowed only 5 weighings and you have 64 coins. With the best weighing strategy, how many coins will remain untested, that is uncertain? What is that uncertainty in bits?

7. In Exercise 6, suppose it is not known whether the counterfeit coin is over weight or under weight. How does this affect your procedure? How much information is contained in the answer?

8. In general, if you have a decision that contains n bits, what is the minimum number of bits in the algorithm by which this decision was made?

3

The principle of maximum entropy

It is now evident how to solve our problem; in making inferences on the basis of partial information we must use that probability distribution which has maximum entropy subject to whatever is known. This is the only unbiased assignment we can make; to use any other would amount to arbitrary assumption of information which we do not have.

E. T. Jaynes (1957a)

3.1 Establishing probabilities in the case of insufficient knowledge

3.1.1 Bernoulli's 'principle of insufficient reason'

Consider the situation, which is frequently the case, where we have some knowledge of the problem under consideration but our knowledge is faulty and incomplete. A specific problem of this character in molecular biology is in the study of the replaceable amino acids in evolution. As the evolutionary tree develops we need to know the probability of each of the functionally equivalent amino acids at each site. A best estimate of the probability of the acceptable amino acids at each site is needed to calculate the complexity, or the information content, of a homologous protein family.

We wish to organize our knowledge without, at the same time, inadvertently assuming knowledge we do not have. The method we seek must comply with all known constraints and give positive weight to all possibilities that are not excluded by prior knowledge. The problem is one of statistical inference and is as old as the theory of probability. In Chapter 1 we assumed that coins are equally likely to fall on either side. We assumed that a die will fall with probability 1/6 on each face. We made this assumption because we throw the die and the coin in such a way that each event is 'equally possible'. This is essentially the 'principle of insufficient reason' due to Jacob Bernoulli (Jaynes, 1979), which says that events are assumed to be equally likely if one has no reason to believe otherwise. Except in games of chance that are specifically designed to generate events that are equally possible, the notion of 'equally possible' does not survive a change of variable. For example, if we are studying the probability distribution in an experiment in which we drop a small shot on a circle we assume that the probability of striking various zones in the circle is proportional to the area. If later, we find that what we thought was a circle is in fact a sphere we may wish to change the variable to the subtended area of the zones on the sphere. This has been a fertile ground for inventing paradoxes (Feller, 1971). The 'principle of insufficient reason' applies only to the case where the events are equally probable and is silent for the general case, a deficiency of which Bernoulli was well aware (Jaynes, 1979).

Although one is justified in taking the *a priori* probability of the appearance of each face of a die to be $1/6$, a real die may be expected to have slightly different probabilities for each face. This follows from errors in manufacture and the weighting of the material that forms the spots. Thus the six face may be very slightly heavier than other faces. Since the sum of the numbers on opposite faces is seven, this may increase the probability of a one. The means by which the die is cast may also influence the outcome. Jaynes (1979) and Fougère (1988) examined the data of some 20 000 tosses of two dice. By means of maximum entropy they found deviations from perfection that caused the small variations from the predictions obtained from the *a priori* probabilities, namely, $1/6$ for each face.

3.1.2 Is the choice of a probability distribution that maximizes entropy the most unbiased?

The fact that our knowledge is incomplete reflects uncertainty and since we found in Chapter 2 that Shannon entropy is a unique measure of uncertainty, perhaps the ideal method of statistical inference should be one that selects that probability distribution that maximizes the Shannon entropy under those constraints of which we know. Maximizing the uncertainty thus avoids the arbitrary introduction of knowledge that we do not have and is therefore unbiased. It supplies the missing objective criterion that removes the apparent arbitrariness of the principle of insufficient reason and shows how that principle should be generalized in cases where the events are not 'equally probable'. This idea was first stated explicitly by Jaynes (1957a, b) as a general method of statistical inference and applied by him in statistical mechanics. Kullback's book applying information theory to statistics followed soon afterwards (Kullback, 1968). Since then there have been numerous applications in many diverse fields. Among them are the dramatic restoration of smeared or grainy photographs (Gull & Daniell, 1978; Skilling, 1984). See Shore & Johnson (1980, 1983) for references to other applications.

In Chapter 2 one of our approaches to the derivation of Shannon entropy was from the point of view of the number of possibilities (regarding them as all equally probable). According to Boltzmann's intuition, the most probable distribution was the one that could be realized in the greatest number of ways, or, in modern terminology, with the maximum degeneracy. From equation (2.30) it is easily seen that the number of possibilities, W, varies exponentially with the entropy. The maximum is extremely sharp, especially for a large number of possibilities. It is therefore quite *unreasonable* to choose any value of W but the maximum, which, of course, determines the entropy to be a maximum. The known restrictions are that the total number of particles is constant and that the collisions are elastic, that is, the molecules do not radiate energy, so that the total kinetic energy of the particles is constant. Applying the Lagrange method of undetermined multipliers to maximize the entropy, it was possible to derive, in a few lines, the Maxwell–Boltzmann expression for the energy distribution in a perfect gas. This was the beginning of the principle of maximum entropy.

On the other hand, Maxwell used the laws of dynamics to describe in detail the collisions of the molecules in a long and devious derivation of that expression. But Boltzmann's derivation seems to ignore the dynamics entirely. The answer seems to come too easily. These are among the criticisms directed at the maximum entropy formalism by Rowlinson (1970) and by Friedmann & Shimony (1971). See the response by Jaynes (1979) and by Tribus & Motroni (1972). The fact is that Boltzmann included only the relevant dynamics, namely, thermal equilibrium, the conservation of energy and the conservation of particles. All other factors cancelled out of Boltzmann's expression. The same criticism, that we are neglecting biology, may be directed at applications in molecular biology. So, by the same token, when the principle of maximum entropy is applied in biology, it will be necessary to present a thorough justification of the principle and to show that all the pertinent biology is indeed taken into account.

Without a rigorous proof, the reasonableness and the plausibility of the principle of maximum entropy, even with the derivation of well-known results, leaves the justification for general application incomplete, as Jaynes admits (Jaynes, 1968). Proof of the principle has been published by several authors. Skilling (1984) has summarized the current state, summarizing the work of Shore & Johnson (1980; see corrections, 1983) and that of Tikochinsky, Tishby & Levine (1984). Both papers start from considerations other than entropy and prove that the unique functional that satisfies the quite reasonable and general criteria for an algorithm for selecting the most unbiased probability distribution is equivalent to maximizing the entropy. This complements the remarks in Chapter 2 where I showed that entropy plays a stellar and often unanticipated role in the theory of probability.

3.1.3 The generalization to continuous probability distributions

In molecular biology we are normally interested only in discrete probability distributions. Since there are important applications in statistical inference, it would be as well to consider the generalization from discrete to continuous distributions and to allow for prior knowledge as reflected by prior probabilities (Kullback, 1968; Jaynes, 1963, 1968). The principle of maximum entropy, in this form, was employed by Borstnik & Hofacker (1985) in their study of the physico-chemical properties of replacement amino acids in the evolution of proteins. We will put their results to use in Chapter 6.

If the generalization is carried out as Shannon did originally, one has

$$H(x) = -\int f(x) \log f(x) \, dx \tag{3.1}$$

where $f(x)$ is a probability density, that is, a probability per unit of x. This expression is not invariant under a change of variables $x \to y = y(x)$. A function or an object should appear the same no matter how we look at it if it is to be taken as real. To say this with more mathematical sophistication it is generally possible to deal with probability theory by use of Lebesgue–Stieltjes integrals. There is no Lebesgue–Stieltjes integral that reduces equation (2.24) to equation (3.1) (Hobson, 1969). Kullback (1968) defined the function $I(2:1)$ in statistical inference as the mean information per observation for discrimination in favor of observation H_2 against observation H_1.

$$I(2:1) = \int f_2(x) \log_e \left(\frac{f_2(x)}{f_1(x)} \right) d\lambda(x) \tag{3.2}$$

where $\lambda(x)$ is a probability measure. From this equation Kullback (1968) derived rigorously the correct equation for the channel capacity $I(\Delta\omega, \Delta t)$ of a continuous analogue signal, where $\Delta\omega$ is the band width, Δt is the duration, S is the signal power and N is the noise power

$$I(\Delta\omega, \Delta t) = \Delta\omega\Delta t \log_2\left(1 + \frac{S}{N}\right) \qquad (3.3)$$

This equation will have important consequences in mathematical molecular biology as discussed in section 11.3.5.

3.1.4 *Applications of the maximum entropy criterion to decision theory*

Suppose a decision-maker has a finite number of courses of action from which he must choose one. He may set up a reward or loss system together with a probability of success for each course of action. The problem is to establish a utility function and maximize it under the uncertainties that are measured by the probabilities of success. The most objective method of carrying out this maximization is by the maximum entropy criterion (Smith, 1974).

3.1.5 *Applications in molecular biology*

The principle of maximum entropy was applied to the derivation of equation (5.22) (Yockey, 1977b). Holmquist & Cimino (1980) used the principle to find the best values of the empirical amino acid property eigenvectors (section 6.3.2). A number of applications to general scientific problems can be found in Erikson & Smith (1988).

4

Coding theory and codes with a Central Dogma

It is further postulated that the specific arrangement of amino acid residues in a given peptide chain is derived from the specific arrangement of nucleotide residues in a corresponding specific nucleic acid molecule, and that in addition the nucleic acid molecule in question is concerned in transferring energy necessary for peptide bond synthesis.

Dounce (1952)

There is now a mass of evidence which suggests that the amino-acid sequence along the polypeptide chain of a protein is determined by the sequence of the bases along some particular part of the nucleic acid of the genetic material. Since there are twenty common amino-acids found throughout Nature, but only four common bases, it has often been surmised that the sequence of the four bases is in some way a code for the sequence of the amino-acids.

Crick *et al.* (1961)

4.1 Introduction to coding theory

The idea of an hereditary *code-script* carried by an *aperiodic solid* in the chromosomes was expressed by Schrödinger (1944) (but see comments by Perutz 1987). Schrödinger pointed out that the enormous diversity in biology could be expressed by the transcomputationally large number of sequences that can be formed from the letters of a code such as the Morse code. Before the discovery of the structure of DNA, Dounce (1952) proposed the idea of a template between nucleic acid chains and protein chains and a duplicating mechanism for nucleic acids without quite coming to the notion of a code between these sequences. The discovery of the DNA molecular structure by Watson & Crick (1953a) and by Wilkins, Stokes & Wilson (1953), together with its identification as the substance carrying the genetic message that directs the formation of protein, again brought forward the following proposal: 'It follows that in a long molecule many different permutations are possible, and it therefore appears likely that the precise sequence of the bases is the code that carries the genetical information' (Watson & Crick, 1953b). The question immediately arose of how the sequence of four nucleotides of DNA could specify the 20 amino acids in protein. George Gamow was the genius who made the suggestion that the nucleotides of DNA were mapped onto the amino acids of protein by a code (Gamow, 1954; Gamow & Ycas, 1958). Gamow immediately set about to find arrangements of four letters that would yield 20 code words. He put it this way in his inimitable fashion: There are four ways to select three cards of one suit. If two cards are of the same suit and one is different, one has four choices for the cards of the same suit and three for the card that is different, giving 12 arrangements. And finally there are four possibilities for choosing the cards all of different suits. This makes the required 20 code words (Gamow, 1970). His overlapping diamond code was soon shown to be incorrect but the idea was fixed and the search was on. The means by which the genetic code was found are well known and will not be discussed here.

It will help to understand the genetic code if we compare it with codes in general, even though we don't have the flexibility of the communication engineer to devise codes that take advantage of the theorems that we learned about in the previous chapters. In order to avoid the confusing terminology of the early papers, it is essential to make a clear distinction between the sequences of nucleotides or amino acids and the code that relates one sequence to another. Regrettably, this distinction is not always observed even in the literature of today. Computer programs are sometimes called codes but this terminology is disappearing. Therefore we must now have a formal definition of a code.

Definition of a code: Given a source with probability space $[\Omega, A, \mathbf{p}_A]$ and a receiver with probability space $[\Omega, B, \mathbf{p}_B]$, then a mapping of the letters of alphabet A onto the letters of alphabet B is called a *code* (Perlwitz, Burks & Waterman, 1988). Here $\mathbf{p}_A$ is the probability vector of the elements of alphabet A and $\mathbf{p}_B$ is the probability vector of the elements of alphabet B.

Thus the genetic code is a mapping of the mRNA code words onto the code words of the protein space. The suggestion of Schrödinger (1944), Watson & Crick (1953b) and of Crick *et al.* (1961) in this chapter's epigraph, is called *the sequence hypothesis* in this book.

As we learned in section 2.2.1, alphabet A and alphabet B will not be isomorphic unless they are bijective, that is, there must be a one-to-one mapping from one alphabet to the other. In the important case of error detecting and error correcting codes and indeed, the genetic code, the two alphabets are not isomorphic. The binary alphabet, with the letters 0, 1, is the natural choice of an alphabet in communication engineering. This alphabet is manifested by a hole punched or not punched in a card or tape, a relay closed or open, or a magnetic domain magnetized North–South or South–North. By the same token, because of the structure of DNA and mRNA, the natural choice for the genetic alphabet is four letters that correspond to the four nucleotides typical of DNA or mRNA.

In section 1.2.2 I discussed the raising of probability spaces to integral powers. If the elements of a probability space are the letters of an alphabet then one has the following definition.

Definition: Given a probability space $[\Omega, A, \mathbf{p}_A]$ the elements of which are the letters of an alphabet A, the first, second and higher integral powers of this probability space are called *extensions* of the first, second, third etc. kind of the alphabet A.

The elements of these successive extensions are the elements of the alphabet A, the ordered pairs, the ordered triplets, the ordered quadruplets etc. For example, the numbers on a die are an alphabet. The pairs of numbers are the second extension of that alphabet. The code by which they are mapped on a second alphabet is the sum of those numbers.

4.2 Shannon–Fano coding and Shannon's noiseless coding theorem

4.2.1 *The properties of instantaneous block codes*

A number of useful codes have code words of variable length, including the Morse code, which was used for many years in telegraphy. The genetic code is a member of a class of codes in which all code words have the same number of letters. This class of codes has many interesting properties, a number of which are important in studying and understanding the genetic code.

Definitions:

A set of M sequences *all* of length n composed of an alphabet of q symbols that maps onto a second alphabet is called a *block code*.

A code is called *distinct* if each code word is distinguishable from any other code word.

A code is called *uniquely decodable* if every code word is identifiable when contained in a sequence of other code words. A uniquely decodable code is said to be *instantaneously decodable* if no code word is the prefix of another. A code is said to be *optimal* if (a) it is instantaneously decodable and (b) if the code words have a minimum average length.

We shall now prove that the necessary and sufficient condition for the existence of an instantaneous code is that it satisfies the Kraft inequality (Kraft, 1949). In order to do so we need to establish further relations between the source entropy and the length of code words (equation (2.61)).

Theorem 4.1 (The Gibbs inequality): Suppose we have two probability vectors, x and y, of q elements. Then the Gibbs inequality states that

$$\sum_{i}^{q} x_i \log_e(y_i/x_i) \leqslant 0 \tag{4.1}$$

Proof: It is well known that for all $x > 0$,

$$\log_e x \leqslant x - 1 \tag{4.2}$$

and

$$\log_2 x \leqslant (x-1)\log_2 e \tag{4.3}$$

Therefore, substituting expression (4.2) in expression (4.1):

$$\sum_{i}^{q} x_i \log_e(y_i/x_i) \leqslant \sum_{i}^{q} x_i [(y_i/x_i) - 1] \tag{4.4}$$

$$\sum_{i}^{q} x_i \log_e(y_i/x_i) \leqslant \sum_{i}^{q} (y_i - x_i) = 0 \tag{4.5}$$

Since y_i and x_i are components of the probability vectors **y** and **x**, they are normalized and the right-hand side of (4.5) vanishes. Therefore the theorem is proved. And of course the expression is true when the logarithm is to base two.

Next we need the Kraft inequality (Kraft, 1949).

Theorem 4.2 (The Kraft inequality): A necessary and sufficient condition for the existence of an instantaneous code S of q symbols s_i with code words (codons) arranged according to length as follows, $l_1 \leqslant l_2 \leqslant \ldots \leqslant l_q$, is

$$K = \sum_{i}^{q} 1/r^l i \leqslant 1 \tag{4.6}$$

where the radix r is the number of symbols in the source alphabet.

Proof: In the case of molecular biology the radix $r = 4$ corresponds to the four nucleotides U C A G. The q symbols are the words: the 64 codons of equal length in the second extension of the source alphabet. It is easy to see by trial that the expression is true for the first and second extensions. The proof is done by induction. Assume the statement is true for the probability space Ω^{n-1}. At the next extension l_i is increased by one letter so that

$$\sum_i^q 1/r^{l_i+1} \leqslant \sum_i^q 1/(r)(r^{l_i}) = 1/r \sum_i^q 1/r^{l_i} \leqslant 1 \tag{4.7}$$

There are at most r branches in a decoding tree starting with the first nucleotide, so the inequality is true. Since it can be shown to be true for $r = 4$, it is true in general.

Next we need the relation of the entropy of the probability vector **p** to the average code word length. We define the numbers Q_i as follows:

$$Q_i = r^{-l_i}/K \tag{4.8}$$

where the Q_i are normalized. We may substitute Q_i in the Gibbs inequality, expression (4.1):

$$\sum_i^q p_i \log_2(Q_i/p_i) \leqslant 0 \tag{4.9}$$

$$\sum_i^q p_i(\log_2 Q_i - \log_2 p_i) = -\sum_i^q p_i \log_2 p_i + \sum_i^q p_i \log_2 Q_i \leqslant 0 \tag{4.10}$$

Equation (4.10) may be expressed in terms of the source entropy $H_2(A)$:

$$H_2(A) + \sum_i^q p_i \log_2 Q_i \leqslant 0 \tag{4.11}$$

Substituting the value of Q_i, the expression (4.11) becomes

$$H_2(A) \leqslant \sum_i^q p_i(\log_2 K - \log_2 r^{-l_i}) = \log_2 K + \sum_i^q p_i l_i \log_2 r \tag{4.12}$$

If, as the extension of code words proceeds, every terminal node is a code word, as it is in the genetic code, then $K = 1$. The $\log_2 K$ term vanishes and we have

$$H_2(A) \leqslant \sum_i^q p_i l_i \log_2 r = L \log_2 r \tag{4.13}$$

where L is the expectation value of the length of the code word. It will be convenient to convert to logarithms to the base r. This expression then becomes

$$H_r(A) \leqslant L \qquad (4.14)$$

The entropy is a lower bound for the average code word length for any instantaneous code.

4.2.2 Shannon's noiseless coding theorem

Let us now arrange the code words in ascending order according to probabilities. The code word lengths are integers so we have the following inequality, taking the logarithms to the base r:

$$-\log_r p_i \leqslant l_i < -\log_r p_i + 1 \qquad (4.15)$$

Removing the logarithms we have

$$p_i \geqslant 1/r^{l_i} > p_i/r \qquad (4.16)$$

Let us sum this inequality over all i; because the p_i are normalized,

$$1 \geqslant \sum_i^q 1/r^{l_i} > 1/r \qquad (4.17)$$

This satisfies the Kraft inequality, expression (4.6), and so there exists an instantaneously decodable code having these Shannon–Fano lengths.

The entropy of this distribution may be found by multiplying expression (4.15) by p_i and summing:

$$H_r(A) \leqslant \sum_i^q p_i l_i = L \leqslant H_r(A) + 1 \qquad (4.18)$$

The entropy serves to define an upper bound as well as a lower bound on the length of the average code word.

The entropy of the nth extension of the probability space Ω^n is $H_r(A^n) = nH_r(A)$. If L_n is the length of the average code word in that probability space then

$$H_r(A) \leqslant L_n/n \leqslant H_r(A) + 1/n \qquad (4.19)$$

Since we can make n as large as we wish, the average code word, $L = L_n/n$ can be made as nearly equal to the source entropy as we wish.

This is Shannon's noiseless coding theorem, which relates the entropy of the source to the length of the average code word at any extension. Thus a code always exists that satisfies (4.19).

4.3 Error detecting and error correcting codes

4.3.1 The definition of genetic noise

The source that generates the message is regarded mathematically as a stochastic Markov process characterized by a probability space $[\Omega, A, \mathbf{p}]$ where A, the alphabet of the source, is a random variable. Ideally, there will be a close relation between the sequences generated by the source and those recorded in alphabet B at the receiver. In the real case there is always a second random Markov process that acts on the original message and interchanges the letters of the alphabet. In molecular biology this process is caused by point mutations. Because of the basically unpredictable nature of random processes, no fixed code will recover the original message. Let us borrow a term from communication engineering and call this effect *genetic noise* (Yockey, 1956, 1958, 1974, 1977b). This effect is lumped in Figure 5.1, which describes a general communication system, but in fact it operates at each of the components of the communication system. If the interchange of letters is made with equal probability, the stochastic process is called *white noise*. I shall call this phenomenon *white genetic noise* if the probability of a random change between any two nucleotides is equal. This is analogous with the practice in communication engineering.

   White noise is an idealization that will seldom be found in the laboratory, so in general, the effect of point mutations will be called simply *genetic noise*. Nevertheless, white noise has important applications in coding theory. In practice the transition matrix of the Markov process that operates on the message from the source is not known. Some forms of genetic noise are not white noise for example, burst errors such as may be caused by a flaw in a computer tape or by a particle passing through a segment of DNA. Although burst-error correcting codes have been constructed in communication theory, they are highly specialized. Insertions and deletions cause a frame shift that results in a *catastrophic error propagation*. This phenomenon is also known to communication engineers and is to be avoided in the construction of communication codes (Peterson & Weldon, 1972). Almost all considerations of noise in coding theory and information theory are directed at white noise.

Table 4.1. *Error correcting code: the decoding table for q = k = 2 and* n = 5

Code words in receiver alphabet *B*	U	A	G	C
Sense code words in source alphabet *A*	11000	00110	10011	01101
Code words received at Hamming distance 1	11001	00111	10010	01100
	11010	00100	10001	01111
	11100	00010	10111	01001
	10000	01110	11011	00101
	01000	10110	00011	11101
Code words received at Hamming distance 2	11110	00000	01011	10101
	01010	10100	11111	00001

Source: Peterson & Weldon (1972), with permission.

4.3.2 Error detecting and error correcting properties of block codes

Today a great deal of communication is in the form of data transmission rather than by natural languages. Errors in data transmission or in computing devices are virtually intolerable. Furthermore, one cannot rely on the statistics or meaning of the text to correct errors. Since the genetic code is a block code let us consider how block codes, constructed from the *n*th extension of the source code letters, are used to construct error detecting and correcting codes. In the *n*th extension of the alphabet *A* there are a total of q^n sequences or code words. Of these $q^k(k < n)$ are a subset that is used by the source to send messages to the receiver. These sequences are called *sense code words*. For example, suppose the source alphabet is binary and the fifth extension is used so there are five digits in each code word and 32 possible code words. Let the alphabet, *B*, at the receiver be a quaternary code, U, A, C, G. Of the 32 possible code words, a subset of at least four sense code words in alphabet *A* is needed to be mapped onto the receiver alphabet *B*. Suppose the source and the receiver agree that the source sense code words are those listed at the head of each column of Table 4.1. A receiver of primitive design would decode only the source sense code words to the alphabet *B*. For this primitive receiver, the 28 code words that are not sense code words are *non-sense* code words.

A distinction is made in this book between *non-sense* and *nonsense*, the latter term being synonymous with *foolishness*. Thus in the genetic code, the codons UAA, UGA and UAG are *non-sense* codons. Referring to them as *nonsense* codons is not correct and is unnecessarily pejorative. Since they are not ordinarily assigned code words in the receiver alphabet they act as chain terminators. As a matter of fact, as I shall show in Chapter 7, they do have assignments in mitochondria. Furthermore, I shall not use the term *missense* to indicate a transition or transversion from a codon for one amino acid to a codon for another.

In Chapter 6, I discuss the fact that many sites in a protein sequence may be occupied by a number of functionally equivalent amino acids. Mutations between functionally equivalent amino acids do not affect the activity of the protein. Like the term *nonsense*, the term *missense* is unjustifiably pejorative. A more sophisticated design of the decoder would enable it to decode all the code words in the respective column of the sense word as shown in Table 4.1. Source code words that are assigned to the same destination code word are called *synonymous*. Notice that there are no *ambiguous* codes. That is, each of the source code words is always assigned to only one destination code word.

4.3.3 The effect of the Hamming distance in error correcting and error detecting codes

The number of positions in which synonymous source words differ is called the *Hamming distance* (Hamming, 1950). It is possible to decode all patterns of t or fewer errors if and only if the minimum Hamming distance is at least $2t + 1$. The sense code words listed at the head of each column in Table 4.1 were selected such that they have the largest possible Hamming distance between them and each code word in the other columns. Five of the synonymous code words in each column are separated from the sense code words by only one Hamming unit. The two code words at the bottom of each column are separated by two Hamming units. In each column, all the source sense code words with only one error, due to white noise, and two that have two errors, are received correctly. All other cases result in the sense code word being decoded incorrectly. Of course, the source could be perverse and send any of the synonymous code words in each column. This would compromise the error detecting and correcting property for which the code was designed. For example, referring to the code in Table 4.1, it is clear that any four of the 32 source code words could have been chosen as sense code words. It is undecidable at the destination which of the set of correct code words was actually sent. If the source does stick to the sense code words, the decoder may obtain a measure of the noise on the line and still decode correctly if the noise does not become too great.

In this example the decoder could also be designed to choose a second extension of the quaternary alphabet to include all pairs of U, A, C, G and thus 16 of the source code words would be sense code words. However, then $k = 4$, and only two other source code words would be received correctly. The cost of this would be a greater vulnerability to white noise. Furthermore, in the limit, each of the 32 code words could be designated as a sense code word. In that case, which of the set of correct code words was actually sent is decidable but there is no protection at all from white noise. If 32 code words were needed, the designer of the code would go to a sixth and presumably a seventh extension to construct an error correcting and error detecting code.

If all the source code words are mapped onto the destination alphabet, so that the source entropy and the destination entropy are equal, the code is said to be *complete* or *saturated*. Thus the DNA–mRNA code is complete or saturated, whereas the mRNA–protein code is not saturated unless we regard termination as a coded function.

4.3.4 The error correction effect of the number of codons assigned to each amino acid

In the example of a code in Table 4.1, all source code words that were not sense code words are assigned in equal numbers to the destination alphabet. One of the ways to make a code more secure from white noise is to assign more alternate source code words according to the frequency of code words in the destination alphabet. Such a code uses the redundance in a nearly optimal manner to reduce the effect of genetic noise (Figureau & Labouygues, 1981). The best-protected codons are Leu CUA, CUG, and Arg CGA, CGG. The least-protected is Trp, which has two transitions to a termination codon. Modern codon usage follows the ideal selection of codons only in the case of the CUG codon of Leu. The CUA codon for Leu is used very little in *E. coli*. The CGG and CGA codons of Arg are used very little in *E. coli* and in yeast (Wong & Cedergren, 1986; Bulmer, 1988). It has been pointed out by many authors over several years that the genetic code is arranged to minimize the effect of genetic noise (Sonneborn, 1965; Woese, 1965b; Goldberg & Wittes, 1966; MacKay, 1967; King & Jukes, 1969; Alff-Steinberger, 1969; Jukes, Holmquist & Moise, 1975; Batchinsky & Ratner, 1976; Wolfenden, Cullis & Southgate, 1979; Inarine, 1981; Figureau & Labouygues, 1981; Labouygues & Gourgand, 1981; Dufton, 1983; Cullmann & Labouygues, 1987). In the first place, these authors recognized that having assigned one codon to each of the 20 amino acids, some assignment of the remaining 44 codons, other than non-sense or stop, was necessary to reduce the vulnerability to mutation, that is, to genetic noise. As this process evolved, the genetic code became nearly *complete* or *saturated* in the form we find it today. This is clearly illustrated by analogy with the error correcting binary code in Table 4.1. Thus redundance in the genetic code plays a role in contributing to error protection. King & Jukes (1969) suggested that the codon redundancy of the modern genetic code corresponded to the frequency of amino acids in protein. Later Jukes, Holmquist & Moise (1975) found a frequency of amino acids that differed in most, but not all, cases from that predicted according to the number of codons assigned. The evidence is very plausible, both from the biochemistry of protein and the mathematical requirements of the code for error correction, that the present code assignment resulted from selection pressure very early in the history of life (Chapter 7). The modern genetic code is optimal (section 7.4.2) for the 20 amino acids for which codons have been assigned and it could not be improved without an extension to four or more letters in each code word (Cullmann & Labouygues, 1985). Thus the fact that more than one codon is assigned to 18 of the more common amino acids in protein is seen as very natural, and indeed necessary to achieve a moderately error correcting capability in the genetic code. The *Angst* over this matter in the early investigations is unjustified and the designation of the genetic code as *degenerate* is inappropriately pejorative. There are much more effective error correction mechanisms available in the modern protein formation.

4.4 Codes with a Central Dogma

The binary error correcting code in Table 4.1 has some close parallels with the genetic code. In section 4.3.3 I discussed the error detecting and error correcting properties of the code in Table 4.1. Thus, like that code and any error correcting and error detecting code, the genetic code is *redundant*.

The source need not use the sense code words for transmission because any code word in the column will be received correctly. However, if the less protected code words are used, one error will cause them to be read incorrectly. (Further discussion of the application of this issue to genetic noise and protein error in molecular biology is given in section 11.4.)

The alphabets A and B in Table 4.1 are not bijective, since there are 32 code words in A and four in B. Therefore, after the decoding it is undecidable which of the source code words was actually sent. Two probability spaces are isomorphic if and only if they have the same entropy (Kolmogorov, 1958; Billingsley, 1965; Shields, 1973; Ornstein, 1974). The alphabets A and B do not have the same entropy, and consequently are not isomorphic. This matter is discussed in detail by Petersen (1983). Neither the code in Table 4.1 nor the genetic code is invertible (Billingsley, 1965). Accordingly, no code exists such that the destination can send messages from alphabet B to the source in alphabet A. Thus, the genetic code, like all codes between probability spaces that are not isomorphic, has a Central Dogma.

There is more information in DNA than can be translated to protein. Information is discarded in the transmission from mRNA to protein. The discarding of information by any means, including genetic noise, makes the genetic logic operation irreversible because the transition function lacks a single-valued inverse.

It is not possible to reconstruct the reading of each die if one knows only the sum of the two numbers on each face. By the same token, it is not possible to reconstruct the mRNA sequence from a single protein. Shannon (1948) showed in his channel capacity theorem that a message may be transmitted with an arbitrarily small error frequency if and only if enough redundance is included in the code to correct for the information discarded by noise (section 5.2). I have given further discussion of the irreversibility of the genetic logic system in the derivation of equations (5.22) and (5.23) and in section 11.3.5. Crick (1970) stated that: 'It [the Central Dogma] was intended to apply only to present-day organisms, and not to events in the remote past, such as the origin of life or the origin of the code.' Nevertheless, one sees that the Central Dogma is a fundamental theorem that applies to any communication system in which the entropy of the sending alphabet exceeds that of the receiving alphabet.

However, the information in DNA that cannot be transmitted to a single protein need not be lost. In section 6.5, my explanation of the phenomena of overlapping genes shows that the information that cannot be translated to one protein can be translated to a second and in some cases to a third.

I first pointed out in 1974 that the Central Dogma is a property of any code in which the source alphabet is larger than the destination alphabet (Yockey, 1974, 1978) although not with the mathematical rigor in this section. The Central Dogma is not a first principle or axiom of molecular biology (section 1.1.3). Placing the Central Dogma alongside the Pythagorean theorem enhances rather than diminishes the former's importance. Although many people feel that the Central Dogma belongs only to biology we must, nevertheless, render unto biology that which is biology's and to mathematics that which belongs to mathematics.

There is, of course, no restriction on the DNA to mRNA code being invertible (Baltimore, 1970; Temin & Mizutani, 1970). Crick (1970) pointed out that the Central Dogma is often misunderstood; for example, an unsigned article in *Nature* (1970) interpreted the work of Baltimore (1970) and that of Temin & Mizutani (1970) as requiring a critical reappraisal. The argument given above shows that the transfer of information from protein to DNA or mRNA is *strictly forbidden* because of the mathematical properties of *all* codes in which the entropy of the source is greater than that of the receiver. It does *not* forbid the protein–protein transfer of information, which is forbidden by the Central Dogma as stated by Crick (1970): 'On the other hand, the discovery of just one type of present day cell that could carry out any of the three unknown transfers would shake the whole intellectual basis of molecular biology, and it is for this reason that the Central Dogma is as important today as when it was first proposed' (Crick 1970).

Crick (1970) mentioned the disease scrapie specifically as a possible exception to the restriction of the protein–protein transfer of information. If a mechanism exists to implement a protein–protein genetic code, that is allowed by the entropy theorems and would be no violation of the Central Dogma as expressed in this section. The chemical agent of scrapie, called a *prion*, is a proteinaceous infective agent devoid of nucleic acid (Griffith, 1967; Prusiner, 1982; Kimberlin, 1982; and Wills, 1986, 1989). Griffith (1967) proposed three ways in which the self-replication of proteins could occur. Wills (1986) proposed that the replication is effected through 'pirating' rather than the expression of information stored in the host genome. He has little good to say of the Central Dogma. The difficulties that Wills (1986, 1989) points out with regard to prion infections can be understood if the reproduction of prions is carried out by a protein–protein genetic code.

Root-Bernstein (1983) has proposed two mechanisms of amino acid pairing by which the protein–protein transfer of information might be accomplished. He points out that the question of the existence of the protein–protein transfer of information is important to the problem of scrapie and to the scenarios on the origin of life. If, in the origin of life, proteins came first then there is a requirement for replication by a protein–protein code. Research on the problem of the reproduction of prions and the possibility that this is accomplished by a protein–protein genetic code is important in both the practical and the fundamental spheres of molecular biology and deserves further work.

Exercises

1. The frequency of English letters is as follows in descending order: e, t, a, o, n, r, i, s, h, d, l, f, c, m, u, g, y, p, w, b, v, k, x, j, q, z. Construct an efficient code of dots, dashes, letter spaces and word spaces.
2. There are nine ways to change a codon letter. Show that the error protection of the genetic code is such that the average number of ways in which this change can be made so that the amino acid is changed is 6.509. What is the average Hamming distance? (see Yockey, 1974).
3. Suppose you have an urn from which you draw one ball at a time with replacement. If you draw a red or a white ball you are rewarded with one denarius. If the ball is blue you must forfeit one denarius. You find, after playing the game a suitable number of times, that you estimate that the frequency of red balls is about 0.3, the frequency of white balls is about 0.2 and the frequency of blue balls is about 0.5. What is the estimate of the entropy at the receiver? What is the estimate of the entropy at the source? Is this game isomorphic at the destination with any other gambling game that you know? Does this game have a Central Dogma, that is, does information flow only from the source to the destination?

5

The source, transmission and reception of information

*Yes, Simmias, replied Socrates, that is well said: and I may add that first principles,
even if they appear certain, should be carefully considered; and when they are
satisfactorily ascertained, then, with a sort of hesitating confidence in human reason,
you may, I think, follow the course of the argument; and if that be plain and clear,
there will be no need for any further inquiry.*

Plato *Phaedo*, English translation by George Kimball.
Plochmann, Dell Publishing Co. with permission

5.1 Introductory ideas on the components of a communication system

5.1.1 The standard communication system

Let us now consider, from the first principles discussed in the previous
chapters, the model shown in Figure 5.1 of a general communication
system commonly used by communication engineers. The object of such
systems is to accept messages from the source and to transmit them
through a channel to the destination as free from errors as the specifications
given to the design engineer require. The source generates an ensemble of
messages selected by a Markov process (section 1.3.5) and written in the
finite source alphabet, A, with probability space $[\Omega, A, \mathbf{p}]$. The message is
encoded from the source alphabet to the channel alphabet for transmission
through the channel. At all stages of the communication the message is
acted upon by a second stochastic Markov process that interchanges some
letters in a random and non-reproducible fashion. The result of this
process is called *noise*. It occurs in all blocks, but in Figure 5.1 it is lumped
for clarity because engineers are in the habit of lumping circuit elements
such as resistance, capacitance and inductance. The ensemble of messages,
modified by noise, is received and decoded to the alphabet B in the
probability space $[\Omega, B, \mathbf{p}]$ of the destination.

For the most part we consider only stationary Markov processes. The
sequences we consider are ergodic (section 1.3.6) and have an entropy by
which we may measure the amount of information using either the
ensemble concept or the algorithmic one (section 2.4.1). In general, the
source may not be a prime cause of the messages; rather, these messages

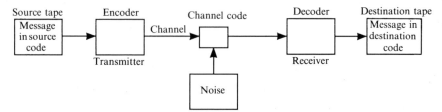

Figure 5.1 The transmission of information from source to destination as conceived in electrical engineering. Noise occurs in all stages but is shown lumped according to accepted practice in electrical engineering.

may be impressed on a tape or record of some sort in the manner of the Turing machine (section 2.4.6).

The characteristics of the ensemble of messages at the source are part of the specifications given to the design engineer. Presumably the characteristics of the noise in the system are also known. The encoder, the channel alphabet, the channel code and the decoder are subject to design. The design engineer chooses a channel alphabet and a channel code from among those error detection and error correction codes that will enable him to recover the message from a noisy channel to meet the accuracy and communication bit rate (bits per second) required in the specifications of the system. Although Shannon's channel capacity theorem (Shannon, 1948; Shannon & Weaver, 1949) is not constructive, nevertheless, it assures the design engineer that a channel code exists such that, by introducing sufficient redundance and thereby accepting a reduced communication rate, he can recover the message with as few errors as specifications require (section 5.2). On the other hand, as in the case of natural languages, the ensemble of messages from the source may have considerable intrinsic redundance. If the channel is very quiet the engineer may choose a data compaction or data compression code for use in the channel (Theorem 2.3; Blahut, 1987). This will allow a higher bit rate and a more economical utilization of the communication system.

5.1.2 The DNA–mRNA–protein communication system

Figure 5.2 shows the DNA–mRNA–protein communication system in simplified form in order to demonstrate its isomorphism with the standard communication system of the communication engineer. The genome or ensemble of genetic messages is generated by a stationary Markov process as described in section 5.1.1 and recorded in the DNA sequence, which is isomorphic with the tape in a Turing machine. The encoding of the genetic

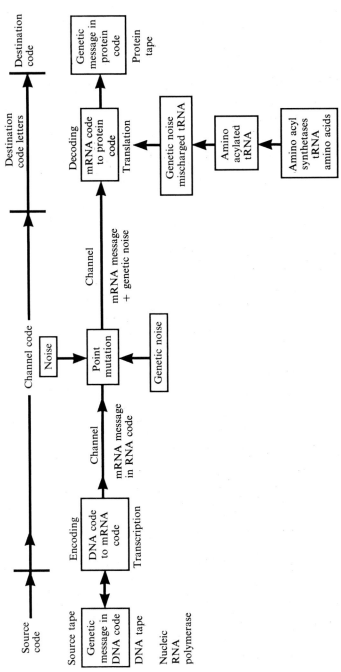

Figure 5.2 The transmission of genetic messages from the DNA tape to the protein tape as conceived in molecular biology. Genetic noise occurs in all stages but is lumped in the figure to fix the idea.

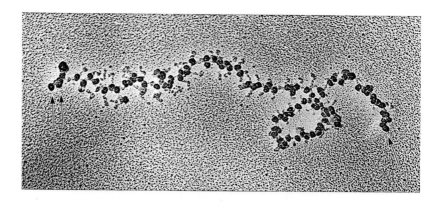

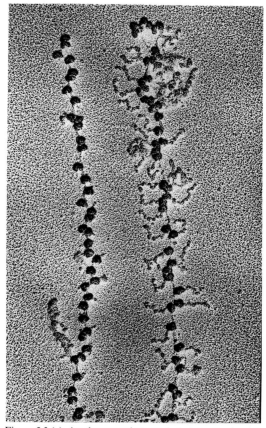

Figure 5.3 (*a*). An electron micrograph showing the spread contents of the salivary glands of *Chironomus thummi*. Single and double arrows indicate the 5' and 3' ends of mRNA. The individual polyribosomes and endoplasmic membrane-bound polyribosome complexes are isolated from the secretory cells. (*b*) An electron

message from the DNA alphabet to the mRNA alphabet is called *transcription*. mRNA plays the role of the channel that communicates the genetic message to the ribosomes, which serve as the decoder. A second stochastic Markov process operates on the message and introduces point mutations. It is shown lumped in the mRNA channel and also in the process of aminoacylation of tRNA in the decoding process although genetic noise is introduced in all stages. Genetic noise is corrected by well-known repair processes in DNA and by the proofreading processes in the charging of tRNA. The means by which the genetic message is recovered in the presence of genetic noise is discussed in Chapter 11.

The genetic message is decoded by the ribosomes from the 64-letter mRNA alphabet to the 20-letter protein alphabet. The decoding is accomplished by the well-known genetic code. This decoding process is called *translation* in molecular biology. The protein molecule, which is the destination, is also a tape. Figures 5.3(*a*) and (*b*) show actual electron micrographs of the ribosomes as they move along the mRNA sequence decoding from the mRNA alphabet to the protein alphabet and thereby producing protein (Kiseleva, 1989). The ribosomes are acting like the reading head on a tape machine. Thus the one-dimensional genetic message is recorded in a sequence of amino acids, which folds up to become a three-dimensional active protein molecule. This process has been designed by evolution to meet the requirements of all organisms that have ever lived, those that are alive today and all organisms yet to evolve.

The direction of flow of information is governed by the entropy of the alphabets where encoding and decoding take place (section 4.4). Thus if an error detecting code is used in any system described in Figures 5.1 or 5.2, that system has a Central Dogma. In the retroviral case, where the two alphabets have the same entropy, information may flow in either direction in the DNA–mRNA encoding if the process is catalyzed by a reverse transcriptase.

5.1.3 Mathematical properties of the source and channel

A *discrete memoryless source* is defined as one in which there are no restrictions or intersymbol influence between letters of the alphabet. If the response of the channel to the input letters is independent of previous

micrograph showing the ribosomes acting like the reading head on a tape machine as they move along the mRNA sequence decoding from the mRNA alphabet to the protein alphabet and thereby producing protein. From Kiseleva (1989) with permission.

letters, the channel is called a *discrete memoryless channel.* A channel that allows input symbols to be transmitted in any sequence is called an *unconstrained channel.* A source that transmits messages written in natural languages is not a memoryless source, since natural languages do have intersymbol influence. Monod (1972), in his philosophy of chance and necessity, says that protein sequences are due to chance and justifies this on the ground that one cannot predict missing amino acids from the properties of their neighbors. That reflects only the fact that there is no intersymbol influence in proteins as there is in natural languages. We shall see in Chapters 6 and 9 that this view of Monod (1972) is fatal to that part of his philosophy.

Thus the DNA–mRNA–protein system is *discrete, memoryless* and *unconstrained.* It transmits the same message to the destination repeatedly, as in a tape recorder. The particular message recorded in the DNA is independent of the genetic information apparatus.

In communication theory the messages that have meaning, or in molecular biology what is called *specificity,* are imbedded in the ensemble of random sequences that have the same statistical structure, that is, the same entropy. We know the statistical structure of the ensemble but not that of the individual sequences. For that reason, the output of any information source, and, in particular, DNA in molecular biology, is regarded as a random process that is completely characterized by the probability spaces $[\Omega, A, \mathbf{p}]$, $[\Omega, B, \mathbf{p}]$. The alphabets A and B are random variables.

Shannon considered only sources with the character of stationary Markov processes. This was generalized to include other sources by McMillan (1953). For our purposes, at least at first, we may regard the particular message transmitted as being one member of a stochastic ensemble generated by a stationary Markov process. The code words of the source, the channel and the destination are the states of a Markov process.

5.2 Shannon's channel capacity theorem

Let us now take up the question of the proper mathematical description of the effect noise has on the ability of a communication system to operate. Although the genetic system is remarkably accurate, genetic noise (Yockey, 1956, 1958, 1974, 1977b) does cause some codons to be translated or decoded incorrectly. At first sight, in order to eliminate errors we might ask for repeated transmissions of the message. If the errors are produced by a stochastic process, they will be scattered along the message and will not be likely to occur repeatedly in the same place. If we use an alphabet (0, 1) and have at least three transmissions, we can use majority logic to determine the correct letter. Let p be the probability of an error, then the probability of two errors at the same position in the three sequences is $3p^2(1-p)$. If p is a number $\ll 1$ the frequency of errors is greatly reduced. The method of majority logic is used to make reliable mechanical or electrical systems from unreliable components (Moore & Shannon, 1956; Kochen, 1959). Majority logic is wasteful of communication system time and does not provide a measure of the effect of noise or a clear procedure to find the maximum performance of a communication system in the presence of noise.

A measure of the effect of noise was provided by Shannon's (1948) discovery of a number of theorems on the behavior of communications systems. The channel capacity theorem is of particular importance. It states that for a noisy channel a code exists between alphabet A and alphabet B such that the communication system can transmit information as close to the channel capacity as one desires with an arbitrarily small error rate. This is done at the cost of incorporating redundance in the original message by means of an error correcting code. The other part of the theorem is that it is not possible to transmit at a rate greater than the channel capacity. The theorem is not constructive and is no help in finding such codes. However, it and other theorems establish from first principles *conditional entropy* (section 5.2.1) as a measure of the effect of noise and, in general, *mutual entropy* as a measure of the relationship between any two sequences one wishes to compare (section 5.5.2). Although, unlike the communications engineer, we are not free to change the genetic code, there are several ways in which Nature has incorporated redundance for error correction in the genetic code and in the protein message. I shall discuss these in Chapters 6, 7 and 11. We may therefore proceed to apply Shannon's channel capacity theorem to problems in molecular biology, secure in the knowledge that we are not just inventing an *ad hoc* procedure.

5.2.1 Conditional entropy is the proper measure of genetic noise

We need a mathematical definition and measure of genetic noise. Shannon (1948) gave an elementary and very lucid explanation of the proper measure of noise and how that measure should be used to calculate the amount of information that can be transmitted through a noisy channel. This explanation is repeated here with permission and a change of notation to conform with that in this book.

Suppose there are two possible symbols 0 and 1, and we are transmitting at a rate of 1000 symbols per second with probabilities $p_0 = p_1 = \frac{1}{2}$. Thus our source is producing information at a rate of 1000 bits per second. During the transmission the noise introduces errors so that on the average one symbol in 100 is received incorrectly (0 as 1, or 1 as 0). What is the rate of transmission of information? Certainly less than 1000 bits per second since about 1 % of the received symbols are incorrect. Our first impulse might be to say that the rate is 990 bits per second, merely subtracting the number of errors. This is not satisfactory since it fails to take into account the recipient's lack of knowledge of where the errors occur. We may carry it to an extreme case and suppose the noise is so great that the received symbols are entirely independent of the transmitted symbols. The probability of receiving 1 is $\frac{1}{2}$, whatever was transmitted, and similarly for 0. Then about half of the received symbols are correct due to chance alone, and we would be giving the system the credit for transmitting 500 bits per second while actually no information is being transmitted at all. Equally 'good' transmission would be obtained by dispensing with the channel entirely and flipping a coin at the receiving point.

Evidently the proper correction to apply to the amount of information transmitted is the amount of this information that is missing in the received signal or, alternatively, the uncertainty when we have received a signal of what was actually sent. From our previous discussion of entropy as a measure of uncertainty it seems reasonable to use the conditional entropy of the message, knowing the received signal, as a measure of this missing information. This is indeed the proper definition, as we shall see later. Following this idea the rate of actual transmission, R, would be obtained by subtracting from the rate of production (i.e. the rate of events x at the source) the average rate of conditional entropy (of events y at the receiver).

$$R = H(x) - H(x|y)$$

The conditional entropy $H(x|y)$ will, for convenience, be called the equivocation. It measures the average ambiguity of the received signal.

In the example considered above, if 0 is received, the *a posteriori* probability that 0 was transmitted is 0.99, and that 1 was transmitted is 0.01. These figures are reversed if 1 is received. Hence

$$H(x|y) = -(0.99 \log_2 0.99 + 0.01 \log_2 0.01)$$

$$= 0.081 \text{ bits per symbol}$$

or 81 bits per second. We may say that the system is transmitting at a rate of $1000 - 81 = 919$ bits per second. In the extreme case where a transmitted 0 is equally likely to be received as 0 or 1, and similarly for 1, the *a posteriori* probabilities are $\frac{1}{2}$ and $\frac{1}{2}$ and

$$H(x|y) = -(\tfrac{1}{2}\log_2 \tfrac{1}{2} + \tfrac{1}{2}\log_2 \tfrac{1}{2}) = 1 \text{ bit per symbol}$$

or 1000 bits per second. The rate of transmission is then zero as it should be.

This anecdotal explanation shows that the probabilities of the errors cannot be a measure of the amount of information that is lost by noise. It shows that it is plausible that *conditional entropy* is the proper measure. It is also apparent that in the quotation above R is a measure of the shared or mutual information of two sequences. The measure of this mutual information is called the *mutual entropy*. This conjecture will be reinforced by the discussion of the mathematical properties of R in section 5.2.2. For example, I shall show in Theorem 5.1 that the mutual entropy is symmetric between two sequences and in Theorem 5.2 that it is zero if and only if the two sequences are independent. Applications to specific problems in molecular biology are given in section 5.4 and in Chapter 6.

We are always allowed to make definitions provided we also prove theorems that show that the new concept is not merely a convenient empiricism. Definitions are useful if and only if such theorems exist, otherwise they are empty. Shannon was too good a mathematician to leave the matter as in the quotation given above. He proved that conditional entropy is the proper measure of the effect of noise. In addition, he proved a number of powerful theorems about mutual entropy that I shall discuss in section 5.2.2.

Regrettably, in the literature of error theories of aging the notion prevails that the proper measures of the loss of genetic information in the genome are the probabilities of protein error (Orgel, 1963, 1970, 1973a; Eigen, 1971; Eigen & Schuster, 1977, 1978a, b; Kirkwood, Rosenberger & Galas, 1986; Parker, 1989). The authors of a number of important papers collected by Holliday (1986) also regard error frequency as a measure of the loss of information due to genetic noise. I made a suggestion (Yockey, 1956) clearly stating that the Shannon conditional entropy is the correct measure of the effect of error or noise on the genetic message, rather than the set of error probabilities *per se* (section 4.4 *et seq*). This idea was developed further in later papers (Yockey, 1958, 1974, 1977b). The correct application of the conditional entropy to protein error and to error theories of aging is given in Chapter 11.

5.2.2 Mathematical properties of mutual entropy

Let us now prove certain theorems that provide support for the use of mutual entropy as a measure of the information transmitted from source to receiver. The input of the channel is characterized by a probability space $[\Omega, A, \mathbf{p}]$ with an input alphabet A of elements x. The output of the channel is characterized by a probability space $[\Omega, B, \mathbf{p}]$ with an output alphabet B of elements y. There is a conditional probability matrix $\mathbf{P}$, with matrix elements $p(j|i)$ that give the probability that if letter y_j appears at the output then the letter x_i was sent. The probabilities p_i and p_j are related by the following equation:

$$p_j = \sum_i^r p_i p(j|i) \tag{5.1}$$

Definition: The *mutual entropy* describing the relation between the input of the channel and the output is defined as follows:

$$I(A;B) = H(x) - H(x|y) \tag{5.2}$$

where $H(x|y)$ is the conditional entropy of x_i given that y_j has been received. The conditional entropy is written in terms of the components of $\mathbf{p}$ and the elements of $\mathbf{P}$ as follows:

$$H(x|y) = -\sum_{i,j} p_j p(i|j) \log_2 p(i|j) \tag{5.3}$$

Theorem 5.1: The value of the mutual entropy is symmetric between the source and the receiver, that is, $I(A;B) = I(B;A)$.

Proof: First express the two entropies in terms of the probabilities, where $p(i,j)$ is the probability of the pair (i,j):

$$I(A;B) = -\sum_i p_i \log_2 p_i + \sum_{i,j} p(i,j) \log_2 p(i|j) \qquad (5.4)$$

$$p_i = \sum_j p(i|j) \qquad (5.5)$$

Substitute p_i in the first term of equation (5.4) (but not in the logarithm).

$$I(A;B) = -\sum_{i,j} p(i,j) \log_2 p_i + \sum_{i,j} p(i,j) \log_2 p(i|j) \qquad (5.6)$$

$$I(A;B) = \sum_{i,j} p(i,j) \log_2 [p(i|j)/p_i] \qquad (5.7)$$

The probability of the pair (i,j) is

$$p(i,j) = p_i p(j|i) = p_j p(i|j) \qquad (5.8)$$

Substitute from equation (5.8) to the argument of the logarithm in equation (5.7):

$$I(A;B) = \sum_{i,j} p(i,j) \log_2 [p(j|i)/p_j] = I(B;A) \qquad (5.9)$$

which proves the theorem.

Theorem 5.2: The mutual entropy $I(A;B)$ is zero if and only if the sequences in alphabet A and those in alphabet B are independent.

Proof: From equation (5.8) we have, assuming that $p_j \neq 0$,

$$p(i,j)/p_j = p(i|j) \qquad (5.10)$$

Substitute $p(i|j)$ in the argument of the logarithm in equation (5.7):

$$I(A;B) = -\sum_{ij} p(i,j) \log_2 \frac{p(i|j)}{p_i p_j} \qquad (5.11)$$

The sequences in alphabet A and those in alphabet B are independent *if and only if* $p(i|j) = p_i p_j$. In that case the argument of the logarithm in equation (5.11) is unity and the logarithm vanishes for all pairs (i,j).

If the sequences in alphabet A and those in alphabet B are identical then $H(x|y)$ vanishes. Therefore, all values of the mutual entropy lie between $H(x)$ and zero. The mutual entropy, or mutual information, is a measure of the information in one set of sequences about another set of sequences (Chaitin, 1975). One can think of it as a measure of the information that the output of a channel with alphabet B, gives about the input from the source with alphabet A (both A and B are random variables). Consequently, the mutual entropy is a measure of the *similarity* of the sequences in alphabet A and those in alphabet B and is essentially the mathematical expression of what we mean by similarity. Since $I(A;B)$ has a finite maximum value, it is reasonable, in the case of a communication channel, to call that maximum the *channel capacity*. According to Shannon's channel capacity theorem, there is no code that can be used to transmit information at a rate greater than the channel capacity. This is intuitively reasonable since sequences in alphabet B cannot be more similar to sequences in alphabet A than identically equal to them. We are, in fact, not limited in this interpretation of the mutual entropy and Shannon's channel capacity theorem. We may remove the communication system entirely and consider the mutual entropy between any two sequences in alphabets A and B. Mutual entropy is a measure of their similarity. Thus the irrelevant and uncorrelated details of each set of sequences cancel out, revealing their similarity. These theorems establish mutual entropy as the *only* measure of similarity between sequences. In this connection we are reminded of the principle of maximum entropy, which gives us the most unbiased estimate of a probability distribution and avoids the assumption of knowledge we do not have.

The discussion given above is descriptive and is directed at problems in molecular biology. My objective is to make the principles easy to understand. Once the whole picture is well in mind the reader may wish to refer for rigorous proofs to the following references: Shannon & Weaver (1949), Khinchin (1957), Feinstein (1958), Billingsley (1965), Ingels (1971), Shields (1973), Ornstein (1974), Martin & England (1981), Petersen (1983), McEliece (1984), Hamming (1986), Chaitin (1987b) and others. The proofs are long and require mathematical sophistication beyond that needed to understand the applications to molecular biology.

5.3 The mutual entropy of the genome as a function of genetic noise

It will prove to be more convenient to deal with the message at the source and therefore with the p_i and the $p(j|i)$ (Yockey, 1974). From Bayes' theorem on conditional probabilities we have (section 1.2.7; see also Lindley, 1965; Feller, 1968; Hamming, 1986):

$$p(i|j) = \frac{p_i p(j|i)}{p_j} \tag{5.12}$$

where the $p(j|i)$ matrix elements are the forward conditional probabilities and the $p(i|j)$ matrix elements are the backward conditional probabilities. Substituting this in the expression for $H(x|y)$ in equation (5.2) we have:

$$I(A;B) = H(x) - H(y|x) - \sum_{i,j} p_i p(j|i) [\log_2(p_j/p_i)] \tag{5.13}$$

where

$$H(y|x) = - \sum_{i,j} p_i p(j|i) \log_2 p(j|i) \tag{5.14}$$

$H(y|x)$ vanishes if there is no genetic noise because then the matrix elements $p(j|i)$ are either 0 or 1. The third term in equation (5.13) is the information that cannot be transmitted to the receiver if the source alphabet is larger than the alphabet at the receiver, that is, if the entropy of the source is greater than that of the receiver so that the source and receiver alphabets are not isomorphic. This is true in general and is a manifestation and quantitative measure of the effect of the Central Dogma that was discussed in section 4.4.

Let us first consider the genetic noise caused by mischarged tRNA species. We may divide the mischarged tRNA species into two groups; namely, those in which the codon of the mischarged amino acid differs by only one nucleotide from the appropriate codon in the mRNA and those in which more than one nucleotide differs from the appropriate codon. An example of the first group is the isoleucyl-tRNA synthetase catalyzed misacylation of tRNA[Ile] with Val (Hopfield *et al.*, 1976). An example of the second group is the valyl-tRNA synthetase catalyzed misacylation of tRNA[Val] with Thr. According to Freter & Savageau (1980), the frequency of the first example is 2.2×10^{-4}, while that for the second is 2.0×10^{-7} (see also the discussion on p. 385 of Freifelder, 1987). Thus, at least in these two cases, misreading one nucleotide is much more likely than misreading two.

Table 5.1. *Genetic code transition probability matrix elements* $p(j\,|\,i)$

Codon x_i	Leu	Ser	Arg	Ala	Val	Pro	Thr	Gly	Ile	Term	Tyr	His	Gln	Asn	Lys	Asp	Glu	Cys	Phe	Trp	Met
UUA	$(1-7\alpha)$	α			α				α	2α									2α		
UUG	$(1-7\alpha)$	α			α					α									2α	α	α
CUU	$(1-6\alpha)$		α		α	α			α			α							α		
CUC	$(1-6\alpha)$		α		α	α			α			α							α		
CUA	$(1-5\alpha)$		α		α	α			α				α								
CUG	$(1-5\alpha)$		α		α	α							α								α
UCU		$(1-6\alpha)$		α		α	α				α							α	α		
UCC		$(1-6\alpha)$		α		α	α				α							α	α		
UCA	α	$(1-6\alpha)$		α		α	α			2α											
UCG	α	$(1-6\alpha)$		α		α	α			α										α	
AGU		$(1-8\alpha)$	3α				α	α	α					α				α			
AGC		$(1-8\alpha)$	3α				α	α	α					α				α			
CGU	α	α	$(1-6\alpha)$			α		α				α						α			
CGC	α	α	$(1-6\alpha)$			α		α				α						α			
CGA	α		$(1-5\alpha)$			α		α		α			α								
CGG	α		$(1-5\alpha)$			α		α					α							α	
AGA		2α	$(1-7\alpha)$				α	α	α	α					α						
AGG		2α	$(1-7\alpha)$				α	α							α					α	α
GCU		α		$(1-6\alpha)$	α	α	α	α								α					
GCC		α		$(1-6\alpha)$	α	α	α	α								α					
GCA		α		$(1-6\alpha)$	α	α	α	α									α				
GCG		α		$(1-6\alpha)$	α	α	α	α									α				
GUU	α			α	$(1-6\alpha)$			α	α							α			α		
GUC	α			α	$(1-6\alpha)$			α	α							α			α		
GUA	2α			α	$(1-6\alpha)$			α	α								α				
GUG	2α			α	$(1-6\alpha)$			α									α				α
CCU	α	α	α	α		$(1-6\alpha)$	α					α									
CCC	α	α	α	α		$(1-6\alpha)$	α					α									
CCA	α	α	α	α		$(1-6\alpha)$	α						α								
CCG	α	α	α	α		$(1-6\alpha)$	α						α								

ACU		2α		α		α	(1–6α)		α					α							
ACC		2α		α		α	(1–6α)		α					α							
ACA		α	α	α		α	(1–6α)		α						α						
ACG		α	α	α		α	(1–6α)								α						α
GGU		α	α	α	α			(1–6α)								α		α			
GGA			2α	α	α			(1–6α)		α							α				
GGC		α	α	α	α			(1–6α)								α		α			
GGG			2α	α	α			(1–6α)									α			α	
AUU	α	α			α		α		(1–7α)					α					α		α
AUC	α	α			α		α		(1–7α)					α					α		α
AUA	2α		α		α		α		(1–7α)						α						α
UAA	α	α								(1–7α)	2α		α		α		α				
UAG	α	α								(1–8α)	2α		α		α		α			α	
UGA	α	α	2α					α		(1–8α)								2α		α	
UAU		α								2α	(1–8α)	α		α		α		α	α		
UAC		α								2α	(1–8α)	α		α		α		α	α		
CAU	α		α			α					α	(1–8α)	2α	α		α					
CAC	α		α			α					α	(1–8α)	2α	α		α					
CAA	α		α			α				α		2α	(1–8α)		α		α				
CAG	α		α			α				α		2α	(1–8α)		α		α				
AAU		α					α		α		α	α		(1–8α)	2α	α					
AAC		α					α		α		α	α		(1–8α)	2α	α					
AAA			α				α		α	α			α	2α	(1–8α)		α				
AAG			α				α			α			α	2α	(1–8α)		α				α
GAU				α	α			α			α	α		α		(1–8α)	2α				
GAC				α	α			α			α	α		α		(1–8α)	2α				
GAA				α	α			α		α			α		α	2α	(1–8α)				
GAG				α	α			α		α			α		α	2α	(1–8α)				
UGU		2α	α					α		α	α							(1–8α)	α	α	
UGC		2α	α					α		α	α							(1–8α)	α	α	
UUU	3α	α			α				α		α							α	(1–8α)		
UUC	3α	α			α				α		α							α	(1–8α)		
UGG	α	α	2α					α		2α								2α		(1–9α)	
AUG	2α		α		α		α		3α						α					(1–9α)	

Reformatted from Yockey, H. P. (1974). An application of information theory to the central dogma and the sequence hypothesis. *Journal of Theoretical Biology*, **46**, 369–406. Published with permission.

If we assume this is true in general, we can set up the matrix elements of the transition probability matrix **P**. I have done this in Table 5.1 (Yockey, 1974) where α is the probability of a base interchange of any one nucleotide, all interchanges being equally probable. By lumping all these probabilities into a single parameter, we are calculating the effect of *white genetic noise*. Substitute the matrix elements from Table 5.1 into equations (5.13) and (5.14) and, replacing the logarithm by its expansion and including only terms of the second degree, we have (Yockey, 1974):

$$I(A;B) = H(\mathbf{p}) - 1.7915 + 34.2108\,\alpha^2 + 6.8303\,\alpha\log_2\alpha \qquad (5.15)$$

In the absence of noise, the terms involving α vanish and the number 1.7915 is the difference in entropy between the source and the receiver and is, therefore, the amount of information that cannot be transmitted from a DNA sequence to a protein sequence because of the redundance in the genetic code (remember that $0\log 0 = 0$). There are actually 12 transversion and transition probabilities among the four nucleotides, considering both directions of change. If these transversion and transition probabilities were known, an equation similar to (5.15) can be derived with these additional parameters. However, equation (5.15) is sufficient to describe the general effect of genetic noise on the genome.

Whether or not the amino acids replaced by the action of genetic noise are errors in the protein sequence must be determined at each site in a particular protein. To get the actual decrease in mutual information as a function of α, one must substitute zero for α if the replacement amino acid is functionally acceptable *at each site* in the protein sequence. As α increases, the value of the mutual information falls nearer and nearer to that of the information content of the protein being considered. There will be a distribution of protein sequences that are functionally active and those that are not. Consequently, the population of functional proteins falls gradually below the level needed to preserve the viability of the cell (Yockey, 1958).

5.4 Mutual entropy as a measure of the information content or complexity of protein families

In section 6.3 I shall discuss the functionally equivalent replacement that may be made at each site in iso-1-cytochrome c. If one selects an active iso-1-cytochrome c from the ensemble of all iso-1-cytochrome c sequences, one is uncertain which of the several functionally equivalent amino acids occupies any given variable site. Clearly the measure of this uncertainty is the conditional entropy $H(y|x)$ given in equations (5.13) and (5.14). We may therefore subtract the conditional entropy $H(y|x)$ from the source entropy. This will give us a measure of the information content at that site. If the site is invariant, there is no uncertainty and the conditional entropy vanishes: the alphabet of the source is larger than that of the receiver and, consequently, the entropy of the source is larger than that of the receiver. In order to take this difference into account, it is more instructive to use equations (5.13) and (5.14).

We must now find the conditional probability matrix **P** of the ergodic Markov chain that describes the evolution communication channel (Cullmann & Labouygues, 1987). **P** must be obtained in its equilibrium state. It completely describes the communication channel, and its matrix elements $p(j|i)$ give the probability of the appearance of the codon j of an amino acid residue following the occurrence of a codon i. As shown in Table 6.3, as many as 19 amino acids may appear at certain sites in iso-1-cytochrome c.

The method of calculating the value of the matrix elements follows the discussion of the Perron–Frobenius theorem in section 1.3.3. We now divide the codons into two groups. The codons for invariant amino acids and for those amino acids that are not functionally equivalent at the site in question will obey the genetic code; therefore the matrix elements will be either zero or unity. There is no uncertainty and those terms in equation (5.14) will vanish. We may now allow the functionally equivalent amino acids to mutate among themselves. The course of these mutations may be followed in much the same way as in section 1.3.3. The matrix elements of **P** are the transition probabilities for the codons of the functionally acceptable amino acids at a given site, and **P** must be doubly stochastic and regular. (We recall that a regular matrix is one in which, at some power, all matrix elements are > 0.) Let **p**(0) be the prior probability vector of the functionally equivalent amino acids at a given site in the protein sequence. Let t be the number of steps in which a mutation is fixed in a population. **P** is a square doubly stochastic matrix since the nucleotides interchange among themselves. We may therefore raise **P** to the power t. The probability vector after t steps will be **p**(t):

$$\mathbf{p}(0)\mathbf{P}^t = \mathbf{p}(t) \tag{5.16}$$

As t grows beyond bounds, the matrix elements approach those of the limiting transition matrix **T**. In the limit they become equal to each other and all knowledge of the original probability vector $\mathbf{p}(0)$ is lost. Therefore, as time goes on, eventually the $p(j|i)$ can all be set equal to $1/s$ where s is the total number of codons of all the functionally equivalent amino acids at a given site. We shall divide the amino acids into classes C_6, C_4, C_3, C_2, and C_1, the subscripts indicating the number of codons for each class. At first I shall assume that the probability of each amino acid is proportional to the number of codons. The number of codons for the class of the functionally equivalent residue j is r_j. Recalling equation (5.1), for all functionally equivalent amino acids we have

$$p_j = \sum_j (1/n)(1/s) r_j = \sum_j r_j(1/ns) \tag{5.17}$$

and therefore, for the functionally equivalent amino acids *only*:

$$p_j = r/ns \tag{5.18}$$

where

$$\sum_j r_j = r \tag{5.19}$$

The sum of all codons pertaining to the class of the accepted amino acids is r. That is, if Ser and Tyr are the functionally equivalent amino acids, then $r = 6 + 2 = 8$. We may now substitute in equations (5.13) and (5.14):

$$I = H(\mathbf{p}) + \sum_{i,j} (1/n)(1/s) \log_2 (1/s)$$

$$- \sum_{i,j} \{(1/n)(1/s) \log_2 [(r/ns)\, n] + (1/n) \log_2 r_j\} \tag{5.20}$$

Combine the first terms of the second and third expressions in equation (5.20):

$$I = H(\mathbf{p}) + \sum_{i,j} (1/n)[(1/s) \log_2 (1/s) - (1/s) \log_2 (1/s) - (1/s) \log_2 r]$$

$$- \sum_{i,j} (1/n) \log_2 r_j \tag{5.21}$$

The first two terms in the second term in the brackets of equation (5.21) cancel. There are $s \times r$ terms in the third expression in the bracket, so that after performing the summation that term becomes $-(r/n) \log_2 r$. The last term is summed over the amino acids that are *not* included in the class of the functionally accepted mutations. Upon summation, that term produces terms such that the coefficient of $\log_2 r_j$ is the number of amino acids, a_j, in class C_j *not* included in the accepted ones, multiplied by the number of codons for that amino acid. Then equation (5.21) reduces to the following very simple equation:

$$I = \log_2 n - (r/n)\log_2 r - a_6(6/n)\log_2 6 - a_4(8/n)$$
$$- a_3(3/n)\log_2 3 - a_2(2/n) \quad (5.22)$$

Equation (5.22) takes into account the information not needed when there is more than one functionally equivalent amino acid residue, the probability of those amino acids and the information that cannot be transmitted to protein because of the redundancy of the genetic code.

Let us remind ourselves that $\mathbf{P}^t$ is a λ-matrix (section 1.3.3) in which the elements are polynomials of degree t. We can, if we wish, stop at any step t to calculate the matrix elements and vector components for substitution in equations (5.13) and (5.14). We can, as a matter of fact, follow the progress of the mutual entropy, step by step, to its equilibrium value.

In deriving equation (5.22) I have assumed that the amino acid probabilities are proportional to the number of codon assignments in the genetic code. The probabilities p_j are often not proportional to the number of codons in the genetic code and in those cases this must be taken into account. Jukes, Holmquist & Moise (1975) suggested the following proportions: $\text{Ala}_{5.3}\text{Arg}_{2.7}\text{Asn}_{3.2}\text{Asp}_{3.3}\text{Cys}_{1.4}\text{Gln}_{3.2}\text{Glu}_{3.3}\text{Gly}_{4.9}\text{His}_{1.3}\text{Ile}_{3.0}\text{Leu}_{3.9}$ $\text{Lys}_{3.9}\text{Met}_{1.1}\text{Trp}_{0.8}\text{Tyr}_{2.0}\text{Val}_{4.1}$. One may establish as many as 20 classes and assign a number of fictitious codons to each class, according to these subscripted numbers r_j (Yockey, 1977b). The r_j need not be whole numbers. Equation (5.22) takes the following form then:

$$I = \log_2 n - (r/n)\log_2 r - \sum_j \delta_j a_j r_j \log_2 r_j \quad (5.23)$$

$$r = \sum_j (1 - \delta_j) r_j \quad (5.24)$$

where $\delta_j = 0$ if the jth amino acid is included in the set of functionally equivalent amino acids, and $\delta_j = 1$ if not. This provides the means, in Chapter 6, to calculate what is called the *information content* or the *complexity* of a protein sequence.

Exercises

1. Apply equation (5.22) to the DNA to mRNA genetic code. What happens to the a_j? What is the mutual information in the case of white genetic noise?

2. Calculate the matrix elements of **P** by means of the principle of maximum entropy. See Yockey (1977b).

3. In nature Leu is found more frequently in protein than Trp. Why does equation (5.22) give the same information content to Leu as it does to Trp? Calculate $H = -\Sigma_i p_i \log_2 p_i$ using the same values for p_i as used in the derivation of equation (5.22). Does this explain why the same information content is calculated for each amino acid?

4. Examine the robustness of the assumption that all p_i are equal to each other in the derivation of equation (5.22). Suggestion: expand the functions in a Taylor or Cauchy series, substituting $p_i' = p_i + \mathrm{d}p_i$ under the condition that $\Sigma_i \mathrm{d}p = 0$.

5. Why is the term in α^2 in equation (5.15) positive?

II

Applications to problems in molecular biology

6

The information content or complexity of protein families

6.1 The information content or complexity of a homologous protein family

6.1.1 Functionally equivalent amino acids and the neutral theory

In early attempts to calculate the information content of proteins, it was supposed that the message had to specify the exact position of each atom in the protein molecule in the folded condition. According to the sequence hypothesis, however, it is necessary only to specify the sequence of amino acids in the genetic message. Thus, far less information is needed to specify a protein than was thought at first. A further decrease in the estimate of information content is due to the fact that proteins are surprisingly tolerant of some substitutions at certain sites (Anfinsen, 1973; Lesk, 1988). The more amino acids that are functionally equivalent at a given site in a homologous protein family, the less information is needed to specify at least one such residue. The functionally equivalent amino acids may play a very specific role in the active pocket or they may play a more passive role in other parts of the folded protein. Research on the protein-folding pathways has shown that some substitutions at certain sites may have a destabilizing effect on intermediates. Thus, the selectivity of amino acids is determined by the primary role played in the protein-folding process as well as by the requirements of the activity of the completed and folded molecule (Goldenberg *et al.*, 1989). Mutations between functionally equivalent amino acids are predicted by the neutral theory in homologous proteins. It is now widely accepted that such mutations are selectively neutral (Zuckerkandl, 1987).

6.1.2 The conditional entropy of functionally equivalent amino acids

Communication systems and information theory are concerned with sending messages from here to there, from past to present, or from the present to the future. Let us consider evolution as a communication system from past to present. At some time in the history of life the first cytochrome c appeared. As a result of drift, random walk and natural selection, this ancestor genetic message was communicated along the dendrites of a fractal (Mandelbrot, 1982) representing a phylogenetic tree. Most organisms that lived once are now extinct and, of course, their protein sequences are lost. Some dendrites lead to modern organisms, the sequence having changed with time. Thus the original genetic message of the common ancestor specifying cytochrome c, regarded as an input, has many outcomes that nevertheless carry the same specificity. The evolutionary processes can be considered as random events along an ergodic Markov chain (Cullmann & Labouygues, 1987) that have introduced uncertainty in the original genetic message. This uncertainty is measured by the conditional entropy in the same manner as the uncertainty of random genetic noise is measured (section 5.3). Since the specificity of the modern cytochrome c is preserved, although many substitutions have been accepted, this conditional entropy may be subtracted from the source entropy, $H(x)$, to obtain the mutual entropy or information content needed to specify at least one cytochrome c sequence or at least one sequence of any other protein for which a list of functionally equivalent amino acids is available.

Since the sample of cytochrome c sequences available is only a tiny fraction of all the organisms that have ever lived or even of those that live today, it is wise to include in the list of known functionally equivalent amino acids those amino acids that have similar properties. Some of these amino acids may be found in protein sequences in the future. If they are not included in the estimate of the information content or the complexity, the result will be too large. I have shown how to predict functionally acceptable amino acids at given sites by means of a prescription (Yockey, 1977a) that will here be brought up to date.

The final result will be obtained by using equation (5.22) to calculate the information content of a message that determines at least one among the functionally equivalent amino acids at any site in the cytochrome c molecule. The information content of the sequence that determines at least one cytochrome c molecule is the sum of the information content of each site. The total information content is a measure of the complexity of cytochrome c. This will be applied to the solution of certain problems in molecular biology.

6.2 The mutual entropy of homologous protein families

6.2.1 The distinction between 'homologous' and 'similar'

The study of natural proteins shows that they are grouped in homologous families, for example, myoglobin, the α and β hemoglobin chains and cytochrome c. These sequences perform similar functions in the organisms in which they are found (Dayhoff *et al.*, 1972; Anfinsen, 1973). The term 'homologous' means having a common evolutionary origin as evidenced by common function and structure (Jukes & Bhushan, 1986; Reeck *et al.*, 1987; Lewin, 1987). Thus 'homologous,' like 'unique,' cannot be qualified. Sequences are either homologous or they are not. The term 'similar,' with which 'homologous' is often confused, means 'being identical except for a number of sites in the chain'. Human and gorilla cytochrome c are both very similar and homologous, but human and yeast cytochrome c are homologous but less similar. Thus one or more sequences may be more or less similar and consequently the similarity of the sequences may be quantified.

6.2.2 Mutual entropy must replace 'per cent identity' as a measure of the similarity of amino acid or nucleotide sequences

There is a need in molecular biology to have a measure of the similarity between sequences or the information in one set of sequences about another. The notion prevails that the appropriate measure of the similarity or sometimes the degree of homology between sequences is an *ad hoc* score, namely, the 'per cent identity' in the amino acid alignment (Doolittle, 1981, 1987a, b, 1988). Per cent identity is not, however, a measure of similarity for the same reasons that error frequency is not a measure of the effect of noise on the transmission of information (section 5.2.1). In addition, the per cent identity score does not take into account the fact that the letters of the alphabet of the sequences are, almost always, not equally probable, nor, in the application to molecular biology, does it take into account the degree of functional equivalence between the amino acid replacements. For example, Tomarev & Zinovieva (1988) compare squid major lense polypeptides with gluthione S-transferase subunits using the *ad hoc* 'per cent homology' (*sic*) method of scoring. This method scores as non-homologous (*sic*) the closely related Gln–Glu and Gly–Glu pairs and the relatively interchangeable Phe–Tyr pair (Table 6.6) as well as the unrelated Met–Ala pair (Table 6.2). Thus two sequences may have a vanishing per cent identity score yet be closely related. I have pointed out previously (Yockey, 1974) that *which* amino acids may occupy a site as well as *how many* is important. Holmquist & Jukes (1981) have made the same observation. It will be clear from the following discussion that the mutual entropy as calculated from equation (5.22) takes the functional equivalence of amino acid replacements and their probability into account without *ad hoc* assumptions and, accordingly, is the only measure of similarity that has any meaning (section 5.2.2).

'Similarity' derives its meaning and mathematical definition from mutual entropy, just as other words find their meaning in mathematical definitions and the theorems that justify the definitions (Chapter 2). Mutual entropy is the measure of the similarity of any two sequences *or set of sequences* of whatever kind, whether or not they are associated with communication systems and however they may occur (section 5.2.2). In contrast the *ad hoc* per cent identity score cannot be applied to more than two sequences. This is a more general statement than that used in the first application of mutual entropy in establishing Shannon's channel capacity theorem.

Mutual entropy is symmetrical between a sequence from a source and a sequence at a receiver (Theorem 5.1). If the sequences are identical, the mutual entropy has its maximum value. If the sequences are independent, the mutual entropy is zero (Theorem 5.2). If the sequences are those of nucleotides in DNA, RNA or protein, then mutual entropy is a measure of biological similarity. The determination of homology requires other considerations (Jukes & Bhushan, 1986; Reeck *et al.*, 1987; Lewin, 1987). Since homology cannot be qualified (i.e. two or more sequences are either homologous or they are not), the expression 'degree of homology' must be replaced by mutual entropy as a measure and definition of similarity. Two or more sequences may have a large mutual entropy as a result of *convergent* evolution. The criterion for best alignment is that which exhibits the most similarity. Consequently, the best alignment is that for which the mutual entropy is at its maximum. Algorithms similar to that of Needleman & Wunsch (1970) to accomplish proper alignment, as discussed by Feng, Johnson & Doolittle (1985), are based on an *ad hoc* method of scoring. No *ad hoc* corrections need be made to allow for chance coincidences since the equations that define mutual entropy account for chance coincidences.

6.3 A prescription that predicts functionally equivalent amino acids at a given site in protein sequences revisited

6.3.1 *The functional equivalence of cytochrome c in the electron transfer pathway*

In this section I shall revisit a prescription (Yockey, 1977a) for predicting functionally equivalent amino acids in homologous protein families, bring it up to date and evaluate its usefulness. I call it a *prescription* or an *Ansatz* because it does not at this time meet the requirements for a *theory*. The sequences of cytochrome c, a small globular heme-containing protein, are ideal for the application of this prescription. They are included in the aerobic respiratory electron transfer pathway between cytochrome oxidase and cytochrome b_5. The molecular dynamics of the cytochrome c–cytochrome b_5 electron transfer complex is a subject of active research (Wendoloski *et al.*, 1987). Cytochromes of the c type are found in eubacteria, eukaryotes and archaebacteria (Mikelsaar, 1987). With proper control, all cytochromes c are found to be functionally identical in spite of considerable differences in the amino acid sequences (Margoliash, Barlow & Byers, 1970; Dickerson *et al.*, 1971; Margoliash, 1972; Margoliash *et al.*, 1972; Margoliash *et al.*, 1973). Although they are functionally identical in the electron transfer complex, antibodies have been found in rabbits inoculated with a wide variety of cytochromes c, including those from man, *Macaca mulata*, horse, rabbit, kangaroo, turkey and tuna. The studies on the enzymic reactions and the stoichiometry of binding to antibodies points to a relatively small number of areas on the surface of the protein that exhibit this dissimilarity (Smith *et al.*, 1973).

The c-type cytochromes have a long history. Almasy & Dickerson (1978) trace the cytochrome c superfamily to the earliest fermenting bacteria at least 3.8×10^9 years ago (Schidlowski, 1988). Kunisawa *et al.* (1987) suggest that the cytochrome c superfamily can be traced back 3.2×10^9 years. Baba *et al.* (1981) date the origin of eukaryotic cytochrome c at 1.4×10^9 years ago. Wu *et al.* (1986) estimate a figure of 1.2×10^9 years.

6.3.2 Representation of amino acid functional equivalence in an abstract Euclidean space

Sneath (1966) and Epstein (1967) correlated accepted point mutational replacement probabilities (Dayhoff *et al.*, 1972; McLachlan, 1971) with characteristics such as size and polarity to avoid basing predictions of functional equivalence on intuitive notions of chemical similarity that are not generally accepted by all biochemists. This objective approach was improved by Grantham (1974), who proposed a three-dimensional Euclidean space in which the composition, polarity and volume are measured along the three axes. Taylor (1986) prepared a classification of measures of amino acid relatedness in common use and presented it in the form of Venn diagrams for use in protein engineering. Lim & Sauer (1989) have reported that composition, volume and steric interactions are the three most important factors in the physical constraints that determine the functional acceptability of amino acid replacements. The several properties mentioned in this paragraph are not independent and consequently do not form orthogonal eigenvectors on which an abstract Euclidean vector space can be established.

As I discussed in section 1.3.3, such a space can be established only by the use of the orthogonal eigenvectors of the matrix of mutation frequencies. Borstnik & Hofacker (1985) introduced a 20-dimensional Euclidean space of characteristics spanned by eigenvectors of a property preservation matrix closely related to the Dayhoff matrix of mutation frequencies. They showed, using maximum entropy analysis, that regarding protein evolution as a random process, three normalized orthogonal eigenvectors established a three-dimensional flat Euclidean space, in which each amino acid is represented by a point. The position in this space reflects a proper weighting of all the relevant properties of the amino acids including the properties mentioned above. By use of the Pythagorean theorem, we can define a measure of their relatedness as the distance between the points that represent the amino acids. This statement cannot be made if the eigenvectors that define the space are not orthogonal and therefore mutually independent. The procedures of the prescription can be adapted to a space of any finite number of dimensions but it is very interesting that three dimensions are adequate. This work was continued by Borstnik, Pumpernik & Hofacker (1987). The method of presentation of the data given by Borstnik & Hofacker (1985) and by Borstnik *et al.* (1987), given in Table 6.1, is readily adapted to the prescription. Their

Table 6.1. *Three orthonormal eigenvectors* h_4, h_5, h_6 *which are used for the construction of the metric space of polypeptide sequences*

Amino acid	h_4	h_5	h_6
Gly	0.10	0.09	−0.09
Ala	0.08	0.05	−0.90
Pro	0.11	0.10	−0.09
Ser	0.09	0.07	−0.05
Thr	0.08	0.04	−0.09
Gln	0.12	0.13	0.13
Asn	0.10	0.12	0.03
Glu	0.12	0.14	0.02
Asp	0.13	0.15	0.01
Lys	0.11	0.06	0.10
Arg	0.13	0.10	0.21
His	0.08	0.20	0.34
Val	0.03	−0.12	−0.26
Ile	0.02	−0.17	−0.35
Met	0.01	−0.84	0.40
Leu	−0.05	−0.23	−0.22
Cys	0.07	−0.06	0.02
Phe	−0.55	−0.03	−0.45
Tyr	−0.74	0.20	0.42
Trp	0.02	−0.01	0.00

Source: Borstnik & Hofacker (1985), with permission of the Adenine Press.

results differ from that of Grantham (1974) in that they found no tendency in mutations to preserve composition or to form an α-helix. I can find no relation between the distances given in Grantham's table and those in Table 6.2 that are calculated from the coordinates given by Borstnik & Hofacker (1985). For example, the largest distance in Grantham's table is between Cys and Trp, a value of 215. The Cys–Trp distance in Table 6.2 is one of the shortest, 0.07. In Yockey (1977a), I used the data from Grantham (1974) but here I use the coordinates given by Borstnik & Hofacker (1985). The Borstnik–Hofacker (BH) Euclidean eigenvector space and its implementation by the prescription described below elaborates the rationale of the role played by neutral mutations in the Darwinian paradigm and leads to an understanding of the evolution of homologous proteins and the evolution of *de novo* protein functions (Chapter 12).

Table 6.2. *BH amino acid distance matrix (Hamming distance in parentheses)*

Amino acid	Leu	Ser	Arg	Ala	Val	Pro	Thr	Gly	Ile	Tyr	His	Gln	Asn	Lys	Asp	Glu	Cys	Phe	Trp	Met
Leu	0	0.37 (1,2,3)	0.57 (1,2,3)	0.75 (1,2)	0.14 (2,3)	0.39 (2,3)	0.33 (2,3)	0.38 (2,3)	0.16 (1,2)	1.03 (2,3)	0.72 (2,3)	0.53 (2,3)	0.46 (2,3)	0.46 (2,3)	0.48 (2,3)	0.47 (2,3)	0.32 (1,2,3)	0.59 (2,3)	0.32 (1,2)	0.87 (1,2)
Ser	0.37 (1,2,3)	0	0.26 (1,2,3)	0.85 (1,2,3)	0.29 (2,3)	0.05 (2,3)	0.05 (1,2,3)	0.05 (2,3)	0.39 (1,2,3)	0.96 (2,3)	0.41 (2,3)	0.19 (2,3)	0.095 (1,2,3)	0.15 (2,3)	0.11 (2,3)	0.10 (2,3)	0.15 (1,2,3)	0.76 (2,3)	0.12 (1,2)	1.00 (2,3)
Arg	0.57 (1,2,3)	0.26 (1,2,3)	0	1.11 (2,3)	0.53 (2,3)	0.30 (2,3)	0.31 (1,2,3)	0.30 (1,2,3)	0.63 (1,2,3)	0.90 (2,3)	0.17 (2,3)	0.086 (2,3)	0.18 (2,3)	0.12 (1,2,3)	0.21 (2,3)	0.19 (2,3)	0.26 (1,2,3)	0.96 (2,3)	0.26 (1,2)	0.97 (1)
Ala	0.75 (2,3)	0.85 (1,2,3)	1.11 (2,3)	0	0.66 (1,2)	0.81 (1,2)	0.81 (1,2)	0.81 (2,3)	0.60 (2,3)	1.56 (2,3)	1.25 (2,3)	1.03 (2,3)	0.93 (2,3)	1.00 (1,2,3)	0.92 (1,2)	0.93 (1,2)	0.93 (2,3)	0.78 (2,3)	0.90 (2,3)	1.58 (2,3)
Val	0.14 (1,2)	0.29 (2,3)	0.53 (2,3)	0.66 (1,2)	0	0.29 (2,3)	0.24 (2,3)	0.28 (2,3)	0.10 (2,3)	1.08 (2,3)	0.68 (2,3)	0.47 (2,3)	0.38 (2,3)	0.41 (2,3)	0.40 (2,3)	0.39 (2,3)	0.29 (2,3)	0.62 (2,3)	0.28 (2,3)	0.98 (2,3)
Pro	0.39 (2,3)	0.05 (1,2,3)	0.30 (2,3)	0.81 (1,2)	0.29 (2,3)	0	0.07 (1,2)	0.01 (1,2)	0.39 (1,2)	1.00 (2,3)	0.44 (2,3)	0.22 (2,3)	0.12 (2,3)	0.19 (2,3)	0.11 (1,2)	0.12 (1,2)	0.20 (2,3)	0.76 (2,3)	0.17 (2,3)	1.06 (2,3)
Thr	0.33 (2,3)	0.05 (1,2,3)	0.31 (2,3)	0.81 (1,2,3)	0.24 (1,2)	0.07 (1,2)	0	0.05 (2,3)	0.34 (2,3)	0.98 (2,3)	0.46 (2,3)	0.24 (2,3)	0.15 (2,3)	0.19 (2,3)	0.16 (2,3)	0.15 (2,3)	0.15 (2,3)	0.73 (2,3)	0.12 (2,3)	1.01 (2,3)
Gly	0.38 (1,2)	0.05 (1,2,3)	0.30 (1,2,3)	0.81 (1,2)	0.28 (1,2)	0.01 (1,2,3)	0.05 (2,3)	0	0.37 (2,3)	0.99 (2,3)	0.44 (2,3)	0.22 (2,3)	0.12 (1,2)	0.19 (1,2)	0.12 (2,3)	0.12 (2,3)	0.19 (2,3)	0.75 (2,3)	0.16 (2,3)	1.06 (2,3)
Ile	0.16 (1,2)	0.39 (1,2,3)	0.63 (1,2,3)	0.60 (2,3)	0.10 (1,2)	0.39 (1,2)	0.34 (1,2)	0.37 (2,3)	0	1.14 (2,3)	0.79 (2,3)	0.58 (2,3)	0.49 (1,2)	0.51 (1,2)	0.49 (2,3)	0.49 (2,3)	0.39 (2,3)	0.60 (1,2)	0.38 (1,2)	1.00 (1)

Tyr	1.03 (2,3)	0.96 (2,3)	0.90 (2,3)	1.56 (2,3)	1.08 (2,3)	1.00 (2,3)	0.98 (2,3)	0.99 (2,3)	1.14 (2,3)	0	0.82 (1,2)	0.91 (1,2)	0.93 (1,2)	0.92 (2)	0.96 (1,2)	0.95 (2)	0.94 (1,2)	0.92 (1,2)	0.89 (2)	1.28 (3)
His	0.72 (2,3)	0.41 (2,3)	0.17 (2,3)	1.25 (2,3)	0.68 (2,3)	0.44 (2,3)	0.46 (2,3)	0.44 (2,3)	0.79 (2,3)	0.82 (1,2)	0	0.23	0.32 (1,2)	0.28 (2)	0.34 (1,2)	0.33	0.41	1.04 (2,3)	0.40 (3)	1.04 (3)
Gln	0.53 (2,3)	0.19 (2,3)	0.086 (2,3)	1.03 (2,3)	0.47 (2,3)	0.22 (2,3)	0.24 (2,3)	0.22 (2,3)	0.58 (2,3)	0.91 (1,2)	0.23	0	0.10 (2)	0.08 (1,2)	0.12 (2)	0.11 (1,2)	0.23	0.90 (3)	0.22 (2,3)	1.01 (2,3)
Asn	0.46 (2,3)	0.095 (1,2,3)	0.18 (2,3)	0.93 (2,3)	0.38 (2,3)	0.12 (2,3)	0.15 (2,3)	0.12 (2,3)	0.49 (1,2)	0.93 (2)	0.32 (1,2)	0.10	0	0.09 (1)	0.05 (2,3)	0.03	0.18 (3)	0.82 (2,3)	0.15 (2,3)	1.03 (2,3)
Lys	0.46 (2,3)	0.15 (2,3)	0.12 (1,2,3)	1.00 (2,3)	0.41 (2,3)	0.19 (2,3)	0.19 (1,2)	0.19 (2,3)	0.51 (1,2)	0.92 (1,2)	0.28 (2)	0.08	0.09 (1)	0	0.13 (2)	0.11 (3)	0.15 (3)	0.86 (3)	0.15 (2,3)	0.95 (1,2)
Asp	0.48 (2,3)	0.11 (2,3)	0.21 (2,3)	0.92 (1,2,3)	0.40 (2,3)	0.11 (1,2,3)	0.16 (1,2)	0.12 (2,3)	0.49 (1,2)	0.96 (2)	0.34 (1,2)	0.12 (1,2)	0.05 (1)	0.13	0	0.02	0.22 (2)	0.84 (3)	0.73 (2,3)	1.30 (1,2)
Glu	0.47 (2,3)	0.10 (2,3)	0.19 (2,3)	0.93 (1,2)	0.39 (1,2)	0.12 (1,2)	0.15 (2,3)	0.12 (1,2)	0.49 (2,3)	0.95 (1,2)	0.33 (2,3)	0.11 (1,2)	0.03 (2,3)	0.11 (2)	0.02 (1)	0	0.21	0.84 (3)	0.18 (2,3)	1.06 (2,3)
Cys	0.32 (1,2,3)	0.15 (1,2,3)	0.26 (1,2,3)	0.93 (2,3)	0.29 (2,3)	0.20 (2,3)	0.15 (2,3)	0.19 (1,2)	0.39 (2,3)	0.94 (1,2)	0.41 (2)	0.23 (3)	0.18 (2)	0.15 (3)	0.22 (2)	0.21 (3)	0	0.78 (1,2)	0.07 (1)	0.87 (3)
Phe	0.59 (1,2)	0.76 (2,3)	0.96 (2,3)	0.78 (2,3)	0.62 (1,2)	0.76 (2,3)	0.73 (2,3)	0.75 (2,3)	0.60 (1,2)	0.92 (1,2)	1.04 (2,3)	0.90 (3)	0.82 (2,3)	0.86 (3)	0.84 (3)	0.84 (3)	0.78 (1,2)	0	0.78	1.30 (3)
Trp	0.32 (1,2,3)	0.12 (2,3)	0.26 (1,2)	0.90 (2,3)	0.28 (2,3)	0.17 (2,3)	0.12 (2,3)	0.16 (2,3)	0.38 (3)	0.89 (2)	0.40 (3)	0.22 (2,3)	0.15 (3)	0.15 (2,3)	0.19 (3)	0.18 (2,3)	0.07 (1,2)	0.73 (2)	0	0.92 (2)
Met	0.87 (1,2)	1.00 (2,3)	0.97 (1)	1.58 (2,3)	0.98 (2,3)	1.06 (2,3)	1.01 (2,3)	1.06 (2,3)	1.00 (1)	1.28 (3)	1.04 (3)	1.01 (2,3)	1.03 (1,2)	0.95 (1,2)	1.07 (3)	1.06 (2,3)	0.87 (2,3)	1.30 (2)	0.92 (2)	0 (2)

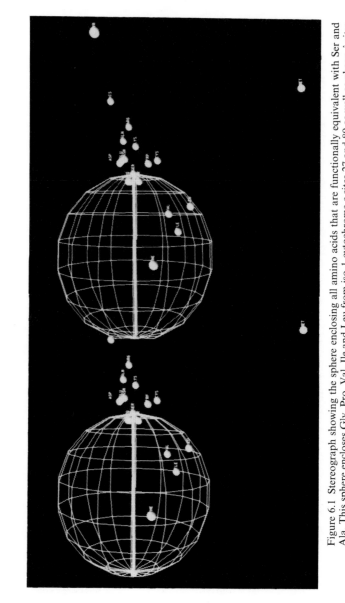

Figure 6.1 Stereograph showing the sphere enclosing all amino acids that are functionally equivalent with Ser and Ala. This sphere encloses Gly, Pro, Val, Ile and Leu from iso-1-cytochrome c sites 27 and 89 as well as phage λ sites 77 and 81 from Reidhaar-Olson & Sauer (1988). See Table 6.7. Stereograph by Clifford A. Pickover, Ph.D. Printed with the permission of International Business Machines Corporation.

The procedure of the prescription for functional equivalence is as follows. One selects all sites with at least two functionally equivalent amino acids in the alignment of the amino acid sequences of a homologous protein family. At each site the pair is chosen that has the largest BH distance of separation. Consider the sphere that has this distance as its diameter and whose center lies on a line between these two amino acids. The prescription asserts that those amino acids that lie on, or are enclosed by, this sphere are functionally equivalent. If they are not already in the list, they are predicted to be found in the future. The success of the prescription is a test of the reliability of the set of eigenvectors, reported by Borstnik & Hofacker (1985), as representing the relative functional equivalence of amino acids. If the BH eigenvectors do not adequately reflect the relative functional equivalence the predictions of the prescription will be found to be erroneous. The only erroneous prediction is that of Pro at sites 17 and 41 in iso-1-cytochrome c. Hampsey *et al.* (1986) report inactive iso-1-cytochrome c sequences that contain Pro at these sites.

Thus the prescription regards all amino acids to be either nearly fully functional or non-functional. A mutation to only one non-functional amino acid in the sequence makes the entire sequence non-functional. Except for the invariant sites there is more than one functionally equivalent amino acid at each site. Thus there is a very large number of functionally equivalent homologous proteins. The method of calculating that number, taking into account the fact that amino acids are not all equally probable, is in section 9.2.2 and the results are given in Table 9.1 for iso-1-cytochrome c. Thus there is a very large number of functionally equivalent 'master sequences' in the high probability set (Theorem 2.2). Except for proteins where nearly all sites are invariant, such as the histones, there is no unique 'master copy,' 'master sequence' or 'consensus sequence' in homologous proteins as some have thought (Eigen, 1971; Eigen & Schuster, 1977, 1978a, b; Eigen, Winkler-Oswatitsch & Dress, 1988).

The stereographs shown in Figures 6.1–6.4 show for four cases the relationships of the amino acids in three dimensions and the application of the prescription.

In order to apply the prescription one needs a table of the distance between all amino acid pairs, the coordinates of the center point between each pair and a means to calculate the radius of each of the 20 amino acids from that center point. According to the Pythagorean theorem, the

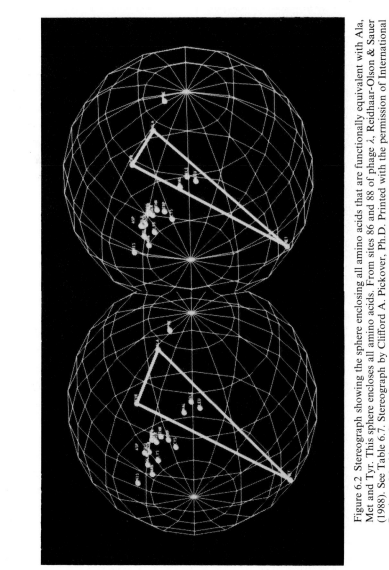

Figure 6.2 Stereograph showing the sphere enclosing all amino acids that are functionally equivalent with Ala, Met and Tyr. This sphere encloses all amino acids. From sites 86 and 88 of phage λ, Reidhaar-Olson & Sauer (1988). See Table 6.7. Stereograph by Clifford A. Pickover, Ph.D. Printed with the permission of International Business Machines Corporation.

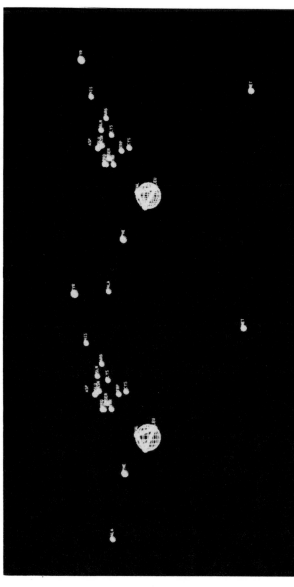

Figure 6.3 Stereograph showing the sphere that encloses Ile, Leu and Val. No other amino acids are enclosed by this sphere. Ala, Met, Phe and Tyr are seen at a distance. The rest of the amino acids form a cluster. From iso-1-cytochrome c sites 102 and 103. See Table 6.4. Stereograph by Clifford A. Pickover, Ph.D. Printed with the permission of International Business Machines Corporation.

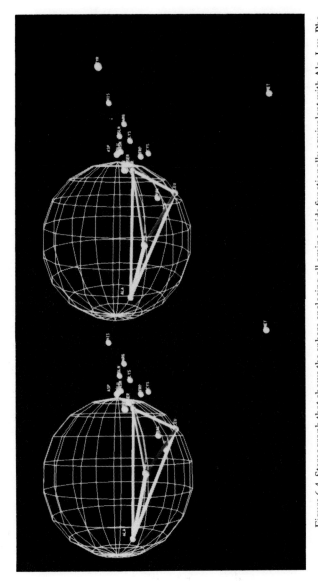

Figure 6.4 Stereograph that shows the sphere enclosing all amino acids functionally equivalent with Ala, Leu, Phe and Thr. Thr is required to complete the Hamming chain to Ala. This sphere encloses Val and Ile. From iso-1-cytochrome c site 43. See Table 6.4. Stereograph by Clifford A. Pickover, Ph.D. Printed with the permission of International Business Machines Corporation.

distance, D, between any two amino acids whose coordinates are, respectively, (x', y', z') and (x'', y'', z''), is given by equation (6.1).

$$D = [(x' - x'')^2 + (y' - y'')^2 + (z' - z'')^2]^{\frac{1}{2}} \tag{6.1}$$

The results are the elements of the distance matrix given in Table 6.2 for each pair of amino acids.

The smallest sphere that encloses the region between the two amino acids with this value of D has its center at (x_2, y_2, z_2):

$$x_2 = (x' + x'')/2 \tag{6.2}$$
$$y_2 = (y' + y'')/2 \tag{6.3}$$
$$z_2 = (z' + z'')/2 \tag{6.4}$$

The radius R of all amino acids from the center is calculated from equation (6.1) substituting the coordinates of each of the 20 amino acids in Table 6.1 for (x', y', z') and (x_2, y_2, z_2) for (x'', y'', z''). The results in the case where Ala–Ser (Table 6.7) determines the sphere are shown by the stereograph in Figure 6.1.

This sphere must enclose all amino acids known at that site. It sometimes happens that three amino acids are nearly equi-distant from each other. One or two amino acids reported at that site may lie outside the sphere defined by any one of the three pairs. In that case the center of the sphere does not lie on a line between any one of the points in the space. Rather, it lies on a plane determined by the coordinates of the three amino acids. The location of the center of the sphere, its radius and the amino acids that lie on its surface or inside can be found by the following procedure.

The general equation of a plane is:

$$x + by + cz = k \tag{6.5}$$

The values of the parameters b, c and k are found by substituting in turn the coordinates of the three amino acids under consideration from Table 6.1 in equation (6.5) and solving the three linear equations in three unknowns by Cramér's rule (section 1.3). Many programs for solving simultaneous equations are available for personal computers. A sophisticated pocket calculator also has this capability.

The general equation of a sphere is

$$x^2 + y^2 + z^2 + 2gx + 2fy + 2hz + d = 0 \tag{6.6}$$

This equation may be put in the following form:

$$(x + g)^2 + (y + f)^2 + (z + h)^2 = g^2 + f^2 + h^2 - d \tag{6.7}$$

The expression on the left of equation (6.7) is that for the square of the distance, in a flat Euclidian space, from the center of the sphere at $(-g, -f, -h)$ to a point (x, y, z). The expression on the right of equation (6.7) is a constant equal to the square of the radius, R, of the sphere:

$$R = (g^2 + f^2 + h^2 - d)^{\frac{1}{2}} \tag{6.8}$$

The coordinates of the three amino acids will satisfy equation (6.7) and this provides three linear equations in four unknowns. The fourth equation is found by the condition that the coordinates of the center of the sphere, $(-g, -f, -h)$, must satisfy equation (6.5). The radius of the sphere and the coordinates of its center are found again by Cramér's rule. One then proceeds as before to calculate the radius of each residue from the center of the sphere. Two examples of this situation are shown in Figure 6.2 by the stereograph for the case of phage λ sites 86 and 88 (Reidhaar-Olson & Sauer, 1988) and in Figure 6.3 by the stereograph for the iso-1-cytochrome c sites 102 and 103.

It may happen that the method discussed in the preceding paragraph will fail to find a sphere that encloses all the reported residues. This is the case when the four most widely separated amino acids are located on the points of a nearly regular tetrahedron. The center of the sphere does not lie on a plane determined by any three of these amino acids. The solution is found by substituting into equation (6.6) the coordinates for each point in turn and solving the four linear equations in the unknown parameters f, g, h, d. An example of this situation is shown in the stereograph of Figure 6.4 for the case of iso-1-cytochrome c site 43. The sphere is determined by Ala, Leu, Phe and Thr.

The closed surface next more complicated than a sphere is the ellipsoid of revolution. This surface is defined by the coordinates of the two foci and a constant that is the sum of the distance from one focus to any point on the surface plus the distance to the other focus. Thus the ellipsoid of revolution is determined by *seven* unknown parameters, which requires coordinates to be known for *seven* points that must be on the surface of the ellipsoid of revolution in a manner similar to the requirement for the coordinates of *four* points to be known in the case of the sphere. Clearly, the simple mathematical procedures described above for the sphere become unwieldy and inappropriate. Referring to Table 6.4, one finds that in cytochrome c there are 13 equivalent amino acids only at site 11. In my analysis this site fits nicely among the Ala–Tyr pairs. Since the Ala–Tyr pairs, taken together, contain all the proteinous amino acids except Met,

Table 6.3. *Information content for iso-1-cytochrome c arranged according to sequence order*

Amino acid site	Radius R	Center coordinates (x_2, y_2, z_2)	Functionally equivalent residues known, plain text / Predicted residues, italics	Information content — All residues	Information content — Known residues
4	0.789	(+0.045, −0.395, −0.25)	(Ala–Met) Gly *Pro* Ser Gln Glu Val Thr Asp Asn *Lys* Ile Leu *Arg Cys Phe Trp*	0.4147581	1.891600784
5	0.528	(−0.39, +0.075, +0.43)	(Phe–Tyr–Asp–Leu) Gly Pro Ser Thr Gln Asn Glu Lys *Cys Trp*	1.702470	2.00146015
6	0.781	(−0.33, +0.125, −0.24)	(Ala–Tyr) *Gly Pro* Ser *Gln* Glu Val Thr Asn *Asp* Lys Ile Leu *Arg His Cys Phe Trp*	0.120681	2.12320374
7	0.789	(+0.045, −0.395, −0.25)	(Ala–Met) *Gly* Pro Ser Gln Glu Glu *Val Thr* Asn *Asp* Lys Arg Ile Leu Cys Phe Trp	0.4147581	2.49615086
8	0.626	(+0.035, +0.115, −0.279)	(Ala–His–Phe) *Gly* Pro Ser Thr Gln Glu Asp *Asn* Lys *Arg* Val Ile Leu Cys Trp	0.208181	2.57039053
9	0.556	(+0.105, +0.075, −0.345)	(Ala–Arg) *Gly* Pro Ser *Gln* Glu Val Thr Asn Asp *Lys* Ile Leu Cys Trp	0.737886	2.79186225
10	0.789	(+0.0455, −0.395, −0.25)	(Ala–Met) *Gly* Pro Ser Gln Glu Asp Thr Asn Leu Ile *Lys* Arg Val Phe Cys Trp	0.4147581	1.82603408
11	0.781	(−0.33, +0.125, −0.24)	(Ala–Tyr) Gly Pro Ser *Thr* Asn Glu *Gln* Asp Val *Arg His* Ile Leu Lys Phe Cys Trp	0.120681	1.62189866
12	0.781	(−0.33, +0.125, −0.24)	(Ala–Tyr) *Gly Pro* Ser Thr Gln Glu Asn Asp Arg His *Val* Ile Leu Lys Phe Cys Trp	0.120681	1.79279439
13	0.556	(+0.105, +0.075, −0.345)	(Ala–Arg) *Gly Pro* Ser Thr *Gln* Asn Glu Asp Lys Val Ile *Ile* Leu Cys Trp	0.737886	2.76892422
14			Gly	4.139192	4.13919287

Table 6.3. (cont.)

Amino acid site	Radius R	Center coordinates (x_2, y_2, z_2)	Functionally equivalent residues known, plain text; Predicted residues, italics	Information content	
				All residues	Known residues
15	0.556	(+0.105, +0.075, −0.345)	(Ala-Arg) Gly *Pro* Ser Thr *Asn* Glu Gln Asp Lys *Val Ile Leu Cys Trp*	0.737868	2.5494943
16	0.155	(+0.105, +0.07, +0.06)	(Thr-Arg) *Gly Pro* Ser Asn Asp Lys *Ala Gln Glu Cys Trp*	1.768044	3.26889868
17	0.626	(+0.035, +0.115, −0.279)	(Ala-His-Phe) *Gly* |Pro|* Ser Thr *Gln Glu Asp Asn Lys Arg Val Ile Leu Cys Trp*	0.672312	2.48852781
18			Phe	4.139192	4.13919287
19	0.257	(+0.065, −0.055, −0.125)	(Ile-Lys) *Gly Pro* Ser Thr *Asn* Glu Asp Val *Leu Cys Trp*	1.701441	3.12645086
20	0.789	(+0.045, −0.395, −0.25)	(Ala-Met) Gly Pro Ser Thr Gln Asn Glu Asp Lys *Arg Val* Ile *Leu Cys Phe Trp*	0.4147581	2.40582668
21	0.059	(+0.12, +0.08, +0.155)	(Arg-Lys) Gln	3.914447	3.91444687
22	0.463	(+0.075, −0.005, −0.44)	(Cys-Ala) *Gly Pro* Ser Thr *Val Ile Leu Tyr (UCU or UCC Ser codons complete Hamming chain)*	2.14982	4.04886869
23	0.517	(+0.10, +0.09, −0.385)	(Ala-Gln) Gly Pro Ser Thr Asn Glu *Asp Lys Val Ile Leu* Leu *Cys Trp*	1.183393	3.00489971
24	0.78	(−0.33, +0.125, −0.24)	(Ala-Tyr) *Gly Pro* Ser Thr *Asn* Glu Gln *Asp Lys Arg His* Val *Ile Leu Cys Phe Trp*	0.121681	2.80171110
25			Cys	4.139192	4.13919267
26			His	4.139192	4.13919287
27	0.425	(+0.085, +0.06, −0.475)	(Ala-Ser) Gly Pro Thr *Val Ile Leu*	2.438366	3.55642443
28	0.78	(−0.33, +0.125, −0.24)	(Ala-Tyr) *Gly Pro* Ser Thr Gln *Asn Glu Asp Lys Arg His* Val *Ile Leu Phe Cys Trp*	0.120681	2.54997885
29	0.78	(−0.33, +0.125, −0.24)	(Ala-Tyr) *Gly Pro* Ser Thr Asn Gln Glu Asp Lys Arg His *Val Ile Leu Cys Phe Trp*	0.120681	3.01576131

30	0.556	(+0.105, +0.075, −0.345)	(Ala–Arg) Gly Pro Ser Thr Gln Glu Asn Asp Val Lys Ile Leu Cys Trp	0.737886	3.17857450
31	0.50	(+0.095, +0.055, −0.40)	(Ala–Lys) Gly Pro Ser Thr Asn Glu Asp Val Ile Leu Cys Trp	1.380022	3.54902894
32	0.463	(+0.10, +0.095, −0.44)	(Ala–Glu) Gly Pro Ser Thr Asp Val Ile Leu Trp (Val required to complete Hamming chain.)	1.845993	3.40265538
33	0.50	(+0.095, +0.055, −0.40)	(Ala–Lys) Gly Pro Ser Thr Asn Glu Asp Val Ile Leu Cys Trp	1.380022	2.85858139
34	0.206	(+0.085, +0.135, +0.145)	(His–Ser) Gln Asn Glu Asp Lys Arg (Only AGU and AGC codons used for Ser.)	3.153886	3.87193410
35	0.518	(−0.37, +0.01, +0.12)	(Gly–Leu–Tyr) Ser Thr Lys Gln Trp Asn Glu Cys	2.377720	2.90374348
36	0.517	(+0.10, +0.09, −0.385)	(Ala–Gln) Gly Pro Ser Thr Val Glu Asn Asp Lys Ile Leu Cys Trp	1.183393	3.35303328
37	0.406	(+0.09, +0.07, −0.495)	(Ala–Gly) Pro Thr Val Ile Leu	3.09277087	4.00804533
38			Pro	4.139192	4.13919287
39	0.466	(+0.09, +0.085, −0.435)	(Ala–Asn) Gly Pro Ser Thr Glu Asp Val Ile Leu Trp (Thr required to complete Hamming chain.)	1.76701428	3.63935312
40			Leu	4.139192	4.13919287
41	0.78	(−0.33, +0.125, −0.24)	(Ala–Tyr) Gly |Pro|* Ser Thr Gln Asn Glu Asp Lys Arg His Val Ile Leu Cys Phe Trp	0.468248	2.07804165
42			Gly	4.139192	4.13919287
43	0.461	(−0.121, +0.134, −0.493)	(Ala–Leu–Phe–Thr) Val Ile (Thr required to complete Hamming chain to Ala.)	3.060877	3.44335746
44	0.572	(−0.36, +0.015, +0.035)	(Ile–Try) Phe Gly Pro Ser Thr Gln Asn Glu Asp Lys Arg His Val Leu Cys Trp	0.468248	3.52899278
45	0.191	(+0.065, +0.00, −0.115)	(Asn–Val) Gly Ser Pro Thr Trp	3.047609	3.63935312
46			Arg	4.139192	4.13919287
47	0.625	(+0.08, +0.125, −0.28)	(Ala–His) Gly Pro Ser Thr Gln Asn Glu Asp Lys Arg Val Ile Leu Cys Trp	0.533821	3.17857450

Table 6.3. (cont.)

Amino acid site	Radius R	Center coordinates (x_2, y_2, z_2)	Functionally equivalent residues known, plain text / Predicted residues, italics	Information content	
				All residues	Known residues
48	0.517	(+0.10, +0.09, −0.385)	(Ala–Gln) *Gly Pro Ser Thr Asn Glu Asp Lys Ile Leu Val Cys Trp*	1.183393	3.49085066
49			*Gly*	4.139192	4.13919287
50	0.12	(+0.10, +0.085, +0.02)	(Gln–Thr) *Gly Pro Ser Asn Glu Asp Lys*	2.744174	3.71635363
51	0.463	(+0.10, +0.095, −0.44)	(Ala–Glu) *Gly Pro Ser Thr Asp Val Ile Leu Trp*	1.845993	3.19202463
52	0.517	(+0.10, +0.09, −0.385)	(Ala–Gln) *Gly Pro Ser Thr Asn Glu Asp Lys Val Ile Leu Cys Trp*	1.183393	2.8017111
53	0.023	(+0.095, +0.08, −0.07)	(Gly–Ser) (*Only AGC codons used for Ser because of Hamming chain.*)	4.055974	4.05597400
54	0.637	(−0.416, +0.458, −0.064)	(Tyr–Phe–Pro) *Gly Ser Asp Asn Glu* (*Ser required to complete Hamming chain.*)	3.015761	3.71635363
55	0.50	(+0.095, +0.055, −0.40)	(Ala–Lys) *Gly Pro Glu Ser Thr Asp Asn Val Ile Leu Cys Trp*	1.380002	3.40265530
56	0.46	(−0.32, +0.13, +0.26)	(Tyr–Lys) *Gln Arg His Trp*	3.563888	4.07361910
57	0.025	(+0.85, +0.055, −0.07)	(Thr–Ser)	3.980021	3.98002064
58	0.50	(+0.095, +0.055, −0.40)	(Ala–Lys) *Gly Pro Ser Thr Asn Glu Asp Val Ile Leu Cys Trp*	1.380002	3.24675743
59	0.406	(+0.09, +0.07, −0.495)	(Ala–Gly) *Pro Thr Val Ile Leu*	3.09277087	4.00804533
60	0.525	(+0.041, −0.39, +0.125)	(Asn–Met–Ile) *Ser Thr Lys Arg Val Leu Cys Trp*	2.180741	3.99567012
61	0.556	(+0.105, +0.075, −0.345)	(Ala–Arg) *Gly Pro Ser Thr Gln Asn Glu Asp Lys Val Ile*	0.737886	3.33748421
62	0.556	(+0.105, +0.075, −0.345)	(Ala–Arg) *Gly Pro Ser Thr Gln Asn Glu Asp Lys Val Ile Leu Cys Trp*	0.737886	2.66071471

No.					
63	0.789	(+0.045, −0.395, −0.25)	(Ala–Met) Gly Pro Ser Thr Gln Asn Glu Asp Lys Arg Val Ile Leu Cys Phe Trp	0.4147581	3.33232212
64	0.466	(+0.09, +0.085, −0.435)	(Ala–Asn) Gly Pro Ser Thr Glu Asp Val Ile Leu Trp	1.76701428	3.69160322
65	0.517	(+0.10, +0.09, −0.385)	(Ala–Gln) Gly Pro Ser Thr Asn Glu Asp Lys Val Ile Leu Cys Trp	1.183393	3.72360551
66	0.517	(+0.10, +0.09, −0.385)	(Ala–Gln) Gly Pro Ser Thr Asn Glu Asp Lys Val Ile Leu Cys Trp	1.183393	2.21689380
67	0.547	(−0.50, −0.155, +0.08)	(Phe–Tyr–Trp–Leu)	3.835193	3.83519332
68	0.517	(+0.10, +0.09, −0.385)	(Ala–Gln) Gly Pro Ser Thr Asn Glu Asp Lys Val Ile Leu Cys Trp	1.183393	2.74417381
69	0.78	(−0.33, +0.125, −0.24)	(Ala–Tyr) Gly Pro Ser Thr Gln Asn Glu Asp Lys Arg His Val Ile Leu Cys Phe Trp	0.120681	2.82900176
70	0.517	(+0.10, +0.09, −0.385)	(Ala–Gln) Gly Pro Ser Thr Asn Glu Asp Lys Val Ile Leu Cys Trp	1.183393	2.74417376
71	0.34	(+0.055, +0.04, +0.04)	(His–Val) Gly Pro Ser Thr Gln Asn Glu Asp Lys Arg Cys Trp	0.891631	2.70564111
72	0.65	(−0.27, −0.435, −0.025)	(Phe–Met) Leu Gly Ser Thr Val Ile Cys Trp	2.455741	3.95854451
73	0.69	(−0.35, −0.31, +0.15)	(Phe–Tyr–Met–His) Gly Pro Ser Thr Arg Gln Glu Asn Asp Lys Val Ile Leu Cys Trp	0.349184	2.92849389
74	0.78	(−0.33, +0.125, −0.24)	(Ala–Tyr) Gly Pro Ser Glu Gln Thr Phe Asn Asp Lys His Val Ile Leu Arg Cys Trp	0.120681	2.316253541
76			Leu	4.139193	4.13919287
77	0.236	(+0.035, −0.045, −0.10)	(Glu–Leu) Thr Lys Gly Pro Ser Asn Val Cys Trp	2.143615	3.71635363
78	0.024	(+0.115, +0.135, +0.02)	(Asn–Asp) Glu	3.983295	4.07361910
79	0.193	(+0.065, −0.035, −0.22)	(Pro–Ile) Gly Val Ser Thr	3.092770	3.35273380
80	0.076	(+0.10, +0.065, +0.025)	(Arg–Ser) Gln Glu Asp Lys Asn (Only AGU and AGC codons used for Ser.)	3.326916	3.84887310
81	0.057	(+0.115, +0.10, +0.06)	(Lys–Glu) Asn	3.983295	4.07361910

Table 6.3. (cont.)

Amino acid site	Radius R	Center coordinates (x_2, y_2, z_2)	Functionally equivalent residues known, plain text; Predicted residues, italics	Information content All residues	Known residues
82	0.543	(−0.363, +0.134, +0.033)	(Phe–Tyr–His–Leu) *Gly Pro Ser Thr* Gln Asn Glu Asp Lys Arg *Cys Trp*	0.949170	3.35970312
83	0.503	(+0.015, −0.505, +0.025)	(Ile–Met) Val *Leu* (*Cys Trp*) (*Leu needed in the Hamming chain; Cys and Trp are not.*)	3.728728	3.95484676
84	0.195	(+0.03, −0.65, −0.155)	(Pro–Leu) *Gly Ser Thr Val Cys Trp*	2.687381	3.98002064
85	0.096	(+0.105, +0.075, 0.005)	(Gly–Lys) *Ser Asn Glu Asp* (*Only AGU and AGC codons used for Ser.*)	3.533286	3.81896041
86	0.073	(+0.09, +0.08, −0.03)	(Thr–Asn) *Gly Pro Ser*	3.402655	4.04886869
87			Lys	4.139193	4.13919287
88			Met	4.139193	4.13919287
89	0.425	(+0.085, +0.06, −0.475)	(Ala–Ser) *Gly Pro Thr* Val Ile *Leu*	2.438366	3.59456655
90			Phe	4.139192	4.13919287
91	0.414	(+0.165, +0.035, +0.102)	(Ala–Gly–Pro–Thr) Val *Ile*	3.167274	3.37790497
92			Gly	4.139192	4.13919287
93	0.650	(−0.27, −0.435, −0.025)	(Phe–Met) *Gly Ser Thr Val Lys Ile Leu Cys Trp*	2.274431	3.78994881
94	0.111	(+0.115, +0.115, +0.02)	(Gln–Pro) *Ser Asn Glu Lys Asp* (*Gln required to complete Hamming chain between Lys and Pro.*)	3.271508	3.71635363
95	0.50	(+0.095, +0.055, −0.40)	(Ala–Lys) *Gly Pro Ser Thr Asn Asp Val Ile Glu Leu Cys Trp*	1.380002	3.71635363
96	0.781	(−0.33, +0.125, −0.24)	(Ala–Tyr) *Gly Pro Ser Thr Asn Glu Gln Asp Lys Arg His Val Ile Leu Cys Phe Trp*	0.120681	2.54934943
97	0.517	(+0.10, +0.09, −0.385)	(Ala–Gln) *Gly Pro Ser Thr Asn Glu Asp Lys Val Ile Leu Cys Trp*	1.183393	2.74417381

98	0.061	(+0.125, +0.14, +0.07)	(Gln–Asp) Glu (*Asx* identified as Asn.)	3.876897	3.98329492
99	0.286	(+0.04, −0.065, −0.005)	(Arg–Leu) Gly *Pro Ser Thr Gln Asn Glu Asp Lys Asp Lys Val Cys Trp*	1.343515	3.72967730
100	0.517	(+0.10, +0.09, −0.385)	(Ala–Gln) Gly *Pro Ser* Thr Asn Glu Lys Asp Val *Ile* Leu *Cys Trp*	1.183393	2.99101090
101	0.169	(+0.105, +0.175, +0.175)	(His–Asp) *Gln* Asn Glu Lys	3.630676	3.98329492
102	0.081	(−0.0097, −0.189, −0.277)	(Ile–Leu–Val)	3.813929	3.81392969
103	0.081	(−0.0097, −0.189, −0.277)	(Ile–Leu–Val)	3.813929	3.81392969
104	0.50	(+0.095, +0.055, −0.40)	(Ala–Lys) *Gly Pro* Ser Thr *Asn Asp Val Ile Glu Leu Cys Trp*	1.380002	3.63935312
105	0.527	(−0.40, +0.067, +0.044)	(Tyr–Leu–Phe–Gln) *Gly Pro Ser Thr Asn Glu Lys Cys Trp*	1.891469	3.78657361
106	0.436	(−0.02, −0.535, +0.09)	(Leu–Met)	4.071295	4.07129599
107	0.625	(+0.08, +0.125, −0.28)	(Ala–His) *Gly Pro Ser* Thr Gln *Asn* Glu *Asp* Lys Arg Val *Ile Leu Cys Trp*	0.533821	2.72628848
108	0.517	(+0.10, +0.09, −0.385)	(Ala–Gln) *Gly Pro* Ser Thr Asn Glu Asp Lys *Val Ile Leu*	1.183393	2.62088490
			Cys Trp		
109	0.50	(+0.095, +0.055, −0.40)	(Ala–Lys) *Gly Pro* Ser Thr *Asn Asp Val Ile* Glu Leu Cys	1.380002	2.90007176
			Trp		
110	0.50	(+0.095, +0.055, −0.40)	(Ala–Lys) *Gly Pro* Ser Thr Asn Glu *Asp Val Ile Leu Cys*	1.380002	3.33708161
			Trp		
111	0.517	(+0.10, +0.09, −0.385)	(Ala–Gln) Gly *Pro Ser Thr* Asp Asn Glu Lys *Val Ile Leu*	1.183393	2.84900176
			Cys Trp		
112	0.517	(+0.10, +0.09, −0.385)	(Ala–Gln) *Gly Pro Ser Thr Asp Asn* Glu Lys *Val Ile Leu*	1.183393	3.26889868
			Cys Trp		
113			Glu	4.139192	4.13919287
Total				233.1288018	373.6288429
Average over 110 sites				2.119353	3.396626

* Pro is predicted by the prescription but found by Hampsey, Das & Sherman (1986, 1988) not to be functionally equivalent.

Source: Hampsey, Das & Sherman (1986, 1988).

Table 6.4. *Information content for iso-1-cytochrome c arranged according to functionally equivalent pairs, triplets and quadruplets*

Amino acid site	Center Radius R	Center coordinates (x_2, y_2, z_2)	Functionally equivalent residues known, plain text Predicted residues, italics	Information content — All residues	Information content — Known residues
23	0.517	(+0.10, +0.09, −0.385)	(Ala–Gln) Gly Pro Ser *Thr* Asn Glu *Asp Lys Val Ile* Leu *Cys Trp*	1.183393	3.00489971
36	0.517	(+0.10, +0.09, −0.385)	(Ala–Gln) *Gly Pro Ser* Thr Val Glu *Asn Asp Lys* Ile *Leu Cys Trp*	1.183393	3.35303328
48	0.517	(+0.10, +0.09, −0.385)	(Ala–Gln) *Gly* Pro Ser Thr Asn Glu Asp Lys Ile *Leu Val* *Cys Trp*	1.183393	3.49085066
52	0.517	(+0.10, +0.09, −0.385)	(Ala–Gln) *Gly* Pro Ser *Thr* Asn Glu Asp Lys Val *Ile Leu* *Cys Trp*	1.183393	2.8017111
65	0.517	(+0.10, +0.09, −0.385)	(Ala–Gln) *Gly Pro Ser Thr* Asn Glu *Asp Lys* Val Ile Leu *Cys Trp*	1.183393	3.72360551
66	0.517	(+0.10, +0.09, −0.385)	(Ala–Gln) Gly *Pro Ser* Thr Asn Glu Asp Lys Val Ile Leu *Cys Trp*	1.183393	2.21689380
68	0.517	(+0.10, +0.09, −0.385)	(Ala–Gln) *Gly* Pro Ser *Thr* Asn Glu Asp Lys *Val Ile Leu* *Cys Trp*	1.183393	2.74417381
70	0.517	(+0.10, +0.09, −0.385)	(Ala–Gln) Gly Pro Ser *Thr* Asn Glu Asp Lys *Val Ile Leu* *Cys Trp*	1.183393	2.74417376
97	0.517	(+0.10, +0.09, −0.385)	(Ala–Gln) Gly *Pro* Ser Thr Asn Glu Asp Lys *Val Ile Leu* *Cys Trp*	1.183393	2.74417381
100	0.517	(+0.10, +0.09, −0.385)	(Ala–Gln) Gly *Pro* Ser Thr Asn Glu Lys *Asp* Val *Ile* Leu *Cys Trp*	1.183393	2.99101090

108	0.517	$(+0.10, +0.09, -0.385)$	(Ala–Gln) *Gly* Pro Ser Thr Asn Glu Asp Lys *Val* Ile *Leu Cys Trp*	1.183393	2.62088490
111	0.517	$(+0.10, +0.09, -0.385)$	(Ala–Gln) *Gly* Pro Ser *Thr* Asp Asn Glu Lys *Val* Ile *Leu Cys Trp*	1.183393	2.84900176
112	0.517	$(+0.10, +0.09, -0.385)$	(Ala–Gln) *Gly* Pro Ser *Thr Asp Asn* Glu Lys *Val* Ile Leu *Cys Trp*	1.183393	3.26889868
6	0.781	$(-0.33, +0.125, -0.24)$	(Ala–Tyr) *Gly Pro Ser Gln* Glu Val Thr Asn *Asp* Lys Ile Leu *Arg His Cys Phe Trp*	0.120681	2.12320374
11	0.781	$(-0.33, +0.125, -0.24)$	(Ala–Tyr) *Gly* Pro Ser Thr Asn Glu *Gln* Asp Val *Arg His* Ile Leu Lys Phe *Cys Trp*	0.120681	1.62189866
12	0.781	$(-0.33, +0.125, -0.24)$	(Ala–Tyr) *Gly* Pro Ser Thr Gln Glu Asn Asp Arg His *Val* Ile Leu Lys Phe *Cys Trp*	0.120681	1.79279439
24	0.78	$(-0.33, +0.125, -0.24)$	(Ala–Tyr) *Gly* Pro Ser Thr *Asn* Glu Gln *Asp* Lys *Arg His* Val *Ile* Leu *Cys Phe Trp*	0.120681	2.80171110
28	0.78	$(-0.33, +0.125, -0.24)$	(Ala–Tyr) *Gly* Pro Ser Thr *Gln Asn* Glu *Asp* Lys *Arg His* Val *Ile* Leu Phe *Cys Trp*	0.120681	2.54997885
29	0.78	$(-0.33, +0.125, -0.24)$	(Ala–Tyr) *Gly* Pro Ser Thr Asn Gln Glu Asp *Lys Arg His* Val *Ile* Leu *Cys Phe Trp*	0.120681	3.01576131
41	0.78	$(-0.33, +0.125, -0.24)$	(Ala–Tyr) *Gly |Pro|** Ser Thr Gln Asn *Glu* Asp Lys *Arg* His *Val Ile* Leu *Cys Phe Trp*	0.468248	2.07804165
69	0.78	$(-0.33, +0.125, -0.24)$	(Ala–Tyr) *Gly* Pro Ser Thr Gln Asn Glu Asp Lys *Arg His Val Ile* Leu *Cys Phe Trp*	0.120681	2.84900176
74	0.78	$(-0.33, +0.125, -0.24)$	(Ala–Tyr) *Gly* Pro Ser Glu Gln *Thr Phe Asn* Asp Lys *His Val Ile* Leu Arg *Cys Trp*	0.120681	2.316253541
96	0.781	$(-0.33, +0.125, -0.24)$	(Ala–Tyr) *Gly* Pro Ser Thr *Asn* Glu Gln Asp Lys *Arg His Val Ile* Leu *Cys Phe Trp*	0.120681	2.54934943

Table 6.4. (*cont.*)

Amino acid site	Center		Functionally equivalent residues known, plain text Predicted residues, italics	Information content	
	Radius R	coordinates (x_2, y_2, z_2)		All residues	Known residues
31	0.50	(+0.095, +0.055, −0.40)	(Ala–Lys) Gly *Pro Ser* Thr Asn Glu *Asp Val Ile Leu Cys* *Trp*	1.380022	3.54902894
33	0.50	(+0.095, +0.055, −0.40)	(Ala–Lys) Gly *Pro Ser* Thr Asn Glu *Asp Val* Ile Leu Cys *Trp*	1.380022	2.85858139
55	0.50	(+0.095, +0.055, −0.40)	(Ala–Lys) *Gly Pro Glu Ser* Thr *Asp* Asn Val *Ile Leu Cys* *Trp*	1.380002	3.40265530
58	0.50	(+0.095, +0.055, −0.40)	(Ala–Lys) *Gly Pro Ser* Thr Asn Glu Asp *Val Ile Leu Cys* *Trp*	1.380002	3.24675743
95	0.50	(+0.095, +0.055, −0.40)	(Ala–Lys) *Gly Pro* Ser Thr Asn *Asp Val Ile* Glu *Leu Cys* *Trp*	1.380002	3.71635363
104	0.50	(+0.095, +0.055, −0.40)	(Ala–Lys) *Gly Pro* Ser Thr *Asn Asp Val Ile* Glu Leu Cys *Trp*	1.380002	3.63935312
109	0.50	(+0.095, +0.055, −0.40)	(Ala–Lys) *Gly Pro* Ser Thr *Asn Asp* Val *Ile* Glu Leu Cys *Trp*	1.380002	2.90007176
110	0.50	(+0.095, +0.055, −0.40)	(Ala–Lys) *Gly Pro* Ser Thr Asn *Glu Asp Val Ile Leu* Cys *Trp*	1.380002	3.33708161
9	0.556	(+0.105, +0.075, −0.345)	(Ala–Arg) *Gly* Pro Ser *Gln* Glu Val Thr Asn Asp *Lys* Ile Leu Cys *Trp*	0.737886	2.79186225
13	0.556	(+0.105, +0.075, −0.345)	(Ala–Arg) *Gly Pro* Ser Thr *Gln* Asn Glu *Asp* Lys Val *Ile* Leu Cys *Trp*	0.737886	2.76892422

15	0.556	(+0.105, +0.075, −0.345)	(Ala–Arg) Gly Pro Ser Thr *Asn* Glu Gln Asp Lys *Val Ile* Leu Cys Trp	0.737886	2.54934943
30	0.556	(+0.105, +0.075, −0.345)	(Ala–Arg) Gly Pro Ser Thr Gln Glu Asn *Asp* Val Lys Ile Leu Cys Trp	0.737886	3.17857450
61	0.556	(+0.105, +0.075, −0.345)	(Ala–Arg) Gly Pro Ser Thr Gln Asn *Glu Asp* Lys *Val* Ile Leu Cys Trp	0.737886	3.33748421
62	0.556	(+0.105, +0.075, −0.345)	(Ala–Arg) Gly Pro Ser Thr Gln Asn Glu Asp Lys *Val Ile* Leu Cys Trp	0.737886	2.66071471
4	0.789	(+0.045, −0.395, −0.25)	(Ala–Met) Gly Pro Ser Gln Glu Val Thr Asp Asn *Lys* Ile Leu Arg Cys Phe Trp	0.4147581	1.89160784
7	0.789	(+0.045, −0.395, −0.25)	(Ala–Met) Gly Pro Ser Gln Glu *Val Thr* Asn *Asp* Lys Arg Ile Leu Cys Phe Trp	0.4147581	2.49615086
10	0.789	(+0.045, −0.395, −0.25)	(Ala–Met) Gly Pro Ser Gln Glu Asp Thr Asn Leu Ile *Lys* Arg Val Phe Cys Trp	0.4147581	1.82603408
20	0.789	(+0.045, −0.395, −0.25)	(Ala–Met) Gly Pro Ser Thr Gln Asn Glu *Asp* Lys *Arg Val* Ile Leu Cys Phe Trp	0.4147581	2.40582668
63	0.789	(+0.045, −0.395, −0.25)	(Ala–Met) Gly Pro Ser *Thr Gln Asn Glu Asp* Lys Arg *Val* Ile Leu Cys Phe Trp	0.4147581	3.33232212
47	0.625	(+0.08, +0.125, −0.28)	(Ala–His) Gly Pro Ser Thr Gln *Asn* Glu Asp Lys Arg *Val* Ile Leu Cys Trp	0.533821	3.17857450
107	0.625	(+0.08, +0.125, −0.28)	(Ala–His) Gly Pro Ser Thr Gln *Asn* Glu Asp Lys Arg Val Ile Leu Cys Trp	0.533821	2.72628848
27	0.425	(+0.085, +0.06, −0.475)	(Ala–Ser) Gly *Pro Thr Val Ile Leu*		2.438366	3.55642443

Table 6.4. (cont.)

Amino acid site	Center Radius R	Center coordinates (x_2, y_2, z_2)	Functionally equivalent residues known, plain text Predicted residues, italics	Information content	
				All residues	Known residues
89	0.425	(+0.085, +0.06, −0.475)	(Ala–Ser) *Gly Pro Thr* Val Ile Leu	2.438366	3.59456655
32	0.463	(+0.10, +0.095, −0.44)	(Ala–Glu) *Gly Pro Ser Thr Asp* Val Ile Leu Trp (*Val required to complete Hamming chain.*)	1.845993	3.40265538
51	0.463	(+0.10, +0.095, −0.44)	(Ala–Glu) *Gly Pro Ser Thr Asp* Val Ile Leu Trp	1.845993	3.19202463
37	0.406	(+0.09, +0.07, −0.495)	(Ala–Gly) *Pro Thr* Val Ile Leu	3.09277087	4.00804533
59	0.406	(+0.09, +0.07, −0.495)	(Ala–Gly) *Pro Thr* Val Ile Leu	3.09277087	4.00804533
39	0.466	(+0.09, +0.085, −0.435)	(Ala–Asn) *Gly Pro Ser Thr Glu Asp* Val Ile Leu Trp (*Thr required to complete Hamming chain.*)	1.76701428	3.63935312
64	0.466	(+0.09, +0.085, −0.435)	(Ala–Asn) *Gly Pro Ser Thr Glu Asp* Val Ile Leu Trp	1.76701428	3.69160322
14			Gly	4.139192	4.13919287
16	0.155	(+0.105, +0.07, +0.06)	(Thr–Arg) *Gly Pro Ser Asn Asp Lys Ala Gln Glu Cys Trp*	1.768044	3.26889868
18			Phe	4.139192	4.13919287
19	0.257	(+0.065, −0.055, −0.125)	(Ile–Lys) *Gly Pro Ser Thr Asn Glu Asp* Val *Leu Cys Trp*	1.701441	3.12645086
21	0.059	(+0.12, +0.08, +0.155)	(Arg–Lys) *Gln*	3.914447	3.91444687
22	0.463	(+0.075, −0.005, −0.44)	(Cys–Ala) *Gly Pro Ser Thr Val Ile Leu Tyr* (*UCU or UCC Ser codons complete Hamming chain.*)	2.144982	4.04886869
25			Cys	4.139192	4.13919267
26			His	4.139192	4.13919287
34	0.206	(+0.085, +0.135, +0.145)	(His–Ser) *Gln Asn Glu Asp Lys Arg* (*Only AGU and ACG codons used for Ser.*)	3.153886	3.87193410

38		Pro		4.139192	4.13919287
40		Leu		4.139192	4.13919287
42		Gly		4.139192	4.13919287
44	(−0.36, +0.015, +0.035)	0.572	(Ile–Tyr) Phe Gly Pro Ser Thr Gln Asn Glu Asp Lys Arg His Val Leu Cys Trp	0.468248	3.52899278
45	(+0.065, +0.00, −0.115)	0.191	(Asn–Val) Gly Ser Pro Thr Trp	3.047609	3.63935312
46		Arg		4.139192	4.13919287
49		Gly		4.139192	4.13919287
50	(+0.10, +0.085, +0.02)	0.12	(Gln–Thr) Gly Pro Ser Asn Glu Asp Lys	2.744174	3.71635363
53	(+0.095, +0.08, −0.07)	0.023	(Gly–Ser) (Only AGU and AGC codons used for Ser because of Hamming chain.)	4.055974	4.05597400
56	(−0.32, +0.13, +0.26)	0.46	(Tyr–Lys) Gln Arg His Trp	3.563888	4.07361910
57	(+0.85, +0.055, −0.07)	0.025	(Thr–Ser)	3.980021	3.98002064
71	(+0.055, +0.04, +0.04)	0.34	(His–Val) Gly Pro Ser Thr Gln Asn Glu Asp Lys Arg Cys Trp	0.891631	2.70564111
72	(−0.27, −0.435, −0.025)	0.65	(Phe–Met) Leu Gly Ser Thr Val Ile Cys Trp	2.455741	3.95854451
76		Leu		4.139193	4.13919287
77	(+0.035, −0.045, −0.10)	0.236	(Glu–Leu) Thr Lys Gly Pro Ser Asn Val Cys Trp	2.143615	3.71635363
78	(+0.115, +0.135, +0.02)	0.024	(Asn–Asp) Glu	3.983295	4.07361910
79	(+0.065, −0.035, −0.22)	0.193	(Pro–Ile) Gly Val Ser Thr	3.092770	3.35273380
80	(+0.10, +0.065, +0.025)	0.076	(Arg–Ser) Gln Glu Asp Lys Asn (Only AGU and AGC codons used for Ser.)	3.326916	3.84887310
81	(+0.115, +0.10, +0.06)	0.057	(Lys–Glu) Asn	3.983295	4.07361910
83	(+0.015, −0.505, +0.025)	0.503	(Ile–Met) Val Leu (Cys Trp) (Leu needed in the Hamming chain; Cys and Trp are not.)	3.728728	3.95484676
84	(+0.03, −0.65, −0.155)	0.195	(Pro–Leu) Gly Ser Thr Val Cys Trp	2.687381	3.98002064

Table 6.4. (cont.)

Amino acid site	Radius R	Center coordinates (x_2, y_2, z_2)	Functionally equivalent residues known, plain text Predicted residues, italics	Information content All residues	Information content Known residues
85	0.096	(+0.105, +0.075, 0.005)	(Gly–Lys) *Ser Asn Glu Asp* (*Only AGU and AGC codons used for Ser.*)	3.533286	3.81896041
86	0.073	(+0.09, +0.08, −0.03)	(Thr–Asn) *Gly Pro Ser*	3.402655	4.04886869
87			Lys	4.139193	4.13919287
88			Met	4.139193	4.13919287
90			Phe	4.139192	4.13919287
92			Gly	4.139192	4.13919287
93	0.650	(−0.27, −0.435, −0.025)	(Phe–Met) *Gly Ser Thr Val Lys Ile Leu Cys Trp*	2.274431	3.78994881
94	0.111	(+0.115, +0.115, +0.02)	(Gln–Pro) *Ser Asn Glu Lys Asp* (*Gln required to complete Hamming chain between Lys and Pro.*)	3.271508	3.71635363
98	0.061	(+0.125, +0.14, +0.07)	(Gln–Asp) *Glu* (*Asx identified as Asn.*)	3.876897	3.98329492
99	0.286	(+0.04, −0.065, −0.005)	(Arg–Leu) *Gly Pro Ser Thr Gln Asn Glu Asp Lys Val Cys Trp*	1.343515	3.72967730
101	0.169	(+0.105, +0.175, +0.175)	(His–Asp) *Gln Asn Glu Lys*	3.630676	3.98329492
106	0.436	(−0.02, −0.535, +0.09)	(Leu–Met)	4.071295	4.07129599
113			Glu	4.139192	4.13919287
8	0.626	(+0.035, +0.115, −0.279)	(Ala–His–Phe) *Gly* Pro Ser Thr Gln Glu Asp *Asn* Lys *Arg Val Ile Leu Cys Trp*	0.208181	2.57039053
17	0.626	(+0.035, +0.115, −0.279)	(Ala–His–Phe) *Gly |Pro|** *Ser Thr Gln Glu Asp Asn Lys Arg Val Ile Leu Cys Trp*	0.672312	2.48852781

35	0.518	(−0.37, +0.01, +0.12)	(Gly–Leu–Tyr) Ser Thr Lys Gln Trp *Asn Glu Cys*	2.377720	2.90374348
54	0.637	(−0.416, +0.458, −0.064)	(Tyr–Phe–Pro) *Gly Ser Asp Asn Glu* (*Ser required to complete Hamming chain.*)	3.015761	3.71635363
60	0.525	(+0.041, −0.39, +0.125)	(Asn–Met–Ile) *Ser Thr Lys Arg Val Leu Cys Trp*	2.180741	3.99567012
102	0.081	(−0.0097, −0.189, −0.277)	(Ile–Leu–Val)	3.813929	3.81392969
103	0.081	(−0.0097, −0.189, −0.277)	(Ile–Leu–Val)	3.813929	3.81392969
5	0.528	(−0.39, +0.075, +0.43)	(Phe–Tyr–Asp–Leu) Gly Pro Ser Thr Gln Asn Glu Lys Cys Trp	1.702470	2.00146015
43	0.461	(−0.121, +0.134, −0.493)	(Ala–Leu–Phe–Thr) Val Ile (*Thr required to complete Hamming chain to Ala.*)	3.060877	3.44335746
67	0.547	(−0.50, −0.155, +0.08)	(Phe–Tyr–Trp–Leu)	3.835193	3.83519332
73	0.69	(−0.35, −0.31, +0.15)	(Phe–Tyr–Met–His) *Gly Pro Ser Thr Arg Gln Glu Asn Asp Lys Val Ile Leu Cys Trp*	0.349184	2.92849389
82	0.543	(−0.363, +0.134, +0.033)	(Phe–Tyr–His–Leu) *Gly Pro Ser Thr Gln Asn Glu Asp Lys Arg Cys Trp*	0.949170	3.35970312
91	0.414	(+0.165, +0.035, +0.102)	(Ala–Gly–Pro–Thr) *Val Ile*	3.167274	3.37790497
105	0.527	(−0.40, +0.067, +0.044)	(Tyr–Leu–Phe–Gln) *Gly Pro Ser Thr Asn Glu Lys Cys Trp*	1.891469	3.78657361
Total				233.1288018	373.6288429
Average over 110 sites				2.119353	3.396626

* Pro is predicted by the prescription but found by Hampsey, Das & Sherman (1986, 1988) not to be functionally equivalent.

Source: Hampsey, Das & Sherman (1986, 1988).

it appears that a sphere is a good approximation to the surface that encloses all functionally equivalent residues. Inspection of the stereographs shown in Figures 6.1, 6.2, 6.3, 6.4 indicates that the prescription is suitable for its purpose without invoking surfaces more complicated than the sphere.

6.3.3 The prediction of functionally equivalent residues in cytochrome c

Hampsey, Das & Sherman (1986, 1988) have provided an alignment of iso-1-cytochrome c sites, together with replacements that from other studies were found to be either functionally equivalent at that site or not functionally equivalent. The alignment is based on 92 eukaryotic cytochromes c and provides more comparative sequences than are available for any other protein. I have used this alignment to apply the prescription to calculate the radius and coordinates of the center of the sphere that encloses all functionally equivalent residues. The amino acids that are found to be on or within the sphere of radius R for each site in the amino acid sequence in cytochrome c are shown in Table 6.3. Presenting the results of the calculations in this way makes it easy to correlate the functionally equivalent amino acids with the cytochrome c structure. Table 6.4 shows the data in Table 6.3 selected according to the alignment of identical pairs, triplets or quadruplets. This arrangement makes it easy to show the similarities among several sites in iso-1-cytochrome c and to make comparisons with other protein sequences by which one may judge the success of the prescription.

Inspection of the alignment of the 13 Ala–Gln pairs shows that each of the 15 predicted amino acids is found at least once except for Trp. In the alignment of 10 Ala–Tyr pairs, all the 19 equivalent amino acids are reported at least once. Similar results are seen in other sets of pairs. Even in the cases where only two pairs are available for comparison, the consistency of the alignment is remarkable. This is especially noteworthy when one considers that 92 sequences is not a large number for statistical evaluation and that many of these sequences are closely related on the phylogenetic tree and are therefore not entirely independent.

One must add another criterion for inclusion in the list of functionally equivalent amino acids, namely, there must be a path by steps of no more than one Hamming distance connecting all members of the list. The Hamming distance between each pair of amino acids is given in Table 6.2. The more amino acids enclosed by the sphere the easier it is to find a

Hamming chain that includes all candidates. Each site has been inspected for assurance that a Hamming chain exists that connects all amino acids. Val is required to complete the chain at sites 32 and 51. Thr is required at sites 39 and 64. The chain between Cys and Ala at site 22 requires the UCU or UCG codons of Ser. Only the AGU and AGC codons of Ser are required at sites 34, 53, 80 and 85. The information content is calculated assuming that the UCN codons of Ser are unassigned so that the total number of assigned codons is 57 rather than 61. At site 83, Leu is needed in the Hamming chain but Cys and Trp are two Hamming distances from the chain that connects the other amino acids. I have selected the *Gln*–Pro pair at site 94 even though Gln is not reported by Hampsey *et al.* in order to complete the Hamming chain between Lys and Pro. The quadruplet Ala–Leu–Phe–*Thr* is used to determine the sphere at site 43 for the same reason.

The prescription may not be falsified in the usual sense since it does not purport to predict amino acids that may be reported in the future if they lie outside the sphere as it is known at present. It may, however, be found not very useful if a number of amino acids that lie well inside the sphere are found not to be functionally equivalent. All the amino acids that are reported to be non-functional by Hampsey *et al.* were found to lie outside the sphere as shown in the lists in Tables 6.3 and 6.4. The only failure of the prescription is that Pro is predicted at sites 17 and 41, whereas Hampsey *et al.* report that iso-1-cytochrome c sequences that contain Pro at these sites are inactive. At the site 34 (my site 37) Das *et al.* (1989) find that the Gly–Ser mutation renders the protein inactive. This is in accordance with the prescription since Ser is not within the sphere of functionally equivalent amino acids. One should note that this mutation is unlikely in Nature since these amino acids are two Hamming steps apart. A mutation Asn–Ile at site 57 (my site 60) restores the function of the protein. At site 38 (my site 41) the mutation His–Pro renders the protein non-functional. Function is restored again by an Asn–Ile mutation.

The prescription may be used to test the invariability of a site. For example, the site Phe-87 (Phe-90 by my numbering) is phylogenetically invariant. To test this and to investigate the electron transfer between cytochrome c and cytochrome c peroxidase, Liang *et al.* (1987) have prepared three mutants at site Phe-87 (Phe-90 by my numbering): namely, Tyr, Gly and Ser. They find that when Tyr or Phe occupy that site, the rate of electron transfer from reduced cytochrome c to the zinc cytochrome c peroxidase π-cation radical is 10^4 times greater than when the site is

Table 6.5. *The conditional information per codon*

Reading frame number, *m*	Reading frame description	$H_m(x|y)$	*I*
0	Primary frame	0.000	4.139
1	Same sense, one to the right	2.144	1.995
2	Same sense, one to the left	2.144	1.995
3	Opposite sense, same frame	1.532	2.067
4	Opposite sense, one to the right	3.424	0.715
5	Opposite sense, one to the left	0.821	3.318

Source: Table 2 of Smith & Waterman (1980), *Mathematical Biosciences* **49**, 27–59.

Table 6.6. *Values of R from the Phe–Tyr center in sites 54, 75 and 90*

Amino acid	*R*	Amino acid	*R*	Amino acid	*R*	Amino acid	*R*
Gly	0.75	Gln	0.78	Arg	0.81	Leu	0.70
Ala	1.14	Asn	0.75	His	0.82	Cys	0.73
Pro	0.76	Glu	0.77	Val	0.75	Phe	0.46
Ser	0.74	Asp	0.78	Ile	0.79	Tyr	0.46
Thr	0.73	Lys	0.76	Met	1.21	Trp	0.67

occupied by the mutant Gly or Ser. Inspection of sites 75 and 90 in Tables 6.3 or 6.4 shows that the Phe–Tyr pair does not include Gly, Ser or, indeed, any other amino acids. In fact, Table 6.6 shows that all other amino acids are so far from the center of the Phe–Tyr pair that one is tempted to predict that the sites 75 and 90 are complete. The Ile–Tyr pair includes Phe at site 44. The other amino acids are accommodated by an increase in *R* and a considerable shift in the center of the sphere, as can be seen from Tables 6.3 or 6.4. Thus the prescription predicts exactly what Liang *et al.* (1987) have reported, namely, that Tyr is functionally equivalent to Phe at cytochrome c site 90 and that this list is complete also for sites 75 and 90. Gardell *et al.* (1985) found that replacing Tyr at site 248 by site-directed mutagenesis in carboxypeptidase A by Phe leaves the catalytic constant toward various peptide and ester substrates unchanged. Noren *et al.* (1989) have utilized recent developments in molecular biology to incorporate unnatural amino acids in proteins by site-specific methods. They studied the conserved site Phe[66] in β-lactimase and replaced Phe by Tyr, Ala and the Phe analogues π-FPhe and π-NO$_2$Phe. Activity was preserved if Tyr and the Phe analogues were incorporated but not in the case of Ala. This is predicted and explained by Table 6.6, which shows that Ala, next to Met, is the most distant of all amino acids from the Phe–Tyr center. Thus the prescription predicts correctly the mutability of the Phe–Tyr pair in several unrelated proteins.

6.3.4 Comparison of cytochrome c sequences with other protein sequences

Site-directed mutagenesis has been used to study the functional equivalence of amino acid replacements. The purpose of these studies is to find the consequences of the chemical changes in order to predict the stability and activity of the altered protein. The prescription draws only on the experimentally determined functionally acceptable amino acids and not on the properties of the host protein itself. Accordingly, site-directed mutagenesis is an interesting test for comparing the properties of sites in different proteins.

Cupples & Miller (1988) have tested changes in activity due to the substitution of 12 different amino acids at four sites in *E. coli* β-galactosidase at sites His[464] and Met[3], and 13 different amino acids at sites Glu[461] and Tyr[503] by site-directed mutagenesis. They find that the substitution of Ser, Gln, Tyr, Lys, Leu, Ala, Cys, Glu, Gly, His, Phe and Pro at Met[3] and His[464] retained most of the activity. The functionally equivalent amino acids at Met[3] and His[464] include Ala, Met and Tyr. The sphere determined by (Ala–Met–Tyr) includes all amino acids (Table 6.7) and so the site has zero information content. This result is in accordance with a remark to that effect by Cupples & Miller (1988). The substitution of any of these amino acids for Glu[461] and Tyr[503] including Asp and Val reduced the activity by two to four orders of magnitude. Thus Glu[461] and Tyr[503] are invariant.

The Cys amino acids at sites 22 and 25 in iso-1-cytochrome c form an S–S bond associated with the heme pocket. Ala is reported at site 22 and this presents the question of the requirement of an S–S bond for folding and formation of the heme pocket. Reference to the need for Cys–Cys S–S bonds in other proteins is helpful. Bovine pancreatic trypsin inhibitor has three disulphide bonds, Cys[14]/Cys[38], Cys[30]/Cys[51] and Cys[5]/Cys[55]. Marks *et al.* (1987) have reported removing the Cys[14]/Cys[38] disulphide bond in

Table 6.7. *Functionally equivalent amino acids in iso-1-cytochrome c, phage T4 lysozyme[1] and phage λ repressor[2,3]*

Source sequence	Site	Radius R	Center coordinates (x_2, y_2, z_2)	Functionally equivalent residues known, plain text / Predicted residues, italics	Information content	
					All residues	Known residues
Phage λ	75	0.517	(+0.10, +0.09, −0.385)	(Ala–Gln) Gly Pro Ser Thr *Asn* Glu Asp *Lys Val Ile Leu Cys Trp*	1.183393	3.33708
Cyto-c	23	0.517	(+0.10, +0.09, −0.385)	(Ala–Gln) Gly Pro Ser Thr Asn Glu Asp *Lys Val Ile Leu Cys Trp*	1.183393	3.00489971
	36	0.517	(+0.10, +0.09, −0.385)	(Ala–Gln) Gly Pro Ser Thr Val Glu *Asn Asp Lys* Ile *Leu Cys Trp*	1.183393	3.35303328
	48	0.517	(+0.10, +0.09, −0.385)	(Ala–Gln) Gly Pro Ser Thr *Asn Glu Asp Lys Ile Leu Val Cys Trp*	1.183393	3.49085066
	52	0.517	(+0.10, +0.09, −0.385)	(Ala–Gln) Gly Pro Ser Thr *Asn* Glu Asp Lys Val *Ile Leu Cys Trp*	1.183393	2.8017111
	65	0.517	(+0.10, +0.09, −0.385)	(Ala–Gln) Gly Pro Ser Thr *Asn Glu Asp Lys* Val Ile *Leu Cys Trp*	1.183393	3.72360551
	66	0.517	(+0.10, +0.09, −0.385)	(Ala–Gln) Gly Pro Ser Thr Asn Glu Asp Lys Val Ile *Leu Cys Trp*	1.183393	2.21689380
	68	0.517	(+0.10, +0.09, −0.385)	(Ala–Gln) *Gly* Pro Ser *Thr* Asn Glu Asp Lys Val *Ile Leu Cys Trp*	1.183393	2.74417381
	70	0.517	(+0.10, +0.09, −0.385)	(Ala–Gln) Gly Pro Ser *Thr* Asn Glu Asp Lys *Val Ile Leu Cys Trp*	1.183393	2.74417376
	97	0.517	(+0.10, +0.09, −0.385)	(Ala–Gln) Gly Pro Ser Thr Asn Glu Asp Lys *Val Ile Leu Cys Trp*	1.183393	2.74417381
	100	0.517	(+0.10, +0.09, −0.385)	(Ala–Gln) Gly Pro Ser Thr Asn Glu Lys *Asp* Val *Ile Leu Cys Trp*	1.183393	2.99101090

	108	0.517	$(+0.10, +0.09, -0.385)$	(Ala–Gln) Gly *Pro* Ser Thr Asn Glu Asp Lys *Val* Ile Leu Cys Trp	1.183393	2.62088490
	111	0.517	$(+0.10, +0.09, -0.385)$	(Ala–Gln) Gly *Pro* Ser Thr Asp Asn Glu Lys *Val* Ile Leu Cys Trp	1.183393	2.84900176
	112	0.517	$(+0.10, +0.09, -0.385)$	(Ala–Gln) Gly *Pro* Ser Thr Asp *Asn* Glu Lys *Val* Ile Leu Cys Trp	1.183393	3.26889868
Phage λ	77	0.425	$(+0.085, +0.06, -0.475)$	(Ala–Ser) Gly *Pro Thr Val Ile* Leu	2.438366	3.9800
Phage λ	81	0.425	$(+0.085, +0.06, -0.475)$	(Ala–Ser) Gly *Pro Thr Val Ile* Leu	2.438366	3.9800
Cyto-c	27	0.425	$(+0.085, +0.06, -0.475)$	(Ala–Ser) Gly *Pro* Thr *Val Ile* Leu	2.438366	3.55642443
	89	0.425	$(+0.085, +0.06, -0.475)$	(Ala–Ser) Gly *Pro* Thr Val Ile Leu	2.438366	3.59456655
Phage λ	83	0.575	$(-0.153, -0.315, +0.155)$	(Met–Gly–Glu–His) Ser Thr Gln Asn Lys Arg Val Leu *Cys*	1.554	1.858
Phage λ	85	0.781	$(-0.33, +0.125, -0.24)$	(Ala–Tyr) Gly Pro Ser Thr Gln *Asn* Glu *Asp Lys* Arg *His* Val Ile Leu Cys *Phe* Trp	0.120681	1.733
Cyto-c	6	0.781	$(-0.33, +0.125, -0.24)$	(Ala–Tyr) *Gly Pro* Ser *Gln* Glu Val Thr Asn *Asp* Lys Ile Leu *Arg His* Cys *Phe* Trp	0.120681	2.12320374
	11	0.781	$(-0.33, +0.125, -0.24)$	(Ala–Tyr) Gly Pro Ser Thr Asn Glu *Gln* Asp Val *Arg His* Ile Leu Lys Phe *Cys* Trp	0.120681	1.62189866
	12	0.781	$(-0.33, +0.125, -0.24)$	(Ala–Tyr) *Gly* Pro Ser Thr Gln Glu Asn Asp Arg *His Val Ile* Leu Lys Phe *Cys* Trp	0.120681	1.79279439
	24	0.78	$(-0.33, +0.125, -0.24)$	(Ala–Tyr) *Gly* Pro Ser Thr *Asn* Glu Gln *Asp* Lys *Arg His Val* Ile Leu Cys *Phe* Trp	0.121681	2.80171110
	28	0.78	$(-0.33, +0.125, -0.24)$	(Ala–Tyr) *Gly* Pro Ser Thr Gln *Asn* Glu *Asp Lys* Arg *His* Val Ile Leu Phe *Cys* Trp	0.120681	2.54997885
	29	0.78	$(-0.33, +0.125, -0.24)$	(Ala–Tyr) Gly *Pro* Ser Thr Asn Gln Glu Asp *Lys* Arg *His* Val Ile Leu Cys *Phe* Trp	0.120681	3.01576131
	41	0.78	$(-0.33, +0.125, -0.24)$	(Ala–Tyr) Gly |Pro|* Ser Thr Gln Asn *Glu* Asp *Lys* Arg *His* Val Ile Leu Cys *Phe* Trp	0.468248	2.07804165
	69	0.78	$(-0.33, +0.125, -0.24)$	(Ala–Tyr) *Gly* Pro Ser Thr Gln Asn Glu Asp Lys Arg *His Val* Ile Leu Cys *Phe* Trp	0.120681	2.84900176

Table 6.7. (*cont.*)

Source sequence	Site	Radius R	Center coordinates (x_2, y_2, z_2)	Functionally equivalent residues known, plain text / Predicted residues, italics	Information content All residues	Known residues
Cyto-c	74	0.78	(−0.33, +0.125, −0.24)	(Ala–Tyr) Gly Pro Ser Glu Gln *Thr Phe Asn* Asp Lys / *His* Val Ile Leu *Arg Cys* Trp	0.120681	2.316253541
	96	0.781	(−0.33, +0.125, −0.24)	(Ala–Tyr) Gly Pro Ser Thr *Asn* Glu Gln Asp Lys *Arg* / *His* Val Ile Leu Cys Phe Trp	0.120681	2.54934943
Phage λ	86	0.86	(−0.183, −0.178, −0.114)	(Ala–Met–Tyr) Gly Pro Ser Thr Gln *Asn* Glu Asp / Lys *Arg His Val Ile* Leu Cys Phe Trp	0.000	1.90384453
Phage λ	88	0.86	(−0.183, −0.178, −0.114)	(Ala–Met–Tyr) *Gly* Pro Ser Thr Gln *Asn* Glu *Asp Lys* / *Arg His* Val Ile Leu Cys Phe Trp	0.000	2.92883644
Phage λ	87	0.436	(−0.02, −0.535, +0.09)	(Leu–Met)	4.071	4.071
Cyto-c	106	0.436	(−0.02, −0.535, +0.09)	(Leu–Met)	4.071	4.071
Phage λ	89	0.789	(+0.045, −0.395, −0.25)	(Ala–Met) Gly *Pro* Ser Thr Gln *Asn* Glu Asp Lys / *Arg Val* Ile Leu Cys *Phe Trp*	0.4147581	1.599
Cyto-c	4	0.789	(+0.045, −0.395, −0.25)	(Ala–Met) Gly *Pro* Ser Gln Glu Val Thr Asp Asn / Lys Ile Leu *Arg Cys* Phe Trp	0.4147581	1.89160784
Cyto-c	7	0.789	(+0.045, −0.395, −0.25)	(Ala–Met) *Gly* Pro Ser Gln Glu Gln Glu *Val Thr* Asn *Asp* / Lys *Ile* Leu Cys Phe Trp	0.4147581	2.49615086
Cyto-c	10	0.789	(+0.045, −0.395, −0.25)	(Ala–Met) *Gly* Pro Ser Gln Glu Asp Thr Leu Ile / Lys *Arg* Val Phe Cys Trp	0.4147581	1.82603408
Cyto-c	20	0.789	(+0.045, −0.395, −0.25)	(Ala–Met) Gly *Pro* Ser Thr Gln Asn Glu *Asp* Lys / *Arg Val* Ile Leu Cys Phe Trp	0.4147581	2.40582668
Cyto-c	63	0.789	(+0.045, −0.395, −0.25)	(Ala–Met) *Gly* Pro Ser Thr Gln Asn *Glu Asp* Lys / *Arg Val* Ile Leu Cys Phe Trp	0.4147581	3.33232212

Phage λ	90	0.790	(+0.047, −0.354, −0.222)	(Ala–Met–His) Gly *Pro* Ser *Thr* Gln *Asn Glu Asp* Lys *Arg* Val *Ile* Leu Cys *Phe* Trp	0.208	2.694
Lysozyme	86	0.625	(+0.08, +0.125, −0.28)	(Ala–His) Gly Pro Ser Thr *Gln Asn Glu Asp Lys* Arg *Val* Ile Leu Cys *Trp*	0.533821	1.778
Cyto-c	47	0.625	(+0.08, +0.125, −0.28)	(Ala–His) *Gly* Pro Ser Thr Gln Asn Glu Asp Lys Arg *Val Ile* Leu Cys *Trp*	0.533821	3.17857450
Cyto-c	107	0.625	(+0.08, +0.125, −0.28)	(Ala–His) *Gly* Pro Ser Thr Gln Asn Glu Asp Lys Arg *Val Ile* Leu Cys *Trp*	0.533821	2.72628848
Cyto-c	108	0.625	(+0.08, +0.125, −0.28)	(Ala–His) *Gly* Pro Ser Thr Gln Asn Glu Asp Lys Arg *Val Ile* Leu Cys *Trp*	0.533821	2.62088490
Phage λ	91	0.463	(+0.075, −0.005, −0.44)	(Ala–Cys) *Gly* Pro Ser Thr Val Ile Leu Trp *(Gly, Pro, Ser, Thr, Val or UUA, UUG of Leu.)*	2.145	2.842
Cyto-c	22	0.463	(+0.075, −0.005, −0.44)	(Ala–Cys) *Gly* Pro Ser Thr Val *Ile* Leu Trp *(Required by Ala–Cys Hamming chain.)*	2.145	4.049
Lysozyme	157	0.626	(+0.035, +0.115, −0.279)	(Ala–Phe–His) Gly Pro Ser Thr *Gln Asn Glu Asp* Lys Arg Val *Ile* Leu Cys Trp	0.672312	1.388677
Cyto-c	8	0.626	(+0.035, +0.115, −0.279)	(Ala–Phe–His) *Gly* Pro Ser Thr Gln Glu Asp *Asn* Lys Arg *Val Ile* Leu Cys Trp	0.208181	2.57039053
Cyto-c	17	0.626	(+0.035, +0.115, −0.279)	(Ala–Phe–His) *Gly |Pro|* Ser* Thr Gln *Glu Asp* Asn Lys Arg Val *Ile* Leu Cys Trp	0.672312	2.48852781

* Found by Hampsey, Das & Sherman (1986, 1988) not to be functionally equivalent.

[1] Alber *et al.* (1988).

[2] Phage λ sites 75–83 from Bowie *et al.* (1990).

[3] Phage λ sites 84–91 from Reidhaar-Olson & Sauer (1988).

[1] T4 Lysozyme 157 from Alber *et al.* (1987), *Nature* **330**, 41–6; T4 Lysozyme 86 Alber *et al.* (1988), *Science* **239**, 631–5.

[2] Phage λ from Reidhaar-Olson & Sauer (1988), *Science* **241**, 53–7.

bovine pancreatic trypsin inhibitor and replacing it with Ala or Thr. They found that at physiological temperatures bovine pancreatic trypsin inhibitor can fold without Cys^{14}/Cys^{38}. This supports but does not prove that the cytochrome c heme pocket may retain its activity if Ala replaces Cys^{22}. One must remember, however, that the Hamming chain requires either the UCU or UCG codons of Ser at this site.

Alber *et al.* (1987) have used oligonucleotide-directed mutagenesis to replace Thr^{157} with thirteen other amino acids in phage T4 lysozyme in order to measure the thermodynamic stability of these replacements. These amino acids, namely, Asn, Arg, Asp, Cys, Leu, Arg, Ala, Glu, Val, His, Phe and Lys, are enclosed by a sphere determined by the Ala–Phe–His triplet. The same complement of amino acids is found at sites 8 and 17 in cytochrome c. Alber *et al.* (1988) reported the eleven functionally equivalent amino acids at Pro^{86} in the same protein. The amino acids are enclosed in a sphere determined by the Ala–His pair. Three comparable sites in iso-1-cytochrome c are shown in Table 6.7. Reidhaar-Olson & Sauer (1988) have reported functionally equivalent amino acids at sites 85 through 91 in the helix 5 region of phage λ repressor. Bowie *et al.* (1990), from the same laboratory, reported functionally equivalent amino acids in phage λ repressor sites 75, 77 and 79–83. Table 6.7 shows the alignment of these amino acids with comparable sites in iso-1-cytochrome c. The amino acids at sites 86 and 88 in phage λ repressor are enclosed in the triplet Ala–Met–Tyr as are those in *E. coli* β-galactosidase at sites His^{464} and Met^3. Table 6.7 shows that all other amino acids are enclosed by the sphere determined by this triplet. The information content of these four sites is zero since any amino acid may occupy these sites. The amino acids at phage λ site 83 are enclosed by the quadruplet (Met–Gly–Glu–His). Cys and Trp are predicted for site 83.

Reidhaar-Olson & Sauer (1988) point out that several amino acids are under-represented. For example, Pro is not found and they regard this as evidence that Pro is not a functionally acceptable amino acid at these sites. This point is worth investigating further since Reidhaar-Olson & Sauer (1988) state that they may not have sequenced a large enough number of candidates to be confident that all candidates have been identified. The Gly–Pro pair has the smallest BH distance and since Gly is found at sites 85, 86, 89 and 90 it is curious that Pro is not also found. The alignment with cytochrome c shows Pro always predicted, but actually found only in iso-1-cytochrome c sites 11, 69 and 96, which are shown aligned with phage λ repressor site 85. Richardson & Richardson (1988) tabulated amino acid preferences from 215 α-helices from 45 different globular proteins for 16 different positions relative to the helix ends. They found a substantial preference for Gly over Pro at the helix ends. Pro was found to have the lowest preference of any residue at the helix ends. Gly is the only one of the proteinaceous amino acids that is symmetric. This fact is not included in the BH considerations. It may be that, because of its symmetry, Gly is more easily accepted by the requirements of protein folding. A second reason may be that Pro is a cyclic imino (not an amino) acid, so that it differs from the others in having highly restricted torsion angles and no N–H bond. A third reason for the absence of Pro is that the Gly codons and the Pro codons are separated by two or three Hamming distances. Thus, a single base interchange goes to a non-functional amino acid if Gly and Pro are the only functionally acceptable amino acids at that site.

6.4 The information content of the cytochrome c family

We are now prepared for calculating the information content or mutual entropy for cytochrome c in the ideal case, using equation (5.22). Although cytochrome c is a mitochondrial protein, the cytochrome c gene is lacking in the animal and fungal mitochondrial genome (Sherman *et al.*, 1970; Raff & Mahler, 1972; Hampsey, Das & Sherman, 1988); accordingly we shall use the standard genetic code for the nucleus. The results are shown in Tables 6.3 and 6.4 for each site including the amino acids actually reported and those predicted by the prescription but not yet reported.

I remarked in section 6.3.1 that immune responses are known between species (Smith *et al.*, 1973). Nevertheless, it remains the fact that with respect to the only known functions of cytochrome c, electron transport and involvement in oxidative phosphorylation, the cytochromes c of all species are identical (Margoliash *et al.*, 1972). However, what I am interested in calculating is the information content of this fundamental function of the cytochrome c homologous family. If other functions are added it is clear that the information content increases.

There is no perceived relation between succeeding sites in a protein and this simply reflects the fact that there is no intersymbol influence. The functionally acceptable amino acids at a site are determined by the folding and chemical requirements. I assume that the functionally equivalent amino acids at each site are independent of those at any other site. This is not true if there is a linkage between sites that are near to each other in the protein-folded condition or which form a complex associated with the active pocket of the protein. Such a linkage implies that, if a mutation to a non-functionally equivalent amino acid occurs, the activity may be restored by a mutation at another site in the protein sequence (section 6.3.3). Such linkages do exist. For example, Brantly, Courtney & Crystal (1988) report that Lys^{290} and Glu^{342} are contiguous in the folded state of α1-antitrypsin and form a salt bridge, Lys^{290} being charged positively and Glu^{342} negatively. These two amino acids are one Hamming distance apart and so a mutation from either one to the other destroys the $+$, $-$ configuration and results in loss of activity and a high risk of emphysema in the patient. The salt bridge configuration is reformed by a second mutation at either site in the protein chain. This is discussed in more detail in section 12.2.1. In any event, the effect of intersymbol influence on the information content can be accounted for by subtracting the conditional entropy associated with the linkage (Shannon, 1951).

The total information content of the cytochrome c family, using only the reported amino acids, is 373.629 bits. Using the reported plus the predicted amino acids the information content is 233.129 bits. The arithmetic average information content of a site is respectively 3.400 and 2.119 bits.

These results will be needed in the discussions of the origin of life scenarios in Chapters 9 and 10, error theories of aging in Chapter 11 and molecular evolution in Chapter 12. The information content of proteins is also important in explaining the phenomenon of overlapping genes. A DNA sequence contains more information than can be transferred to a protein sequence. Do the two overlapping sequences of protein require more information than can be contained by the DNA sequence? If so, this would be a serious challenge to the basic principles of molecular biology.

6.5 The explanation of overlapping genes given by information theory and coding theory

6.5.1 The impact of the discovery of overlapping genes

The possibility that genes might overlap and that more than one protein could be transcribed from the same section of DNA was generally discounted in molecular biology (Fiddes, 1977). One or the other of such sequences is essentially a very long frame-shift mutation. Standard textbooks describe frame-shift mutations and insertion or deletion mutations as catastrophic error propagation hopelessly scrambling the genetic message and therefore destroying specificity. The genetic code exerts a strong degree of coupling between protein sequences coded in different reading frames. It was thought that this would introduce constraints so severe that two protein sequences of any significant length could not have evolved if their genes overlapped.

However, Nature is often full of surprises that upset such well-established beliefs. Two cases in ΦX 174 were discovered (Barrell, Air & Hutchinson, 1976; Sanger *et al.*, 1977; Smith *et al.*, 1977) where genes for proteins of 120 and 151 amino acids each overlap the sequences for longer proteins on a single strand of DNA. In addition, short sections in G4 in which all three reading frames are transcribed were found by Shaw *et al.* (1978). A similar situation was found in the onconogenic virus SV 40 (Durham, 1978; Fiers *et al.*, 1978; Reddy *et al.*, 1978) but within an intron. Henikoff *et al.* (1986) have found that the *Gart* locus in *Drosophila melanogaster* contains an entire gene encoding a cuticle protein from the opposite DNA strand. Spencer, Gietz & Hodgetts (1986) found overlapping transcription in the dopa decarboxylase region in *Drosophila*. Jankowski *et al.* (1986) found two proteins overlapping for 33 sites in trout DNA. Williams & Fried (1986) found complementary mRNAs transcribed from opposite strands of the same cellular DNA sequence in the mouse. Adelman *et al.* (1987) found that the gene in the rat that encodes gonadotropin-releasing hormone (GnRH) from one strand of DNA also transcribes a second gene from the opposite strand, SH, to produce an RNA of undefined function.

The discovery of overlapping genes caused considerable *Angst* expressed by well-respected commentators in *Science* (Kolata, 1977) and in *Nature* (Szekely, 1977, 1978). Kolata (1977) declared: 'Studies of these simple organisms have recently yielded results that shake the foundations of the theories of molecular biology. The hypothesis of nonoverlapping genes is a keystone for many genetic theories'. A complete description of the background and the impact this discovery had at the time is given by these two authors. Kolata called for a redefinition of the '...very concept of a gene'. The interpretation given by Szekely was that it '...illustrates some new tricks a viral genome is capable of in order to compress more information (meaning genetic messages) into a small DNA molecule'. More recently Adelman *et al.* (1987) expressed the same idea: 'Indeed, this situation may be a significant form of molecular genetic evolution. By using both strands of the same conserved DNA, the information content (regulatory and/or structural) of a particular genetic segment becomes amplified, adding a new complexity to the concept of a eukaryotic gene'.

A Protein

↑ Thr ↑ *Term* ↑   ↑ *Term* ↑ *Ala* ↑ *Phe* ↑
↑A C A↑T AA ↑   ↑T A A↑G C A↑T T T↑
↑ Thr ↑ *Tyr* ↑ Lys ↑ *Term* ↑ *Pro* ↑ *Val* ↑
↑A C C↑T AT ↑AA G ↑T G A↑C C A↑G T T↑
↑ Thr ↑ *Tyr* ↑ Lys ↑ *Ser* ↑ *Ser* ↑ *Ile* ↑
↑A C G↑T AC ↑AA G ↑T C A↑T CA ↑A T T↑
Met ↑ Thr ↑ Gln ↑ Lys ↑ Leu ↑ Thr ↑ Leu ↑ Ser ↑ Asp ↑ Ile ↑ Ser
A T G↑A C G↑C A G↑A A G↑T T A↑A C A↑C T T↑T C G↑G A T↑A T T↑T C T
A T↓G A C↓G C G↓G A A↓G T T↓A A C↓A C T↓T C G↓G A↓T A T↓T T C↓T
↑ Asp ↓ Ala ↓ Glu ↓ Val ↓ Asn ↓ Thr ↓ Phe ↓ Gly ↓ Tyr ↓ Phe ↓
↓G *T A*↓C AA ↓G T C↓AT C ↓A A T↓
↓ *Val* ↓ Gln ↓ Val ↓ *Ile* ↓ *Asn* ↓
↓C *T A*↓T AA ↓ G TG↓A C C↓ AG T↓
↓ *Leu* ↓ *Term* ↓ Val ↓ *Thr* ↓ Ser ↓
↓A *T A*↓A AA ↓G T A↓AG C ↓ AT T↓
↓ *Ile* ↓ *Lys* ↓ Val ↓ Ser ↓ Ile ↓

B Protein

Figure 6.5 A and B protein sequences in ΦX 174 from M. Smith *et al.* (1977), Figure 4. The repeated nucleotides are in boldface. The amino acids reported by M. Smith *et al.* are in plain text; the nucleotides and amino acids produced by mutations are in italics.

Szekely (1977, 1978) and Adelman *et al.* (1987) did not, of course, use the terms 'information' and 'complexity' in the sense in which these words are defined in this book.

6.5.2 A comparison of the genetic message in the DNA of ΦX 174 and in comparable sites in cytochrome c

Figure 6.5 shows part of the genetic messages in the overlapping A protein and B protein sequences of the DNA of ΦX 174 from Figure 4 of Smith *et al.* (1977). If, in the A protein, the third nucleotide G for Thr goes through the sequence $N = $ G, T, C, A of the cylinder codons ACN, Thr is preserved in the A protein. Smith *et al.* (1977) reported that the corresponding amino acid codon CGA for Ala, in the B protein, mutates to GTA for Val. Further mutations of GTA to Leu CTA and Ile ATA are compatible with the ACN cylinder codons of Thr. Mutation of the third nucleotide in CAG of the Gln, Tyr, Term series in the A protein causes a Glu, Gln, Term, Lys series in the B protein as reported by Smith *et al.* (1977). Moving to the right to where Val, Asn and Thr in the B protein are opposite Leu, Thr and Leu in the A protein, we note that if the cylinder codons for Val in the B protein mutate through a sequence of all the third nucleotides this causes Leu in the A protein to mutate to Ser TCA and the Term codons TGA, TAA. If the first nucleotide in the ACA codon of Thr which lies between two Leu codons in the A protein mutates, this produces Ser TCA, Pro CCA and Ala GCA in the A protein. It also produces, pairwise, Ile ATC, Thr ACC and Ser AGC in the B protein. If, in the B protein, the second nucleotide of Thr ACT mutates we find Asn AAT, Ser AGT and Ile ATT. These mutations in the B protein cause the second Leu in the A protein to go to Ile ATT, Val GTT and Phe TTT. All these mutations produce amino acid sets that are found to be functionally equivalent in cytochrome c. It is not going too far to say that the functionally equivalent amino acids found at certain sites in cytochrome c are also functionally equivalent in A and B protein. This is another illustration of the phenomenon that sites characterized by the same functionally equivalent amino acids (i.e. enclosed in the same volume in BH space) are found in unrelated protein sequences (Table 6.7).

6.5.3 Considerations from information theory and coding theory showing that overlapping genes are consistent with the foundations of molecular biology

Upon seeing the paper of Barrell *et al.* (1976) I pointed out in a paper that was submitted immediately but not published until 1979 (Yockey, 1979) that the maximum information content of a DNA or mRNA sequence of N codons is $N \log_2 61 = 5.931\,N$ bits. The information content of an 'average protein' sequence is $4.139\,N$ bits. Thus $1.792\,N$ bits are lost because of the Central Dogma, that is, the source alphabet in DNA has 61 letters assigned and the alphabet of the receiver only 20, thus reducing the channel capacity below that of the source (equation (5.15)). This lost information serves a function of partial error correction as discussed in Chapter 5. The value of the information content of cytochrome c that I had at the time (Yockey, 1977b) was $2.953\,N$, slightly less than half $5.931\,N$. The explanation that I proposed (Yockey, 1979) for overlapping genes was that the degeneracy redundance of the genetic code and the redundance due to the information content of a protein family would allow, in some cases, the coding of two, and perhaps more, genetic messages in one of the three reading frames of single-stranded DNA and in the six reading frames of double-stranded DNA. The results of the calculations, shown in Tables 6.3 and 6.4, strengthen this proposal.

Let us now apply what we have learned about the mutual entropy of sequences. As I remarked in section 6.2.2, the mutual entropy may be used as a measure of the relatedness of two or more sequences regardless of the origin of the sequences. We may calculate the mutual entropy of events x and y in any two reading frames. Smith & Waterman (1980) calculated the value of $H_m(x|y)$ (equation (5.14)) for each of the six reading frames in double-stranded DNA. Their results are shown in Table 6.5. The third column in Table 6.5 shows the mutual entropy $I(0|m)$. $H_0(x|y)$ vanishes so that $I(0|0)$ is the information content of an invariant site, 4.139 bits. The other values of $I(0|m)$ are 4.139 bits less the value of $H_m(x|y)$. As I have shown in Tables 6.3 and 6.4, sites that have functionally equivalent amino acids have information content lower than 4.139 bits. Accordingly, this amounts to a flexibility in information content shared between the primary and other reading frames, as I have illustrated in Fig. 6.1.

$I(0|m)$ shows quantitatively what one expects qualitatively from the genetic code. $H_m(x|y)$ is the conditional entropy, which measures the information content per protein site in reading frame m, given a sequence in the primary reading frame. The values of $H_m(x|y)$ in all reading frames compare quite well with the information content found at various sites in cytochrome c. In general, the central bases are the most restrictive, the first bases are next in importance and the third are the least determining. In reading frame 4, on the second strand, the primary frame third base becomes the middle base. This means that reading frame 4 is the least related to the primary frame and therefore protein sequences read in that frame may have the highest information content.

In the case of the single strand DNA of ΦX 174, reading frames $m = 1$, 2 are used (Fiddes, 1977). The same is true of gene K in bacteriophage G4 (Shaw *et al.*, 1978). Overlapping genes in yeast use frame 1 (Clare & Farabaugh, 1985). Spencer *et al.* (1986) report overlapping genes in *Drosophila* on the opposite strand of DNA using frame 3. Williams & Fried (1986) report the first occurrence of transcription on the opposite strand of DNA in the mouse on frame 3. Adelman *et al.* (1987) report overlapping genes in the rat using frame 3. Jankowski *et al.* (1986) report overlapping genes in trout using frame 1.

Overlapping genes are a manifestation of the use of the full information content of the DNA sequence to record and transcribe genetic messages. This phenomenon shows that the source can drive two or even three channels of transcription with appropriate entropy. The study of over-lapping genes gives valuable knowledge about the functional equivalence of amino acids in protein sequences. Although first discovered in viruses, the phenomenon appears in insects and in vertebrates. In spite of the alarm following the initial discovery, the phenomenon of overlapping genes is well on its way to be found to be widespread, as Kolata (1977) feared. The discovery of overlapping genes does not 'shake the foundations of molecular biology'. It illustrates the need for an understanding of the mathematical foundations of molecular biology. The only redefinition of 'the very concept of the gene' needed is to remind ourselves that the gene is the genetic message and not the material DNA or mRNA. It receives a full quantitative explanation from first principles by the discussion given above.

7

Evolution of the genetic code and its modern characteristics

7.1 Early speculations on the evolution of the genetic code

7.1.1 The difficulty of determining the origin of the genetic code

In this chapter I shall apply the principles of information theory and coding theory to find reasonable speculations on the first events in the evolution of the genetic code at the time of the early Earth and show how these events lead to explanations of its modern structure.

Many papers have been published with titles indicating that their subject is the origin of the genetic code, whereas the content actually deals only with its evolution. This is because the authors assume that the origin of the genetic code is inevitable once they have created a scenario that provides the components of an informational molecule. One of the purposes of this chapter is to focus attention on the distinction between the origin of the genetic code and its evolution in order to determine the plausibility of various origin of life scenarios by their ability to account for the assembling of the first effective genetic code.

Looking backward in time, how near to the origin of the genetic code can we see? According to Woese (1970), the origin of the coding process and the sequence control of protein specificity are 'practically inscrutable'. As I have pointed out, 'The calculations presented ... show that the origin of a rather accurate genetic code, not necessarily the modern one, is a *pons asinorum* that must be crossed to pass over the abyss that separates crystallography, high polymer chemistry and physics from biology' (Yockey, 1981). The same comment was made by Delbrück (1986), ' ... a vast gulf separates all creatures large and small from inanimate matter, a gulf that seems much wider today than it seemed only a few decades ago'. Likewise, Orgel (1986), in a review of RNA catalysis, finds the central issue of the origin of the genetic code baffling. Therefore, the reason for the difficulty in speculating on the origin of the genetic code is that there seems to be no place to begin. There is no trace in physics or chemistry of the control of chemical reactions by a sequence of any sort or of a code between sequences. Thus when we make the distinction between the origin and the evolution of the genetic code and find the origin of the genetic code is inscrutable, we are aware that we must take its existence as following from the axiom of the existence of life (Bohr, 1933). The existence of life is based on the sequence hypothesis and consequently there must be a code between each of the several sequences such as those in DNA, mRNA and protein. Accordingly, in this chapter I shall discuss the evolution of the genetic code, not its origin.

It was once thought that the genetic code did not evolve because any change would totally scramble the genetic message and would therefore be lethal. Crick & Orgel (1973) challenged this idea, saying that there was no good argument for the universality of the genetic code and that it is surprising that organisms with slightly different genetic codes did not exist. This proved prophetic when the paper of Barrell, Bankier & Drouin (1979) was published, showing that mitochondria use some codon assignments that differ from the standard code. We have seen in Table 6.7 that the structure of proteins admits the replacement at some sites of up to all amino acids. Since these amino acids do not all have the same codon assignment this reflects *de facto* a non-lethal change of codon. Therefore the discoveries of the alternate mitochondrial codons, other variations from the standard code, the overlapping genes and the readthrough tRNAs of the termination codons show that there is flexibility in the genetic code and that it did evolve, at least in the early period before saturation in the second extension.

It is frequently pointed out that the genetic code was solved by experiment, not by theory, in spite of considerable attention by George Gamow and other extremely capable theoreticians. Although philosophers of science regard prediction as an important support of theory, nevertheless, they also regard the explanation of a phenomenon after it has been discovered to be significant, provided that the theory is not *ad hoc*. The reason that the character of the genetic code was not predicted is that the theories that were used are inapplicable. After examining certain alternatives, I shall present in this chapter a scenario by which the genetic code may have evolved to its present structure by applying what the reader has learned of the mathematical principles of coding theory in Chapter 5 to several primitive conjectures about protein synthesis in early primitive organisms. But first let us consider scenarios for the evolution of the genetic code that assume it originated as a second extension of a quadruplet alphabet and consequently was composed of triplet codons.

7.1.2 The 'frozen accident' theory of the evolution of the genetic code

Did the genetic code reach its current form by a 'sequence of happy accidents [as] ... the result of trying all possible codes and selecting the best' (Crick, 1968) as the frozen accident theory states? Although events in evolution can be considered an ergodic chain, the scenario that the genetic code is a 'frozen accident' evades the question by stating that the genetic code happened without saying how. This evasion is comparable to my point in section 1.1 that we put decisions in the hands of the fickle goddess Fortuna when we wish to avoid responsibility in important matters. Since Crick (1968) proposed the 'frozen accident' it has been learned that the standard genetic code is not universal, fixed or frozen, so the 'frozen accident' scenario explains nothing and should be replaced.

7.1.3 The stereochemical theory of codon assignment

The stereochemical theory of codon assignment (Woese, 1967) stated that the codon assignments are required thermodynamically because there is a stereochemical relation between the amino acid and the codons. This was a reasonable assumption from the point of view of biochemistry and deserved consideration. The theory purported to explain the universality of the genetic code, which was accepted as fact; however, this fact is no longer available since the discovery of the mitochondrial genetic codes and other non-standard genetic codes (section 7.2.5).

7.1.4 Estimates of the number of possible original genetic codes based on triplet codons

If the modern genetic code is the result of Nature's trying a number of possible codes and selecting the best, the plausibility of that proposal can be tested by comparing estimates of the number of codes to be tested with the time available for the origin of life and therefore of the genetic code. If there are more codes to be tested in the paradigm than the time available allows, then one must look for another means by which the genetic code originated. It will also help plausibility if the scenario selected the modern code by a process that did not require trying all possible codes.

First let us go to the geological record to find estimates of the time available for the origin of life. The Earth was formed about 4.6×10^9 years ago. It is well established that life existed in abundance at or before 3.8×10^9 years ago (Schidlowski, 1988). There are 2.52×10^{16} seconds in the 8×10^8 years between these two events. However, during most of these 8×10^8 years Gaea was busy with other things, such as getting heavily bombarded by meteorites, comets and other objects, losing her original atmosphere, outgassing and cooling off to form the oceans (Maher & Stevenson, 1988). A better estimate of the time between the formation of the oceans and the origin of life is about 2×10^8 years or 6.3×10^{15} seconds. However, this time may be too generous since Raup & Valentine (1983) suggest that there were multiple origin of life events but that the Earth was sterilized at intervals of 10^5 to 10^7 years by the climatic consequences of bolide impacts (section 8.1.8).

Having established how much time is available, let us now consider how to calculate the number of genetic codes to be tested. The codons are formed from the letters of the genetic alphabet by taking all *permutations* of those letters with replacement. That is, the order in which the letters are arranged distinguishes the codons. In the case of r_k-tuples with replacement, the number of permutations of n objects is n^k. Once the code has been established we are not concerned in which order, say, the six codons of Leu appear. The approach to the problem is to calculate the number of ways n objects can be arranged, in an unrestricted fashion, without replacing any of them. That is, we need to calculate the number of *combinations* of n objects taken in r_k-tuples without replacement. If we have n objects to be arranged in n permutations without replacement, there are n objects available for the first state. Once the first object is placed there are $(n-1)$ objects left for the second state. There are $(n-2)$ states unoccupied for the third object and so on. Therefore there are $n!$ permutations of n objects in n states. Suppose we consider dividing the states into two subpopulations, one of t states and the other of $(n-t)$ states. By the argument just given, there are $t!$ permutations in the first population and $(n-t)!$ in the second one. To determine how many arrangements or combinations there are of the total population we must divide $n!$ by both $t!$ and $(n-t)!$.

$$C(n, t) = \frac{n!}{(n-t)!\,t!} \tag{7.1}$$

By extending this argument we see that the number of arrangements of n objects in k subpopulations of r_k-tuples without replacement is given by

$$C(n, k) = \frac{n!}{r_1!\,r_2!\dots r_k!} \tag{7.2}$$

The expression in (7.1) is the same as the binomial coefficients and the expression in (7.2) is the same as the multinominal coefficients (section 1.2.9).

Suppose we have a second set of objects each one of which we wish to identify by a mapping with the r_k-tuples of the subpopulations of the set of the n objects. The correspondence is between each object in the second set of objects and the subpopulations in the first set, not with each object in the subpopulation. In general, there is more than one object in each of the subpopulations of the first set of objects. This mapping relationship is what I have defined as a code in section 4.1. It is now clear that the reason we must calculate the number of combinations without replacement is that a codon would then be assigned to more than one amino acid and that situation would not constitute a code.

We have now put forward the mathematical ideas necessary to calculate the number of genetic codes under certain specified conditions. Suppose the 20 amino acids and the 64 codons were all present at the origin of the genetic code awaiting assignment. Let us consider the number of codes in the case where each of the 20 amino acids is assigned one codon and the others are non-sense. This is two subpopulations. The first is the 20 codons assigned to 20 amino acids and the second is the 44 non-sense codons. The number of codes is calculated from equation (7.1) to be 1.96×10^{16}. Codes of this type are very vulnerable to genetic noise since most mutations are to non-sense codons.

Let us calculate the number of genetic codes with the codon–amino acid assignment typical of the modern standard genetic code. Leu, Ser and Arg are assigned a 6-tuple subpopulation of codons each. Ala, Val, Pro, Thr and Gly are assigned a 4-tuple subpopulation of four codons each and so on including the three non-sense codons. Thus these r_k-tuple subpopulations may be arranged in 6!, 4!, 3! and 2! different ways without replacement. Substituting these numbers in equation (7.2) we have:

$$\frac{64!}{(6!)^3(4!)^5(3!)(2!)^9 \times 1 \times 1} = 1.40 \times 10^{70}$$

Any other arrangement in which all but three codons are assigned to at least one of the 20 amino acids also results in a very large number of this order of magnitude. Clearly this is an implausibly large number of genetic codes from which the modern standard genetic code is presumed to have been selected by evolution in the 6.3×10^{15} seconds of the Earth's early history during which the origin of life events occurred.

The number of codes is reduced considerably if it is possible to start with fewer than 20 sense codons. For example, Crick (1968) suggested that the code began with only seven amino acids. This idea was further developed by Wong (1976) who also proposed that, according to his co-evolution scenario, only seven amino acids were necessary for the origin of the genetic code. He allocates the codons in this way: 16 for Glu, 14 for Asp, 10 for Val, 9 for Ser, 4 for each of Phe, Ala and Gly and 3 for Term. The idea of Crick (1968), Wong (1976) and Lehman & Jukes (1988) of starting with fewer than 20 amino acids reduces the number of codes enormously, but still leaves an unbelievably large number from which the modern genetic code was to be chosen by the slow processes of natural selection. In addition, as the vocabulary is increased the number of genetic codes increases dramatically as can be seen by substituting the appropriate numbers into equation (7.2).

One must conjecture that the genetic code did not originate as 20 amino acids and 64 codons awaiting assignment. A successful candidate scenario for the evolution of the genetic code must avoid a selection process involving an enormous number of codes; it must show how it was that the modern standard genetic code evolved without testing all possible codes; and it must show how the several mitochondrial codes and other codes that differ from the standard genetic code originated.

7.2 Did the genetic code evolve from a first extension of a four-letter alphabet?

7.2.1 *Was the genetic code ever binary?*

The possibility that the genetic code began with a binary alphabet must be considered first before proposing a four-letter alphabet, because a binary alphabet is the simplest one. Computers and electronic communications use binary alphabets exclusively. The only hint that the genetic code may have begun with a binary alphabet is that the compounds that form the letters of the DNA and RNA alphabet are chemically of two kinds, namely, purines, A and G, and pyrimidines, C, T and U.

There are a number of objections to the belief that the genetic code was binary at some time in its early history. A binary alphabet might have been made from one purine and one pyrimidine. Those who believe in a prebiotic soup (Chapter 8) are aware of the difficulties in the prebiotic formation of pyrimidines. For example, cytosine is obtained in a yield of about 5% in an aqueous solution of 1.0 M potassium cyanate and 0.1 M cyanoacetylene held at 100 °C for 24 hours. This is hardly a reasonable prebiotic synthesis because it is obvious that such controlled conditions require a *deus ex machina* in the form of an expert biochemist. This is not to be expected on the primeval Earth (Joyce, 1989). Even though Joyce (1989) cites this and other pathways to the formation of cytosine, he regards none as convincing.

Wächtershäuser (1988a) has suggested that the pyrimidines in the present nucleic acids are post-enzymatic substitutes for their isoelectronic and isogeometric position 3-bonded purine analogues xanthine and isoguanine. A primeval genetic alphabet could have been made of these two purine compounds.

Crick (1968) regarded a primeval binary alphabet as plausible. Eigen (1971) also proposed a binary genetic code based on an alphabet A, U, although he believed that a binary alphabet on the functional side would not provide enough specificity. Eigen & Schuster (1978b) withdrew the suggestion that the first codons were recruited exclusively from an A, U alphabet because of their views on the prebiotic synthesis of nucleotides, which I discuss in Chapter 10. Had the genetic code ever been binary the transition to a 4-tuple alphabet must have happened very early or there would now be traces, which one does not find, of an octal, hexadecimal or

32-fold extension of a binary letter code, as discussed in section 4.1 (Hamming, 1986).

7.2.2 The proposal that the genetic code may have six letters

Piccirilli *et al.* (1990) have found a new Watson–Crick base pair that may be incorporated in DNA and mRNA and have proposed that the genetic code may be based on a six-letter alphabet. If that had been the case there would be evidence of a 36-letter first extension and, of course, a code between the letters of the first extension and the 20-letter amino acid alphabet. They suggest that 'the extra letters in the nucleotide alphabet could be used to expand the genetic code, increasing the number of amino acids that can be incorporated translationally into proteins'. We learned in Chapter 5 that it is not necessary to increase the number of letters in the nucleotide alphabet to expand the genetic code. The computer on which this book was written is based on a binary code. There are obviously no difficulties in expanding the number of words used here. Furthermore, the standard genetic code, which is the second extension of a quaternary alphabet, can accommodate up to 61 amino acids as it is. The incorporation of amino acids additional to the standard 20 is discussed in section 7.4.4 where it is seen that the capability to add supernumerary amino acids to the standard now exists. There is no evidence that the genetic code ever had six letters. See the comment by Orgel (1990) on his views of the significance of the discovery of the new base pair.

7.2.3 Jukes' proposal that the genetic code evolved from a doublet code

Jukes (1965, 1966, 1973, 1974, 1981, 1983a, b) first suggested that the current standard triplet code evolved from a doublet code with a four-letter alphabet. Crick (1968) concurred, provided that only the first two nucleotides are read and the third letter is effectively a spacer. He believed this to be necessary because the size of the codon is dictated by the diameter of the double helix. Let us accept the condition that the third nucleotide was a spacer and apply the principles of coding and information theory to examine a scenario for the evolution of the genetic code from a first extension of a quaternary alphabet.

The number of amino acids that can be assigned code words by any code based on the first extension of a four-letter alphabet is no more than 15, plus a non-sense or termination code word. So the first test of Jukes' proposal is: could highly specific proteins have been composed of fewer than 20 amino acids? It is a significant support of this scenario that not all 20 amino acids appear in some modern proteins. The ferredoxins, for example, all have fewer than 20 kinds of amino acid. As Lehman & Jukes (1988) have pointed out, there are only 13 amino acids in the clostridial ferredoxins of *Clostridium butyricum*. The seven missing ones are Arg, Leu, His, Lys, Met, Tyr and Trp (Dayhoff, 1978; George *et al.*, 1985). They also pointed out that a comparison of nine different clostridial-type ferredoxins shows that there are only six different amino acids that occupy the invariant sites. As I pointed out (Yockey, 1977b), this is consistent with the fact that the cytochrome c message can be written with 14 amino acids, as inspection of Table 6.4 shows. We may therefore retain for consideration the idea that the first code had fewer than 20 sense codons (Crick, 1968; Orgel, 1968; Jukes, 1965, 1966, 1973, 1974, 1983a, b; Wong, 1976).

I showed in section 7.1.4 that the multiplicity of triplet codes made implausible the idea that the modern genetic code originated directly as a second extension of a quaternary alphabet. Consequently, the second test of Jukes' suggestion must be to show that the multiplicity of codes in the first extension of a quaternary alphabet is very much smaller than the number of codes found in section 7.1.4. It is reasonable that the direction of evolution was toward organisms with genetic codes that had the largest vocabularies and the least number of mutations to non-sense codons. Let us now consider the scenario that a number of different independent origin of life events occurred (Raup & Valentine, 1983), each with its own code of a given vocabulary size τ and with γ mutations to non-sense codons. Only those mutations at a Hamming distance of one are considered because the mutations at two and three Hamming distances are of second and third order and are therefore very improbable. The following argument is substantially due to Figureau & Labouygues (1981), Cullmann (1981), Labouygues & Figureau (1982, 1984) and Cullmann & Labouygues (1983). The code C is composed of sense code words that have a specific assignment. Other doublets of the alphabet are not part of code C. We must first calculate the number of codes that have γ mutations to non-sense codons. In general, the number of exchanges in code C of q symbols K_i times in the first position is qK_i. We must subtract the number of times symbol i replaces itself, so that the total for symbol i is $qK_i - K_i^2$. The same is true for the number of exchanges of q symbols K_j times in the second position. These two expressions are added and a summation is made over only the sense code words in code C. That is, $i \in C$ and $j \in C$.

$$\gamma = \sum_{i \in C} (q - K_i) K_i + \sum_{j \in C} (q - K_j) K_j \qquad (7.3)$$

The resistance of mutations to non-sense code words in code C is measured by the number of code words, D_1, in code C, that are separated by a Hamming distance of unity. The number of pairs of symbols of sense code words and non-sense code words is the square of the number of times the symbol i appears in the first position. The first symbol in the doublet code word is not one Hamming distance from itself, so the number of pairs of symbol i in the first position must be subtracted from K_i^2, giving $K_i^2 - K_i$. By the same token, the number of pairs of letter j in the second position is $K_j^2 - K_j$. These expressions must be added and summed over all i and all j. We have counted each pair twice and so we must multiply by $1/2$:

$$D_1 = 1/2 \left[\sum_i \in c(K_i^2 - K_i) + \sum_j \in c(K_i^2 - K_j) \right] \qquad (7.4)$$

All other codes of the same size, τ, can be occupied by removing code word ij and replacing it with code word hl. The variations ΔD_1 and $\Delta \gamma$ can be obtained by the following equations, where in each case $i, j, k \in C$:

$$\Delta D_1 = -\{(K_i - 1) + (K_j - 1) - (K_h + K_l)\} \qquad (7.5)$$

$$\Delta \gamma = 2\{(K_i - 1) + (K_j - 1) - (K_h + K_l)\} \qquad (7.6)$$

This is easily understood when we see that the removal of code word ij transforms K_i to $K_i - 1$ and K_j to $K_j - 1$. The introduction of doublet hl changes K_h to $K_h + 1$ and K_l to $K_l + 1$.

The number of doublet codes for each value of γ and τ is calculated from the number of arrangements of the rows and columns of the 4×4 square array of the 16 boxes in which code words may be found (Cullmann, 1981). As we saw in equation (7.1), the total number of codes for each value of τ is given by the number of arrangements of q^2 things taken τ at a time

$$C(q^2, \tau) = \frac{q^2!}{(q^2 - \tau)! \tau!} \qquad (7.7)$$

where q is the number of letters in the alphabet; in this case $q = 4$. These calculations have been carried out by Cullmann & Labouygues (1983) as described by Cullmann (1981). The number of doublet codes for each value of γ and τ is shown in Table 7.1.

Table 7.1. *The number of doublet codes as a function of τ and γ*

The sum of the large print numbers in each column is the number of combinations of 16 items taken τ at a time. These numbers are given in the last row. The small print number below each number in large print is the number of different configurations generating those codes.

τ	4	5	6	7	8	9	10	11	12	13	14	15	16
γ													
0													1
													1
6												16	
												1	
10											48		
											1		
12	8								8	32	72		
	1								1	1	1		
14	0								0	144			
	0								0	1			
16	324	96			12			96	324	288			
	2	1			1			1	2	1			
18	384	432	288	112	0	112	288	432	384	96			
	2	2	3	2	0	2	3	2	2	1			
20	792	864	1008	864	1008	864	1008	864	792				
	3	2	3	2	4	2	3	2	3				
22	288	1392	1152	1728	1344	1728	1152	1392	288				
	1	4	2	4	4	4	2	4	1				
24	24	1152	2752	2304	3168	2304	2752	1152	24				
	1	2	7	5	6	5	7	2	1				
26		432	1584	3024	2304	3024	1584	432					
		2	4	6	5	6	4	2					
28			1224	2592	3792	2592	1224						
			4	4	8	4	4						
30				816	1152	816							
				1	2	1							
32					90								
					2								
	1820	4368	8008	11440	12870	11440	8008	4368	1820	560	120	16	1

Source: Cullmann & Labouygues (1983), with permission.

The total number of all codes for each value of τ is calculated from equation (7.7) and is shown in the bottom row of Table 7.1. For example, the total number of codes for eight amino acids is 12870. This is still a large number, although it is much more manageable than the numbers discussed in section 7.1.4. However, the number of codes that need to be considered can be reduced still more. In the case where $\gamma = 20$ and $\tau = 8$ there are 1008 codes. One sees immediately that there is a most substantial reduction in the number of doublet codes over the number of triplet codes. Jukes' suggestion passes the second test, which is that the multiplicity of codes in the first extension of a quaternary alphabet is very much smaller than the number of codes found in section 7.1.4.

7.2.4 Evolution of the genetic code by random walk between Markov states

The third test of Jukes' suggestion is to show how these doublet codes may have evolved from codes with a vocabulary of, say, six to nine amino acids to the second extension standard triplet modern genetic code with a vocabulary of 20 amino acids and to the other modern genetic codes that differ from the standard genetic code. It is too dark to see all the way back to the *pons asinorum*, but, to take a specific example, let us consider the genetic code at a time when it had a vocabulary of τ amino acids. For example, perhaps these were the amino acids, appearing in the first column of Table 7.2, which, in the second extension, have cylinder codons (section 1.2.3). In the modern standard genetic code the third position in these triplet codons has no discriminatory function. The tRNAs recognize all four cylinder codons (Barrell *et al.*, 1980; Heckman *et al.*, 1980). In this scenario we presume several different independent origin of life events, each with a doublet genetic code characterized by a pair of numbers (τ, γ). Let the pairs of numbers (τ, γ) in Table 7.1 be a set of Markov states. Mutations will occur that change one code word to another. The mechanism for this is the AT to GC pressure or GC to AT pressure driven by the functional importance or requirements of the protein specificity as recorded in the genome, a reduction in the mutations to non-sense codons and the increase in vocabulary (Jukes & Bhushan, 1986; Jukes *et al.*, 1987; Osawa *et al.*, 1987). The progress to a saturated code, which has a vocabulary of 15 amino acids and one non-sense codon, can be regarded as a stationary Markov process in which a step to a second Markov state results in a decrease in γ or an increase in τ (Cullmann & Labouygues, 1983).

The minimization of genetic noise has been suggested as a controlling influence in the evolution of the genetic code by numerous authors, namely, Sonneborn (1965); Woese (1965a); Goldberg & Wittes (1966); MacKay (1967); Crick (1968); King & Jukes (1969); Alff-Steinberger (1969); Jukes, Holmquist & Moise (1975); Batchinsky & Ratner (1976); Eigen & Schuster (1977); Wolfenden, Cullis & Southgate (1979); Inarine (1981); Figureau & Labouygues (1981); Labouygues & Gourgand (1981); Dufton (1983); and Cullmann & Labouygues (1987). To minimize the effect of genetic noise, D_1 will progress to a maximum value and γ to a minimum value. To illustrate the process by which a Markov chain leads to a minimum in γ and a maximum in D_1, suppose the initial Markov states are the four codons {UU, AA, CC, GG}. The value of γ is 24 and the value of D_1 is zero. If one removes the code word GG and substitutes UG then $\gamma = 22$ and $D_1 = 1$. If the Markov chain proceeds in this way to the code {UU, UC, UA, UG}, then $\gamma = 12$ and $D_1 = 6$. If τ is increased to 6 by the additional assignments to UC and CC, $\gamma = 10$ and $D_1 = 9$. If the step to a second Markov state is accomplished by a decrease in γ, the vulnerability of the genetic message to genetic noise is reduced.

7.2.5 Expansion of the genetic vocabulary and decrease in the number of possible genetic codes

As this process proceeds toward the Markov states to the right-hand side of Table 7.1, the number of codes decreases naturally, for mathematical reasons, as the evolution proceeds. Origin of life events that start at any Markov state in Table 7.1 change up and to the right. All possible doublet codes that may have originated in any Markov state (τ, γ) will coalesce into one of the smaller number of possible Markov states as the code word vocabulary increases. In the case of the 14 sense doublet codons where $\gamma = 10$, there are only 48 codes available. This is reduced to 16 codes as the fifteenth amino acid is admitted. If more than one organism followed a random walk, the different codes of those organisms were squeezed through a bottleneck upon arriving at 14 sense codons and tended to assume similar but not necessarily identical genetic codes. It is at this stage in the scenario that the transition to the second extension must occur. In support of this scenario, one must not fail to notice that the number of known mitochondrial codes, together with other genetic codes that differ from the standard code and the standard code itself, is nearly the same as the number of codes allowed in the last two Markov states in Table 7.1 (Fox, 1987).

It is important to note that this scenario does not require that even a substantial fraction of the available codes be tested and therefore Jukes' suggestion passes the third test. The evolutionary process follows an ascending path, so to speak, to the doublet code saturation bottleneck at 14 to 15 sense code words.

This process is irreversible, since all backward steps increase the vulnerability to mutations to non-sense codons. For the same reason the paradigm assumes that once an assignment is made it does not change. Each code evolves naturally in this way to a larger vocabulary and greater protection from mutations to non-sense code words. Furthermore, after emergence from the bottleneck, the situation reverses and then codon sense reassignments are indistinguishable from genetic noise, as Crick (1968) and Orgel (1968) have pointed out. For these reasons one code cannot evolve from another except as a bottleneck is approached where the number of codes decreases and new codon assignments decrease genetic noise. It is, however, possible to increase the vocabulary by assigning specificity for additional amino acids for up to three of the cylinder codons. The triplet code would then approach saturation at 63 amino acids and one stop or release factor. Further amino acids could be added by a third extension to a quadruplet code.

7.3 Proposed doublet codon assignments and evolution to the modern genetic codes

7.3.1 The formation of triplet codons from doublet codons

The next step in the evaluation of Jukes' proposal is to consider the actual codon assignments as they are found in the modern standard code and in the several mitochondrial codes. In Yockey (1977b), Table 4, following Jukes' work, I made a proposal for codon assignments, which is given here with several modifications. The proposed codon assignments in the first and second extension of the quaternary alphabet are shown in Table 7.2. Wong (1988) suggested that Gly, Ala, Ser, Asp, Glu, Val, Pro, Thr, Leu and Ile were the first amino acids to be assigned codons. I have assumed that the amino acids in the left-hand group of columns in Table 7.2 were the first amino acids to be assigned codons. They have cylinder codons in the second extension of the genetic code (section 1.2.2). Six additional amino acids were assigned doublet codons later in the first extension as shown in the second group of columns. The doublet UA became the stop codons UAA and UAG in the second extension. As the second extension

Table 7.2. *A proposed first extension and evolution to the second extension of the genetic code*

First extension

Amino acid	Doublet codon	Triplet codon	mtDNA codon
Gly	GG	GGN	GGN
Pro	CC	CCN	CCN
Leu	CU	CUN**	CUN*
Arg	CG	CGN	CGN
Thr	AC	ACN	ACN
Val	GU	GUN	GUN
Ala	GC	GCN	GCN
Ser	UC	UCN	UCN
Phe	UU	UUU, UUC	UUU, UUC
Cys	UG	UGU, UGC	UGU, UGC
Gln	CA	CAA, CAG	CAA, CAG
Asn	AA	AAU, AAC	AAU, AAC
Glu	GA	GAA, GAG	GAA, GAG
Ile	AU	AUU, AUC, AUA	AUU, AUC
Non-sense	UA	UAA^, UAG^	UAA, UAG
Non-sense	AG		

Second extension

Amino acid	Triplet codon	mtDNA codon
Leu	UUA, UUG	UUA, UUG
Trp	UGG	UGG
Non-sense	UGA++	UGA#
His	CAU, CAC	CAU, CAC
Lys	AAA, AAG	AAA, AAG
Asp	GAU, GAC	GAU, GAC
Met	AUG	AUG+, AUA+, AUU+, AUC+
Tyr	UAU, UAC	UAU, UAC
Arg	AGA, AGG	AGA##, AGG##
Ser	AGU, AGC	AGU, AGC

* Codes for Thr in yeast mtDNA, Li & Tzagoloff (1979), Bonitz *et al.* (1980), Sibler *et al.* (1981) and in Neurospora mtDNA, Heckman *et al.* (1980). # UGA codes for Trp in animal, fungal, yeast and protozoa mtDNA, Anderson *et al.* (1982), Watson *et al.* (1987). + Met code internal is AUG, AUA; for initiation AUG, AUU, AUC, Barrell *et al.* (1980), Montoya *et al.* (1981), Anderson *et al.* (1981), Bibb *et al.* (1981). ^ Codes for Gln in *Paramecium*, Klug & Cummings (1986), in *Tetrahymena*, Horowitz & Gorovsky (1985), Kuchino *et al.* (1985). ## Non-sense in mammalian mtDNA, Barrell *et al.* (1980), Anderson *et al.* (1981), Watson *et al.* (1987). ** CUG codes for Ser in yeast, Kawaguchi *et al.* (1989). ++ Codes for Trp in *Mycoplasma caprolium*, Yamao *et al.* (1985), Muto *et al.* (1985) and in yeast Macino *et al.* (1979). Codes for selenocystein, Leinfelder *et al.* (1988).

emerged the remaining amino acids of the modern 20 were assigned codons as shown in the third group of columns of the table. Cys retains UGU and UGC and Trp is assigned UGG. UGA is a non-sense codon but codes for Trp in *Mycoplasma caprolium* (Yamao *et al.*, 1985; Macino *et al.*, 1979), in *Neurospora crassa* (Heckman *et al.*, 1980) and in yeast (Macino *et al.*, 1979). Meyer *et al.* (1991) report that, in addition to the standard codons UGU and UGC, Cys is translated by UGA in the pheromone 3 mRNA of the ciliate *Euplotes octocarinatus*. According to this scenario the cylinder codons UGN were translating Cys in the first extension and UGA continued to be translated as Cys in the second extension. UAA functions as the stop codon in this organism. This requires no *ad hoc* or contrived explanation.

UAG and UAA are non-sense in the first extension but code for Gln in *Paramecium* (Klug & Cummings, 1986) and Glu in *Tetrahymena* (Horowitz & Gorovsky, 1985; Hanyu, Kuchino & Nishimura, 1986) and in *Acetabularia cliftonii* and *A. mediterranea* (Schneider, Leible & Yang, 1989). UAG codes 1–2 % for Glu in mouse cells infected with Moloney murine leukemia virus (Kuchino *et al.*, 1987). Gln retains CAA and CAG while His is assigned CAU and CAC. Asn retains AAU and AAC and Lys is assigned AAA and AAG in the second extension. Arg and Ser divide between them the previously non-sense codons, AGN. This must have been somewhat arbitrary since in *Drosophila yakuba* mtDNA AGA is assigned to Ser rather than Arg (Clary & Wolstenholme, 1985; Fox, 1987). Furthermore, in the starfish mitochondrial genome AGA and AGG translate Ser (Himeno *et al.*, 1987). This scenario is supported by the following facts: (a) the AGU and AGC are separated by two or three Hamming distances from the UCN cylinder codons of Ser (b) the AGA and AGG are non-sense codons in mammalian mtDNA (Barrell *et al.*, 1980; Anderson *et al.*, 1981). The assignment or non-assignment of the AGA and AGG codons also supports the scenario that they were assigned at the time of the second extension and that Arg and Ser were included in the original set of amino acids with cylinder codons (section 1.2.3).

This paradigm explains the puzzling fact that Arg, which is relatively rare compared to Lys in modern proteins, has six codons while Lys has only two. Lys is one Hamming distance from Asn and the BH distance is 0.09. The assignment requires only the establishment of specificity in the third nucleotide typical of the process of going to the second extension. On the other hand, the Hamming distance between the CGN cylinder codons of Arg and the AAA, AAG codons of Lys is 2,3. The BH distance between

Arg and Lys is 0.12. Jukes (1983a) suggested that Lys was a late entry to the code.

Additional support of the assignments in Table 7.2 comes from the empirical fact that Lys, Tyr, His, Met and Trp are missing or very rare in ferredoxins and are, therefore, perhaps the last to be incorporated into the genetic code. These amino acids are all in the columns headed *Second extension* in Table 7.2. Asp is the only one of those amino acids incorporated in ferredoxin from *Clostridium thermoaceticum*, which is located only in the third group of columns of Table 7.2. The ferredoxins are often regarded as very ancient proteins that were present in ancestral organisms soon after the origin of life (George *et al.*, 1985). They are iron–sulfur electron carriers and may be related to the interesting suggestions of Wächtershäuser (1988a, b, 1990) regarding the role of FeS in the origin of life. Of the amino acid pairs assigned to the first and second extensions in Table 7.2, only Asp and Glu could be exchanged without violating the need for observing the biosynthetic pathways. Met and Trp must have received their single codons after the second extension was well along. The 12 amino acids admitted in saturating the first and second extensions could have been added very smoothly to the vocabulary without violating the principle of continuity (Crick, 1968; Orgel, 1968).

7.3.2 *mRNA editing in the mitochondrial codes*

CGG was reported to be a Trp codon in *Zea maize* mtDNA by Fox & Leaver (1981) and in the mtDNA of *Oenothera berterilana* by Hiesel & Brennicke (1983). No tRNA specific for Trp and recognizing CGG has been found (Maréchal *et al.*, 1985). Although the CGG codon exists in the wheat mtDNA, Gualberto *et al.* (1989) found that the codon UGG is at the corresponding site in the mRNA sequence so that the actual codon read is UGG, the standard codon for Trp. Covello & Gray (1989) found only C → U editing; thus the plant mitochondrion code does not differ from the standard genetic code. Other C → U conversions in the first and second position, which edited His → Tyr, Ser → Leu, Ser → Phe, Leu → Phe and Pro → Leu, were found by Gualberto *et al.* (1989). Covello & Gray (1989) also found C → U editing in maize, rice, wheat, pea, soyabean and *Oenathera*. Hiesel *et al.* (1989) find that the Arg codon CGG is often edited to the Trp codon UGG in higher plant mitochondria. This suggests that the standard genetic code is used in plant mitochondria and this resolves the frequent coincidence of CGG codons and Trp in different plant species.

mRNA editing changes a Gln CAA codon to UAA at site 2152 in apoliprotein B mRNA, which allows the translation of two proteins of different length from the same gene (Powell *et al.*, 1987; Shaw *et al.*, 1988; Tennyson, Sabatos, Higuchi *et al.*, 1989; Tennyson, Sabatos, Eggerman *et al.*, 1989) according to whether or not the editing event occurs. Simpson & Shaw (1989) have reviewed mRNA editing in mitochondria.

It is important to remember that mRNA editing is not the same as proofreading (section 11.3.3).

7.3.3 The biosynthetic pathways

The amino acids assigned cylinder code words (section 1.2.2) in the first group of columns of Table 7.2 are those that are not formed in biosynthetic pathways from other amino acids. The assignment of cylinder codons to amino acids designated as code words in the second and third groups of columns in Table 7.2 was made on the grounds of the analysis by Wong (1975, 1976, 1981, 1988) of the biosynthetic pathways of formation. The six amino acids admitted in the second extension (Table 7.2) have biosynthetic pathways requiring amino acids that were assigned codons in the first extension (Wong, 1988). The biochemistry must be involved in the co-evolution of the genetic code and the amino acids that are added to increase the vocabulary by the increase of τ (Wong, 1975).

This scenario shares much with Wong's co-evolution theory since the scenario also contemplates the possibility that the genetic code expanded its vocabulary as the evolution of the amino acids was occurring (Wong, 1975, 1976, 1981, 1988). It also shares with the stop codon takeover model of Lehman & Jukes (1988) the idea that the evolution of the genetic code must be such that mutations to non-sense codons are minimized. An advantage of this scenario is that it follows the mathematical theory of coding that applies to all codes and thereby avoids *ad hoc* reasoning. Furthermore, it places the burden of evolution of the genetic code on satisfying the specificity needs of the evolving proteins in the primitive organisms rather than, as Weber & Miller (1981) have suggested, on a Procrustean bed of the abundance of amino acids in a phantom primeval soup (section 8.3).

Some authors suggested that editing has a direct bearing on the Central Dogma, that is on the unidirectional flow of genetic messages and the incorporation of genetic specificity in a sequence of nucleotides and amino acids (Simpson & Shaw 1989; Benne, 1989; Blum, Bakalara & Simpson

1990; Simpson, 1990). The major puzzle presented by the discovery of editing is the nature of the mechanism for carrying out these changes and the means by which the genetic messages that control these processes are recorded. Blum *et al.* (1990) proposed that small transcripts that they called *guide RNAs* (gRNA) carry the genetic messages necessary for a post-transcriptional editing process. They report that the genetic message of these gRNA molecules is recorded in the intergenic region of the mitochondrial maxicircle DNA. They present a model in which a partial hybrid is formed between the gRNA and the mRNA before it is edited. Simpson (1990) has given a short review of the status of knowledge about the editing process. Editing is an elaboration of the genetic information mechanism but it is clearly not a challenge to the Central Dogma as I discussed in section 4.4.

7.3.4 The Central Dogma in the first extension

We should note here in passing that some of the doublet codes may not have possessed a Central Dogma because the entropy of some of the source and destination alphabets may have been equal. An inversion code similar to the reverse transcription of the modern DNA to mRNA code could have existed so that there would have been no mathematical stricture against the following of information from protein to mRNA.

7.4 Characteristics of the genetic code

7.4.1 The genetic code is instantaneous

In the course of the evolution of the genetic code, as the genetic code approaches and emerges from the bottleneck that leads to the modern second extension code, specificity must be assigned to the third nucleotide of each codon. The code must be instantaneously decodable (section 4.2.1) at all times in its evolution in order to avoid the confusion that would occur, for example, if the genetic code had the doublet AA for Asn together with the triplet code AAA for Lys. Therefore, the doublet codons in the second group of columns in Table 7.2 must assign specificity, one by one, to the third nucleotide of the codons that are assigned to amino acids that enlarge the vocabulary. However, it is not necessary for additional specificity to be assigned to all doublets. For example, the cylinder codons (section 1.2.2) in the first group of columns of Table 7.2 continue to function as doublet codons.

As the transition to the second extension proceeds, only those newly admitted amino acids must be assigned specificity in the third position. This must happen in such a way that the genetic code is instantaneously decodable at each stage. In order that this be so, the Kraft inequality (Theorem 4.2), which is the necessary and sufficient condition that a code be instantaneous, must be satisfied at all stages of the evolution, both for the first extension and for the second extension. It is easy to see that the first extension code in Table 7.2 satisfies the Kraft inequality:

$$1 \geqslant \sum_{1}^{q} r^{l_i} = \sum_{1}^{16} 4^{-2} = \frac{16}{16} = 1 \qquad (7.8)$$

In addition, the reader can easily verify that the modern second extension triplet code also satisfies the Kraft inequality so that the genetic code is at all times instantaneously decodable.

Instantaneous codes are well known to information theorists (Chaitin, 1975b; Hamming, 1986). Cullmann & Labouygues (1985) first proved that the genetic code is instantaneous. This is an important property since if the genetic code were not instantaneous a decoding device would be needed to record the message before decoding could proceed. The code words of the Morse code are not instantaneous so that the telegrapher must leave a letter space between each letter and a word space between words. Thus the Morse code is effectively a quaternary code.

7.4.2 The genetic code is optimal

It can also easily be shown that the genetic code has the property of being *optimal*, which means that the genetic code has the most economical use of its nucleotides. An optimal code was defined in section 4.2.1 to be one that is both instantaneously decodable and that has the minimum average code word length. In section 2.3.3, Theorem 2.3, it was shown that the maximum compression of a code is:

$$\frac{H(\mathbf{p})}{\log_2 q} \qquad (7.9)$$

where $H(\mathbf{p})$ is the entropy of the probability vector $\mathbf{p}$ and q is the number of letters in the alphabet. In the case where the codons are all equally probable, it is easy to show that as the transition from a first extension to the third extension proceeds the value of the expression (7.9) gradually increases from 2 to 3 so the genetic code continues to be both instantaneous and optimal. This was first proved by Cullmann & Labouygues (1985).

7.4.3 Did the genetic code evolve in such a way that similar amino acids have similar codons?

As the genetic code became available in the literature, numerous authors sought to find its logic or regularities. One of the earliest observations was that some similar amino acids had similar code words. This was based on classifications such as hydrophobicity, side chains etc. (Woese, 1965b; Epstein, 1966; Crick, 1968; Jurka, 1977; Jurka, Kolasza & Roterman, 1982). Such classifications are not independent and therefore do not form a Euclidean space. We now have in this book the means to determine quantitatively whether similar amino acids have similar codons. There are several pairs of amino acids that have a small BH distance and yet are at a Hamming distance 2, 3. For example, the Pro–Gly BH distance is 0.01 but the Hamming distance is 2, 3; the Thr–Gly BH distance is 0.05, with Hamming distance 2, 3; the Asn–Asp BH distance is 0.05 with Hamming distance 2, 3. The Hamming distances and the BH distances for each pair of amino acids are given in Table 6.2 for comparison of those cases where mutationally related amino acids also have related codons and cases where this is not true.

However, all the amino acids that in the second extension were assigned triplet codons derived from the doublet codons of the first extension, according to the Markov evolution scenario described in section 7.2, have codons that are at a Hamming distance of unity. There is also considerable mutational similarity as shown by the BH distances. The BH distance of Cys and Trp is 0.07; the Hamming distance is unity. The BH distance between Asn and Lys is 0.09; the Hamming distance is unity. The BH distance between Gln and His is 0.23; the Hamming distance is 1, 2. The BH distance between Asp and Glu is 0.02; the Hamming distance is unity. Leu captured UUA and UUG from Phe: the BH distance is 0.59 and the Hamming distance is 1, 2. Arg and Ser divide the AGN codons and the BH distance is 0.26. Ile retains three of the AU codons in the second extension and Met is assigned AUG. The BH distance between Met and Ile is 1.00. Note that these BH distance figures are diameters and are rather small compared to the radii R in Table 6.3. Therefore, we see that the amino acids that lie in the columns labeled *Second extension* in Table 7.2 are only one Hamming distance apart from those found in the same row in the third group of columns. The amino acids in the same row in the central and in the third groups of columns have a small BH distance and do indeed have similar codons.

7.4.4 Are there really only 20 amino acids incorporated directly?

Much of the early considerations and speculations about the genetic code regarded the four nucleotides and the 20 amino acids as 'magic numbers' (Gamow, 1954; Crick, Griffith & Orgel, 1957). Subsequent research has shown they are not. Although 20 amino acids are known to participate in the first steps of protein synthesis, some 140 other amino acids are found in various proteins (Uy & Wold, 1977). One is tempted to speculate that supernumerary amino acids were added into the genetic message because, as evolution progressed to more complex organisms, the specificity of protein sequences continued to require an increase in vocabulary. These supernumerary amino acids are believed to be derived by a chemical pathway from one of the 20 amino acids incorporated directly in protein and are thus 'post-translational' or 'derivatized'. The information for the enzymes that catalyze these pathways is incorporated in the genetic message. At least 61 supernumerary amino acids could have been added without going to the 256 codons of a third extension of the alphabet. Since DNA provides plenty of information capacity to accommodate the enzymes needed to change the amino acids directly incorporated to one of the 140 others found in modern proteins, this means of doing so may often be preferred to the more drastic path of revising the genetic code or going to a third extension. This is comparable to a computer programmer who adds a subroutine to a program to incorporate a new capability instead of completely rewriting the original program.

Weber & Miller (1981) have offered an analysis to explain why there are exactly 20 proteinous amino acids. Since that paper was published, Zinoni *et al.* (1986), Chambers *et al.* (1986), Sunde & Evenson (1987), Zinoni *et al.* (1987), Mullenbach *et al.* (1987), Leinfelder *et al.* (1988) and Mizutani & Hitaka (1988) have found that an in-frame UGA non-sense codon can sometimes direct the incorporation in glutathione peroxidase of phosphoserine, which is then changed to selenocysteine. Berry, Banu & Larsen (1991) have found that the mRNA for type I iodothyronine deiodinase, which converts thyroxine to 3,5,3'-triiodothyronine, contains an in-frame UGA that translates selenocysteine. This shows that there are 21 proteinous amino acids and maybe more. The point that 20 is not a magic number is further demonstrated by the fact that certain amino acid analogues are incorporated directly in the protein sequence. For example, Wong (1983) has reported a serial mutation in a tryptophan auxotroph of *Bacillus subtilis* strain QB928 to yield strain HR15, which grows well on 4-fluorotryptophan but marginally on Trp. In the transition from QB928 to HR15, the replacement of Trp by 4-fluorotryptophan changed the growth rate by a factor of 2×10^4 in favor of 4-fluorotryptophan. Noren *et al.* (1989) have developed a site-specific method of incorporating unnatural amino acids into proteins by the use of a chemically acylated suppressor tRNA in response to a stop codon that had been substituted for the codon at the site of interest. This is essentially a method of assigning specificity to a codon previously unassigned. There is nothing in the theory presented in this book that is inconsistent with that and, indeed, the theory may be said to predict that more directly incorporated amino acids will be found, or perhaps could be created by genetic engineering.

7.5 Non-triplet reading in the genetic code

There is good evidence that a transition has occurred from a first extension genetic code to a second extension and, in a few cases, even to a third extension. For example, the reading of the third extension quadruplet codons as suppressors or frame-shifting mutations has been reported by a number of authors (Yourno & Heath, 1969; Yourno & Tanemura, 1970; Yourno, 1971). Yourno & Kohno (1972) report reading the CCCU quadruplet as Pro by a tRNA with a quadruplet anticodon. Riddle & Roth (1972) reported the reading of CCCU and the direct incorporation of Pro in the histodinal gene His D3018 of *Salmonella typhimurium*. Craigen *et al.* (1985) have reported a frame-shift suppressor equivalent to reading UGAC as Asp rather than Term. Riddle & Carbon (1973) report the reading of anticodon quadruplet CCCC to insert Gly in *Salmonella typhimurium*. Atkins & Ryce (1974) have reported a class of UGA suppressor mutations in a tRNA methylase gene, *supK*, of *Salmonella typhimurium*; they favor the interpretation of their data as the reading of a non-triplet codon. For a minireview of frame-shift reading of quaternary codons, see Roth (1981).

In addition to quadruplet codons, doublet codons are also read by the modern protein synthesis system. Yourno & Tanemura (1970) suggested that the Try A91 frame-shift might be of the -1-type mediated by a tRNA with a doublet anticodon. Bruce, Atkins & Gesteland (1986) showed that a frame shift resulting from the tRNA AGC reading the Ala GC doublet results in two proteins at the end of a MS2 coat protein in *Escherichia coli*. The precision of movement of the mRNA is coordinated through the codon–anticodon match. Weiss (1984) proposed a base-pairing scheme (as he called it) between a shifty codon and an 'offset' anticodon in the anticodon loop of each shifty tRNA. His scheme shows how doublet, triplet and quadruplet codons can be read by shifty tRNA. The fact that quaternary codons are read in some instances shows that the evolution of the genetic code may not be static.

7.6 The mitochondrial genetic codes, the endosymbiotic theory and a common genetic code bottleneck

In this section, I shall examine the characteristics of the mitochondrial codes in order to evaluate the conclusion in the preceding paragraph in reference to the endosymbiotic theory of mitochondria (Margulis, 1970). Mitochondria are organelles in the cell that have a protein-synthesizing capability that is physically and genetically separate from the cytoplasmic system. Ten to 20 proteins are in the genetic message in the mitochondrial DNA. According to the endosymbiotic theory, mitochondria were free-living bacteria at an early time in the history of life that were absorbed by eukaryotic organisms and now survive in a symbiotic relationship.

The fact that doublet codons are read in the modern genetic code supports the paradigm that the standard, mitochondrial and other non-standard genetic codes evolved independently from a first extension of the four-letter alphabet. Therefore the conclusion of the discussion of Jukes' suggestion in this Chapter is that the modern genetic code evolved from the first extension of a four-letter alphabet and that the modern standard code, the mitochondrial codes and other non-standard modern genetic codes were all squeezed through a common bottleneck upon the transition from a first extension of a four-letter alphabet to the second extension.

As the codes approached the bottleneck they perforce became more nearly alike and this is reflected in the fact that there are only a few differences between the standard genetic code and the mitochondrial and other non-standard codes. We find this is so in examining the codon assignments, shown in Table 7.2, of mitochondria in *Mycoplasma caprolium*, *Tetrahymena*, *Paramecium*, yeast and perhaps other organisms. The mitochondrial genetic codes differ from the standard genetic code and, indeed, from the codes of other mitochondria. These differences between the standard genetic code and the non-standard code reflect the fact that they are not a property of a stereochemical relation between the codon and the amino acid as was once believed (Pelc & Welton, 1966; Welton & Pelc, 1966; Woese, 1967; Junck, 1978; Lacey & Mullin, 1983). This is demonstrated, for example, by the curious assignment of the CUN codons in yeast mtDNA to Thr rather than to Leu (Li & Tzagoloff, 1979; Heckman *et al.*, 1980; Bonitz *et al.*, 1980; Sibler *et al.*, 1981). As another example, the use of UGA as a Trp codon in yeast mitochondria is reported by Macino *et al.* (1979) and in rabbit reticulocytes by Geller & Rich (1980). Furthermore, UGA codes for Trp in animal, fungal and protozoan mtDNA, whereas UGG codes for Trp in plant mtDNA (Hiesel & Brennicke, 1983). The AUG, AUA and AUU codons are initiator codons in HeLa mtDNA (Anderson *et al.*, 1981; Montoya, Ojala & Attardi, 1981) and in mouse mtDNA (Bibb *et al.*, 1981).

Other deviations from the standard code are exhibited by different assignments of specificity to the third nucleotide in sense codons (Anderson *et al.*, 1982). The readthrough or suppression of non-sense codons (Watson *et al.*, 1987) is seen in this scenario as simply the assignment of specificity to non-sense codons in the same manner as specificity has been assigned to other codons. Let us recall the language example in section 2.1.1 where I pointed out that a German-speaking person reads the word *Gift* with a very different meaning than an English-speaking person. The non-sense codons are not special *per se*.

In addition to the differences from the standard genetic code in mitochondria, one notes that Muto *et al.* (1985) and Yamao *et al.* (1985) have found that UGA codes for Trp in the small prokaryote *Mycoplasma capricolum*. A mechanism for how this occurred was suggested by Jukes (1985), who outlined a series of evolutionary steps that could have led to the replacement of the UGG codon by UGA.

Giardia lamblia is at least one exception to the general rule that eukaryotics have mitochondria (Feely, Erlandson & Chase, 1984). Sogin *et al.* (1989) proposed a phylogenetic tree, based on an analysis of ribosomal sequences, in which *Giardia lamblia* is the first to separate from other eukaryotes. If this tree is correct one must assume that the endosymbiotic event occurred after that separation. Other scenarios have been discussed by Raff & Mahler (1972) and by Uzzell & Spolsky (1973, 1974).

The pervasiveness of the standard genetic code has often been cited as evidence of a single origin of life event (Mayr, 1982). The other side of the coin is that the existence of several different genetic codes is evidence of a number of independent origin of life events. Therefore, those who believe that the origin of life is nearly certain, given the right conditions, must support the scenario that there could have been more than one origin of life event (Raup & Valentine, 1983) (section 7.2 and Chapter 8). Weiner (1987) has suggested that there may have been more than one endosymbiotic event. Schön *et al.* (1988) have suggested that plant and fungal mitochondria appear to be derived from different endosymbiotic events than that from which animal organelles are derived. In that case mitochondria would be polyphyletic and this would account for the fact that mitochondrial genetic codes differ both from the standard code and from other mitochondrial genetic codes.

The proposal discussed in this chapter shows that, assuming several different independent origin of life events, the first extension codes of such independent events would have become similar, but not necessarily identical, as they approached the bottleneck at code saturation. A second suggestion is that there was a single origin of life event but that the random walk took separate paths and several nearly similar genetic codes arrived at the bottleneck. Thus several organisms may have emerged from the bottleneck at (geologically speaking) nearly the same time with genetic codes that differed only in a few assignments, especially in the third position. Notice in Table 7.2 that most of the differences in the mitochondrial codes are, indeed, in the third position (Anderson *et al.*, 1982). This may mean that these differences were established upon emergence from the bottleneck at the extension from the doublet code to the triplet code and that the bacteria that were to become mitochondria had independent genetic codes. For a more detailed review see Fox (1987).

Crick (1981), in one of those marvelous intuitions that have led him to so many discoveries, and without going through the mathematical argument above, has proposed: 'What the code suggests is that life, at some stage, went through at least one bottleneck, a small interbreeding population from which all subsequent-life has descended . . . Nevertheless, one is mildly surprised that several versions of the code did not emerge, and the *fact that the mitochondrial codes are slightly different from the rest supports this*' (italics mine).

In Chapter 8, I discuss the environment on the early Earth at the time during which life originated. This environment was similar to or even more rigorous than that which is believed to have caused the major extinctions. In addition to the possibility that there were a number of origin of life events one must consider, by the same token, a number of extinctions. This

Table 7.3. *The probability that life will survive to the latest Precambriam eon**

Bioclades starting number	Lineage duration $(1/q)$, Myr							
	1	10	20	30	40	50	100	1000
1	0.003	0.03	0.06	0.08	0.11	0.13	0.23	0.75
2	0.01	0.06	0.11	0.16	0.21	0.25	0.41	0.94
3	0.01	0.09	0.16	0.23	0.29	0.35	0.55	0.99
4	0.01	0.11	0.21	0.30	0.37	0.44	0.66	1
5	0.02	0.14	0.26	0.36	0.44	0.51	0.74	1
6	0.02	0.17	0.30	0.41	0.50	0.58	0.80	1
7	0.02	0.19	0.34	0.46	0.56	0.63	0.85	1
8	0.02	0.21	0.38	0.51	0.60	0.68	0.88	1
9	0.03	0.24	0.42	0.55	0.65	0.72	0.91	1
10	0.03	0.26	0.45	0.59	0.69	0.76	0.93	1
20	0.06	0.45	0.70	0.83	0.90	0.94	1	1
50	0.14	0.78	0.95	0.99	1	1	1	1
100	0.26	0.95	1	1	1	1	1	1

* It is assumed that the origination rate p is that value that would predict a final diversity of 10 000 species.
Source: Raup & Valentine (1983), Multiple origins of life, *Proceedings of the National Academy of Sciences USA* **80**, 2981–4, with permission.

point has been examined by Raup & Valentine (1983). They concluded, using several stochastic models with a reasonable duration of survival of the resulting bioclades, that there may have been as many as 10 to 100 origin of life events in order for one to have survived (Table 7.3). This is consistent with the impact history of the early Earth. Maher & Stevenson (1988) and Overbeck & Fogleman (1989) point out the possibility that the larger impacts may have sterilized the surface of the Earth at intervals of 10^5 to 10^7 years. (Table 7.3 is reproduced from Raup & Valentine (1983) with permission.)

7.7 Summary

The proposal for the evolution of the modern genetic codes discussed in this chapter is based on the first principles of coding theory. It provides an explanation of the characteristics of the standard genetic code and other non-standard genetic code assignments. The discussion in section 7.2 is based principally on the mathematics of coding theory and is not anecdotal or *ad hoc*. It enables us to see quite far back in the evolution of the genetic code and therefore of life. As Brenner, Ellington & Tauer (1989) have suggested in connection with their analysis of modern molecular catalysis, the modern state may be regarded as a palimpsest in which the modern features have been superimposed on an earlier state. (A palimpsest is a parchment that has been inscribed two or more times, the previous texts being imperfectly erased and partially legible.) Lys with two codons is more frequent in modern proteins than Arg with six codons. Thus six codons for Arg are selectively unfavored in the modern world. According to the comment in section 7.3.1 this is evidence that these assignments were fixed prior to the incorporation of Lys in the protein alphabet. A change of assignment in the second extension is indistinguishable from genetic noise and is selectively prohibited.

The nature of the genetic code and the assignments in the genetic code are best understood by comparison with other codes under the guidance of coding theory. For example, it is as well to recall the discussion of the construction of block codes in section 7.3.1. Given a binary alphabet 0, 1, we need to construct a code for a destination quaternary alphabet U, G, A, C. This can be done with the first extension, using all pairs of 0, 1, namely, 00, 01, 10, 11. However, this code is saturated and provides no error protection. Therefore the writer of the code may wish to go to the second extension and use all eight triplets of (0, 1) and so on to further extensions as required. In the case shown in Table 5.1, the code is written in the fifth extension and provides a full quaternary alphabet at the destination with considerable error protection. By the same token, it is possible to go to further extensions and provide for a larger alphabet at the destination and as much protection from error as desired. The evidence cited indicates that the evolution of the genetic code has followed the same coding theory procedure. In the period when the genetic code was not saturated there were enough non-sense codons to facilitate evolution from unassigned to assigned codons. When nearly all codons had been assigned amino acids and the codes were nearly saturated, changes in the codon assignments would be indistinguishable from noise. Therefore the possibility of further evolution of the genetic codes was greatly diminished. This may explain why the standard, mitochondrial and other genetic codes have remained unchanged since this early period in the evolution of life.

There may have been more than one origin of life event or there may have been a single origin of life event from which divergences occurred from the first doublet code as the bottleneck was approached where the transition to the second extension occurred. One should remember that, looking backward, a bottleneck appears to be a source. Therefore, the basic form of the Darwinian paradigm that all life is descended from a common ancestor is not challenged by the argument, which I have made in this chapter and in chapter 8, that there may have been a number of origin of life events each with its own first extension doublet genetic code.

Exercise

1. Superman was born on the planet Krypton. Due to the high level of genetic noise caused by natural radioactivity from the mineral kryptonite, life evolved from a doublet genetic code through a second extension to a third extension quadruplet genetic code. This provides more redundance for protection against genetic noise and a much larger vocabulary of amino acids. The evolution of sentient life on Krypton was not possible until that was accomplished. Using what you have learned so far in this book, find out if there is a bottleneck at the transition from a second extension to the third extension genetic code found on Krypton. Some of Superman's extraordinary powers and immunities are due to his third extension genetic code.

8

The early Earth and the primeval soup

Some say the world will end in fire,
Some say in ice.
From what I've tasted of desire
I hold with those who favor fire.
But if it had to perish twice,
I think I know enough of hate
To say that for destruction ice
Is also great
And would suffice.

Robert Frost, 'Fire and Ice', copyright 1923 by Holt, Rinehart & Winston and renewed 1951 by Robert Frost. Reprinted from The Poetry of Robert Frost, edited by E. C. Lathem, by permission of Henry Holt & Co. Reprinted by permission of Jonathan Cape, Ltd in the British Commonwealth.

8.1 The cosmology of the Big Bang theory of the origin of the universe

It is helpful for the molecular biologist to know the elements of current ideas in cosmology in order to understand the modern view of the conditions on the early Earth when life originated. The papers of Oparin (1924) and Haldane (1929), establishing the modern paradigm for research on the origin of life, were written when the accepted cosmology considered the universe to be infinitely large and without a beginning or an end in time. (I use the word *paradigm* in the sense of Kuhn, 1970.) This view of the universe affected their views and those of their followers, even today, on the origin of life. Anything that is not *streng verboten* will happen in an infinite, ageless universe, even events that have a probability measure zero. At some place, at some time, a pot of water placed on a fire will freeze, not boil (Eddington, 1931a). Thus, even if life proves to be improbable, it will happen in such a universe. According to the anthropic cosmological principle, perhaps it happened here where we are fortunately situated to observe it.

Events began to take place, in the decade previous to that of the papers of Oparin and Haldane, which were to force the replacement of the steady state universe cosmology by the amazing notion that the universe began in a huge hydrogen bomb explosion and is small, young and expanding. This trend started with Einstein's general theory of relativity, an example where an impecunious theoretical genius scribbling away by himself changed everyone's view of the world.

8.1.1 Einstein's general theory of relativity

Einstein's General Theory of Relativity regards the universe in terms of a non-Euclidean geometry in which space is distorted by the presence of matter, that is by stars. This may be visualized roughly by a trampoline consisting of a rubber sheet stretched by a rigid frame. A test object such as a steel ball of negligible weight will roll on a straight line. When the athlete steps on the rubber sheet, it is distorted locally by his weight. Steel balls will then be deflected by the distortion. Light serves as the test object in the universe, so that the path of light is predicted to be deflected toward a star as it grazes the limb of the Sun. The first test of the theory came when a total eclipse of the Sun made the sky dark enough so that it was possible to photograph the position of stars in the background. Photographs showing the eclipsed Sun against a background of stars, compared with photographs taken when the Sun was not in that part of the sky, showed an apparent shift in the position of these background stars whose light grazed the limb of the eclipsed Sun. The shift due to general relativity was in the predicted direction and by the predicted amount and was superposed on the classical effect due to refraction in the atmosphere of the Sun.

A theory that captures an appreciable portion of reality finds applications often far outside the considerations from which it was derived. This is in contrast to *ad hoc* or empirical conjectures which often fail to perform even the functions for which they were invented. This has been the case in spectacular fashion for the general theory of relativity. Among the exact predictions or explanations is the advance of the perihelion for planets, or in the case of stars, advance in the periastron of the orbits of binary stars. (The perihelion or periastron is the point in the orbit which is closest to the Sun or to the heavier star in the case of a binary star.) The advance of the perihelion of Mercury is 43.11 ± 0.45 *seconds of arc per century*. The prediction from general relativity is 43.03 *seconds of arc per century*. The advance of the periastron of the binary pulsar star known as PSR1913 + 16 is 4.225 ± 0.002 degrees per year, which is 35 000 times as great (Taylor, Fowler & McCulloch, 1979). The prediction of general relativity is 4.2 degrees per year.

The advent of the exploration of space has resulted in numerous experimental tests of the predictions of both special and general relativity. These effects are small; however, the methods of measurements are extremely accurate. In each case the agreement has been within experimental error. An example is the red shift in the spectrum of light escaping from the intense gravitation fields of large astronomical objects. The gravitational red shift in the Earth's field has been measured by comparing the frequency of a rocket-borne hydrogen maser atomic clock with that of a similar device at a ground station. Frequency can be measured with extreme accuracy. The predictions made according to special and general relativity agreed with experiment to the 70×10^{-6} level (Vessot *et al.*, 1980). General relativity is unique among physical theories in that it has no constants that must be found from experiment. Every test of the theory may make it or break it. Will (1990) has given a review that includes a number of effects not discussed above and the status of their experimental verification.

8.1.2 The application of Einstein's theory to the character and origin of the universe

Friedmann (1922) corrected a mistake in the algebra in Einstein's original papers (Einstein, 1915a, b, c, d, 1916, 1917). Friedmann (1924) showed that, in the correct form, the general theory of relativity predicted that the universe must either expand or contract unless it is empty of matter (Bowers & Deeming, 1984). The seminal theoretical papers on the expanding universe were published by Lemaître (1931a, b, c). The four-dimensional universe in which we live has three space-like dimensions and one time-like dimension. One can visualize the effect of an expanding four-dimensional universe by a model with two space-like dimensions and one time-like. Suppose one blows up a child's balloon decorated with dots. As time increases the balloon expands and each dot recedes from all the others at a rate proportional to the distance. Thus, wherever one is in the expanding four-dimensional universe, all other objects appear to be receding at a rate proportional to the distance (Eddington, 1930, 1931a; Bowers & Deeming, 1984).

Astronomers measure the speed of approach or recession of an object by the shift in wavelength of spectral lines toward the blue or red end of its spectrum. The change in wavelength of sound or of electromagnetic radiation between a source and a receiver in motion is called the Doppler effect. If the sender is approaching the receiver, the apparent wavelength is decreased; by the same token if the sender is receding from the receiver the wavelength is increased.

The seminal experimental papers on the expanding universe were published by Hubble (1929) and Hubble & Humason (1931), who discovered that the amount of red shift was proportional to the distance of stars and galaxies from Earth (Bowers & Deeming, 1984). The Doppler effect gives the speed of this recession. Certain stars, especially the Cepheid variables, pulsate regularly in their luminosity. The distance of galaxies from the Earth is found by measuring the luminosity and the period of these stars. Some Cepheid variables are near enough to Earth so that their distance can be measured by annual parallax as the Earth goes around in its orbit. Thus the relation between the absolute luminosity and the period of variation is known and this can be applied to Cepheid variables in galaxies too far away to have a detectable parallax. The distance of galaxies can be measured since the apparent brightness falls off as the inverse square of the distance. And so one has, immediately, the proportionality between distance and speed of recession of the celestial object. It was difficult to avoid the astonishing conclusion that, extrapolating back in time, the universe had a beginning and is finite in size, but, like the surface of the child's balloon, it has no edges. An expanding universe of finite age and extent with no edges and that had begun in a Big Bang would, perhaps, if the density of matter is above a critical value, because of gravitational attraction, reverse itself to collapse in the Big Crunch (Bludman, 1984). At this time the question of whether or not the universe is closed is not resolved. This is the so-called 'missing mass' controversy. Nucleons fail to close the universe by a factor of 5–7 (Yang *et al.*, 1984).

Gamow (1946), Alpher & Herman (1948) and Alpher, Bethe & Gamow (1948) applied the new knowledge of nuclear physics to put the concept of an expanding universe in a quantitative form. (This is the famous α, β, γ theory. Herman refused Gamow's request to change his name to Delter.) Fundamental physics indicated that, if the universe had once been compressed to a small region, the temperature and density must have been enormously high. Thus, in a popular sense, the universe began in a huge hydrogen bomb explosion. Thermal equilibrium was established at a time $\approx 10^{-2}$ second and the universe was then dominated by a relativistic gas of photons, electron–positron pairs and neutrinos at a temperature of ≈ 10 million electron volts. The fraction of nucleons was $\approx 10^{-10}$ to 10^{-9}. The radiation and matter in this primeval fireball expanded and cooled.

At an age of about 300,000 years, the temperature of the fireball fell to about 3000 K. Radiation ceased to have the energy to ionize atoms so that radiation and matter decoupled and the fireball became transparent. The universe continued to expand and both radiation and matter cooled. Alpher, Bethe, Herman and Gamow concluded that the white hot flash at the time matter became transparent must now have cooled to a cosmic microwave radiation background at a temperature of about 3–10 K. They attempted to persuade radio astronomers to look for the cosmic background radiation but were told that techniques available were not sensitive enough. This radiation was discovered serendipitously by Penzias & Wilson (1965). Measurements are now being taken by the Cosmic Background Explorer satellite of the black body temperature and the spatial distribution of this radiation (Lindley, 1990). At present no intensity fluctuations have been found. The microwave background intensity at 20 frequencies matches the curve for black body radiation derived by Planck from quantum theory within 1 % at each point. The currently accepted value of the black body temperature of the cosmic microwave background is 2.735 ± 0.06 K. This is a textbook example of the agreement of theory and experiment.

A second triumph of the Big Bang theory is that it predicts the correct abundance of ^{1}H, ^{2}D, ^{3}He, ^{4}He and ^{7}Li (Alpher, Bethe & Gamow, 1948; Alpher & Herman, 1948, 1950). Hydrogen is known to be by far the most abundant element in the universe. In the radiation era, before the decoupling of radiation and matter, protons reacted with each other and with the products that were generated to form these elements. The nuclear reactions of the thermonuclear fusion of hydrogen to form helium are as follows, where β^+ stands for positron, v stands for neutrino and γ stands for gamma rays.

$$^1\text{H} + {}^1\text{H} \rightarrow {}^2\text{D} + \beta^+ + v$$

$$^2\text{D} + {}^1\text{H} \rightarrow {}^3\text{He} + \gamma$$

$$^3\text{He} + {}^3\text{He} \rightarrow {}^4\text{He} + 2\,{}^1\text{H}$$

$$^3\text{He} + {}^4\text{He} \rightarrow {}^7\text{Be} + \gamma$$

$$^7\text{Be} + \text{electron} \rightarrow {}^7\text{Li} + v + \gamma$$

$$^7\text{Li} + {}^1\text{H} \rightarrow 2\,{}^4\text{He}$$

$$^7\text{Be} + {}^1\text{H} \rightarrow {}^8\text{B} + \gamma$$

$$^8\text{B} \rightarrow {}^8\text{Be} + \beta^+ + v$$

$$^8\text{Be} \rightarrow 2\,{}^4\text{He}$$

The net effect of these nuclear fusion reactions is to convert four hydrogen atoms into a helium 4 atom plus neutrinos and energy. Both ^{8}B and ^{8}Be are unstable (half life $= 10^{-21}$ second) and so are all nuclei with five nucleons, so that no elements of higher atomic number can be made in this chain. The events involved in the Big Bang paradigm are described in an article by Alpher & Herman (1988) that gives references to other important papers. A detailed exposition of the status of primordial nucleosynthesis is given by Yang *et al.* (1984) who find that no other scenario gives this accurate and detailed account of the abundance of ^1H, ^2D, ^3He, ^4He and ^7Li.

The prediction of an expanding universe is regarded as a necessary consequence of the general theory of relativity even by those who oppose the Big Bang (Arp, Burbidge, Hoyle, Narlikar & Wickramasinghe, 1990). In spite of other successes of the general theory of relativity, the Big Bang, and in particular the idea that the universe had a beginning, was fought bitterly every step of the way. In his Gifford lectures, given in 1927, and published in 1928, Eddington had this to say:

> Scientists and theologians alike must regard as somewhat crude the naïve theological doctrine which (suitably disguised) is at present to be found in every textbook of thermodynamics, namely that some billions of years ago God wound up the material universe and has left it to chance ever since. This should be regarded as the working-hypothesis of thermodynamics rather than its declaration of faith. It is one of those conclusions from which we can see no logical escape – only it suffers from the drawback that it is incredible. As a scientist I simply do not believe that the present order of things started off with a bang; unscientifically I feel equally unwilling to accept the implied discontinuity in the divine nature. But I can make no suggestion to evade the deadlock.

(American readers are reminded that the English billion is 10^{12}; the American billion is 10^9. For this reason I have always given large numbers as powers of ten. Thus at that time Eddington was thinking of a universe 1000 times older than present data indicate.)

Eddington (1930, 1931a) regarded this prediction as an instability which was 'philosophically repugnant' to him. In order to remedy this defect, as it was seen in the paradigm of the time, Einstein put in *ad hoc* a term in his equation with a cosmological constant. Although there are mathematical difficulties with this *ad hoc* step, Eddington (1930) believed that an appropriate value would remove this instability, as he called it. This proved to be a mistake, as experimental evidence showed in the next decades. For these and for theoretical reasons Einstein (Einstein, 1931; Einstein & de

Sitter, 1932) abandoned the cosmological constant. Einstein later told Gamow (Gamow, 1970) that the cosmological constant was 'the biggest blunder of his life'. Hawking (1983) has shown that the cosmological constant is the quantity in physics that is most accurately measured to be zero, namely, $< 10^{-120}$ in dimensionless form. Later Eddington seems to have come to terms with the expanding universe (Eddington, 1931b, 1933).

Among the schemes to rescue the universe with no beginning and no end was the suggestion of Sir Fred Hoyle and of Hermann Bondi & Thomas Gold who proposed separate steady state theories in which matter is continuously created to fill the voids left by the expansion of the universe (Hoyle, 1948, 1949; Bondi & Gold, 1948). Thus, the paradigm of a universe with no beginning and no end was upheld. (According to Alpher & Herman (1988) the term *Big Bang* was coined, in a pejorative sense, by Sir Fred Hoyle in a BBC radio broadcast.) The motivation of both Hoyle (1948) and Bondi & Gold (1948) was based on their philosophical rejection of a universe with a beginning in a Big Bang. Both Hoyle and Bondi & Gold accepted an expanding universe because of the velocity distance relation and also because the universe is clearly not in thermodynamic equilibrium. These facts required astronomers to make a choice between two alternatives:

(1) the universe was much more dense in the past and had a beginning in a condition of enormous density and high temperature, or

(2) matter was being created continuously.

There was some justification for choosing the second alternative at the time when these papers were written. The value of the Hubble constant which was available gave an age for the universe of about 2×10^9 years. The age of the stars in the Milky Way galaxy was thought at that time to be about 5×10^9 years. This discrepancy has been cleared up in recent work. Bondi & Gold (1948) rejected the theory of Hoyle (1948) for various technical reasons. They also rejected general relativity because the field equations demand a conservation of mass. The steady state theory was abandoned, even by its authors, because it could not explain the abundance of the light elements or the measurements of the cosmic background black body radiation (Hoyle & Taylor, 1964; Peebles & Silk, 1990).

A theory of gravitation was proposed by Brans & Dicke (1961) but it was not successful in replacing the general theory of relativity. Results from astronomy, nuclear physics and high energy particle physics have been applied by cosmologists to the extent that it is now believed that the broad outlines of the evolution of the universe are understood back to the time of

10^{-33} second. The laws of physics for times shorter than the Planck time of 10^{-43} second break down because, at that early time, the force of gravity is comparable to the strong nuclear force. Unsolved questions regarding quantum gravity are involved in attempting to describe events at such short times (Hawking & Ellis, 1968; Hawking & Penrose, 1970; Hartle & Hawking, 1983; Hawking, 1984; Hawking, 1988).

In consequence of the success of the Big Bang theory in explaining a number of apparently unrelated phenomena, the universe is now believed to be between 10×10^9 and 20×10^9 years old and of a radius of the same number of light years (Thielemann, Metzinger & Klapdor, 1983; Sandage & Tammam, 1984; Bartel *et al.*, 1985; Tully, 1988). This is based on the consistency of recent measurements of the Hubble time, the age of globular clusters and the chronology of radioactive decay of long-lived nuclei (Cowan, Thielemann & Truran, 1987). The very existence of long-lived isotopes such as ^{187}Re (half life 45×10^9 years) and ^{232}Th (half life 14.05×10^9 years) together with the well-established nuclear physics of the radioactive decay of these and other elements is very convincing evidence that the universe has an age between 10×10^9 and 20×10^9 years (Cowan *et al.*, 1987). The national debts of most nations far exceed in numbers the age of the universe in years. An international fast food chain advertises that it has served 75×10^9 hamburger sandwiches. This is about one sandwich for each 2.4 months of the age of the universe. One sees that, according to the Big Bang theory, the universe, rather than being infinite and without a beginning or an end, is small and young.

However, even today, those who continue to find a universe with a beginning to be philosophically unacceptable have not gone 'gentle into that good night without raging against the dying of the light'. They regard a universe with a 'short age' with the same contempt as scientists worthy of the name regard Archbishop Ussher's date, October 3, 4004 BC, for the creation of the Earth. The distinguished prelate cannot be charged with lack of exactness. Sir Fred Hoyle (1982), with characteristic flamboyance, finds a finite age of the universe 'offensive to basic logical consistency . . . I have always thought it curious that, while most scientists claim to eschew religion, it actually dominates their thoughts more than it does the clergy. The passionate frenzy with which the Big-Bang cosmology is clutched to the corporate scientific bosom evidently arises from a deep-rooted attachment to the first page of Genesis, religious fundamentalism at its strongest' (Hoyle, 1982).

Among Sir Fred's other objections to a universe with a 'short age' is a speculation he made of the probability that an enzyme could have occurred by chance, namely, between 10^{-15} and 10^{-20}. From this he proposed that the probability that a set of enzymes necessary for life could be produced by the random shuffling (*sic*) of amino acids to be $10^{-40\,000}$ (Hoyle, 1980). Since he believed that the origin of life could have happened with such a probability only in a universe with infinite age he regarded that figure as a reason to disbelieve in all universes with a 'short age'. The correct method of making this calculation is given in Chapter 9 where it is found that Sir Fred was wildly optimistic. What this means, of course, is that life did not originate by chance in a primeval soup. It has no bearing on the age of the universe.

Arp *et al.* (1990), who have long regarded the Big Bang paradigm with creative skepticism, revived the steady state creation of matter paradigm. In their article, written in the dramatic style of Sir Fred, they apply probability, in the form of degree of belief without specifically mentioning that interpretation (section 1.2.5) to a number of recently discovered astronomical phenomena that they assert do not conform with predictions of the Big Bang. They challenge the conclusion that the steady state creation paradigm is ruled out by the cosmic background radiation (Peebles & Silk, 1990). Since they agree that the prediction of an expanding universe is a necessary consequence of the general theory of relativity, they propose certain modifications to that theory. Maddox (1989b) finds a universe with an origin thoroughly unsatisfactory. He believes the Big Bang theory will not survive the data to be supplied by the Hubble Space Telescope.

In the history of science, this is *déjà vu* all over again. Philosophical objections were put forward by Leibnitz against Newton's theory of gravitation, namely, the supposed impossibility of 'action at a distance ... and introducing occult qualities and miracles into philosophy' (Darwin 1872). Einstein's arguments on the question of 'simultaneity' in special relativity, and the wave-particle duality of quantum mechanics, were also 'offensive to basic logical consistency' and accordingly were philosophically unacceptable to many people. Nevertheless they are verified experimentally.

This is another example that shows it is important in science to follow the evidence and the mathematics and not one's philosophy, ideology or religion. The Big Bang theory is a revolution as hard to come by and as drastic as the Copernican revolution with which the names Copernicus, Kepler, Galileo and Newton are associated (Kuhn, 1957). The very incomplete sketch of the Big Bang theory given here may be supplemented by a number of quasi-popular books, among which are Weinberg (1977), Davies (1977), Barrow & Silk (1983), Trefil (1983), Audouze & Israël (1985), Gribbin (1986), Hawking (1988) and Penrose (1989). A more technical account of the Big Bang is found in Gibbons, Hawking & Siklos (1983).

8.1.3 The source of stellar energy and the generation of the heavier elements

The source of the energy of the stars was one of the more important problems in the cosmology of the time when the papers of Oparin (1924) and Haldane (1929) were written. In the nineteenth century, Clausius had proposed the entropy concept with respect to the efficiency of heat engines. Von Helmholtz (1854) was the first to apply that idea to the 'heat death' of the universe (Eddington, 1928). It was easy to show that if the Sun were a ball of coal burning in an atmosphere of oxygen it would be burned to a cinder in about 1,500 years. Lord Kelvin calculated that if the Sun's energy came from gravitational contraction that energy would be exhausted in 30×10^6 years (Burchfield, 1975; Bowers & Deeming, 1984). Darwin's theory of evolution and the uniformitarianism of Sir Charles Lyell in geology required a much longer time than the known sources of energy of the Sun could supply (Gould, 1984). Lord Kelvin's value for the upper limit of the age of the Sun was used against the theories of both Darwin and Lyell. Morrison (1981) tells an anecdote about Sir Ernest Rutherford who thought that radioactivity might, in some way, be the source of the energy of the Sun and was able to satisfy the elderly Lord Kelvin by declaring that it was Lord Kelvin himself who had first identified the problem.

Eddington (1926) published a landmark work on the constitution of stars. By applying the basic laws of physics he was able to calculate the temperature and density that must exist in stars, given the temperature of the photosphere, the luminosity and the dimensions. He concluded that the energy of stars is generated near the center and diffuses out as radiation. His estimate of the age of the Sun is nearly that believed to be correct today. He proposed correctly that the source of the energy of stars is from 'subatomic' reactions, that is, what we now call nuclear reactions. He identified the formation of He from protons and electrons as the 'subatomic' reaction, but he was frustrated by the improbability of the simultaneous collision of four protons and two electrons. We now know that this improbable event is not the means by which hydrogen is burned in stars. In 1926 the deuteron and the neutron remained to be discovered. The correct sequence of nuclear reactions is given in section 8.1.2. Eddington's results provided a platform for Bethe, when data from nuclear physics were available, to show that the energy of the stars does, indeed, come from the nuclear burning of hydrogen and heavier elements.

Astronomers believe that galaxies, and the stars of which they are composed, coalesced by gravitational instabilities after the decoupling of the primeval fireball from matter. When a cloud of hydrogen reaches a certain mass, the energy from gravitational contraction increases the temperature and pressure at the center. Although the temperatures in the center of stars, as calculated by Eddington (1926), are very high and the matter is very dense even so there are not enough protons with energy sufficient to overcome the electrostatic repulsion, as calculated by classical electrodynamics, so that nuclear fusion reactions can take place. Here the work of George Gamow came to the rescue. He showed that quantum mechanics require that a probability exists that some protons at much lower energies may penetrate this electrostatic energy barrier. As the temperature in a star increases, the Maxwell–Boltzmann energy distribution, together with the quantum mechanical barrier penetration, results in what are called thermonuclear fusion reactions. Thus the reactions given in section 8.1.2 occur at a rate sufficient to supply the energy of the Sun. These and other nuclear reactions are called *thermonuclear burning.*

Most of the energy of the Sun is produced at its center because that is where the temperature and pressure are highest. The energy then diffuses out in the form of photons to the photosphere where it is radiated to outer space (Eddington, 1926). When the hydrogen at the center nears exhaustion, other thermonuclear fusion reactions take place. Bethe showed that the barrier to the formation of heavier nuclei caused by unstable states for nuclei of atomic weight 5 or 8 is surmounted by the collision of three helium nuclei to form carbon. Such collisions are much too rare to contribute to the generation of heavy elements in the Big Bang because the time available is too short.

Bethe discovered that once some carbon was available, the reaction of protons with carbon led to nitrogen and oxygen and to a cycle in which hydrogen is burned to helium and the carbon is restored. This is called the CNO cycle (Bowers & Deeming, 1984). When hydrogen is exhausted at the centre of a star, the temperature increases, and helium burning to carbon commences. The burning of hydrogen proceeds in a shell about the helium core. Upon the exhaustion of helium the pressure and temperature increase again and still heavier elements form shells like an onion and participate in the nuclear burning. This eventually generates a core of iron and nickel. There is a minimum in the curve of nuclear binding energy at ^{56}Fe so that no more steady state exothermic burning can occur. (Heavier elements are then made from iron and nickel by neutron capture on a timescale long compared to the half lives of the β-decay of the nuclei in question. This is the s-process (slow) of Burbidge, Burbidge, Fowler & Hoyle (1957); see Clayton & Woosley (1974). When the pressure of the escaping radiation no longer supports the outer layers of the star, a catastrophic instability leads to the gravitational collapse and explosion of the star. This takes place in the alarmingly short time of a few seconds. The speed of collapse may reach 50000 km/second. During this time neutron capture (r-process), which is rapid compared with the β-decay (s-process), forms elements heavier than iron (Burbidge *et al.*, 1957; Clayton & Woosley, 1974; Thielemann, Metzinger & Klapdor, 1983; Cowan *et al.*, 1987; Matthews & Cowan, 1990). The neutron flux is provided by ^{22}Ne(α, n) ^{25}Mg or ^{13}C(α, n) ^{16}O reactions. Nuclei participating in the r-process may have from 15 to 40 more neutrons than the stable nuclei at that atomic number. The nuclear physics of the r-process is well established by terrestrial data. Multiple neutron capture in U from extremely high neutron fluxes that produced transuranic elements with super-heavy isotopes has been found in the debris from detonations of thermonuclear weapons. Spontaneous fission of these super-heavy nuclei contributes to the population of nuclei of lower atomic number (Bell, 1967). The exploding star scatters these heavier elements in the form of dust accompanied by hydrogen and helium expelled by the explosion. (Nature seems to have a predilection for thermonuclear explosions.) Such explosions are called *novae* or *supernovae* by astronomers. A supernova was seen by the naked eye on 23 February 1987, in the Large Magellanic Cloud, a satellite of our Milky Way Galaxy. The Sun is a star in the outer arms of the Milky Way galaxy. Our galaxy is a spiral similar to the famous spiral galaxy seen along its axis in Andromeda. Neutrino detectors placed far underground detected the neutrinos from the stellar explosion. An interesting account of this supernova and of stellar explosions in general is given by Waldrop (1988) and by Woosley & Phillips (1988). A popular account of the Big Bang and the cooking of the elements heavier than He along with anecdotes on the personalities who contributed to the story is given by Gribbin (1986).

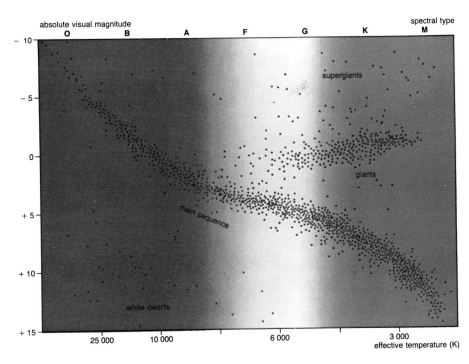

Figure 8.1 The Hertzsprung–Russell diagram. The absolute visual magnitude or luminosity of stars of known distance is shown plotted on the vertical axis. The effective surface temperature is plotted on the lower horizontal axis. The spectral type is given on the upper horizontal axis. The Sun began its life in the gravitational collapse of a cloud of gas and dust. When thermonuclear ignition occurred the Sun was located at the lower right of this figure. As it became hotter and brighter it progressed to the left and up along the main sequence. It is now of the G2 spectral type. The effective surface temperature is about 6000 K. In about 10^9 years the energy received from the Sun will heat the Earth so that the oceans will boil. Eventually the Sun will become a red giant and its surface will extend to the present orbit of the Earth. Reprinted from Audouze, J. & Israël, G. (1985), *Cambridge Atlas of Astronomy*, Cambridge, New York: Cambridge University Press.

Astronomers have convincing evidence that such explosions result in a dense molecular cloud in which solar systems like our own originate from gravitational collapse. The Hertzsprung–Russell diagram (Figure 8.1) is a plot with absolute visual magnitude as the ordinate and surface temperature as the abscissa. The spectral type is shown on the top of the figure. White dwarfs, giants, supergiants and main sequence stars are found in their separate parts of this diagram. The location of each star on this diagram depends on its mass and its stage of evolution. Stars in the main sequence of the Hertzsprung–Russell diagram (Figure 8.1) have sufficiently long lives that carbon may be formed by the collision of three helium nuclei. The events leading to the evolution of the Sun to the main sequence on the Hertzsprung–Russell diagram started at an age of the solar nebula of 10^5 years and were completed in about 50×10^6 years.

Measurements on T-Tauri stars have indicated that these are very young solar-type stars surrounded by large rotating disks of gas and dust (Lada & Shu, 1990). (Astronomers designate stars in the order of the Greek alphabet. T-Tauri is the τ star in the constellation Taurus, the Bull.) Observations indicate that T-Tauri stars are very bright in the ultraviolet region. In a similar phase of its evolution the ultraviolet brightness of the Sun was $> 10^4$ times that of the current Sun. Even after the Sun reached the main sequence it was still 10^2 times its current brightness in the ultraviolet (Sagan, 1965; Canuto *et al.*, 1983). Typical contemporary organisms would acquire a lethal dose of ultraviolet in < 0.3 second (Sagan, 1973). The total luminosity of the 'faint young Sun' when the solar system was formed was about 70% of its present value.

The course of events in an expanding universe powered by nuclear explosions is quite different from the 'heat death' of 19th century physics and cosmology. An expanding universe never reaches equilibrium and a constant temperature and so the Maxwell–Boltzmann–Gibbs entropy, like Achilles in his race with the tortoise, approaches but does not achieve a limit. Some physicists have speculated about the survival of life in the universe as it ages (Dyson, 1979; Frautschi, 1982). However, life on Earth faces a more immediate crisis from the increasing luminosity of the Sun. The progression of the Sun along the main sequence of the Hertzsprung–Russell diagram is shown in Figure 8.1. The abscissa is the effective black body temperature and the ordinate is the ratio of the luminosity with respect to the current luminosity of the Sun. The core of the Sun, at its present position in the Hertzsprung–Russell diagram (Figure 8.1), is gradually being enriched in He. It will begin to contract and increase its temperature. The rate of thermonuclear reactions is very sensitive to temperature and this will cause the insolation to increase about 1% per hundred million years (Gough, 1981). As this proceeds, together with an increasing greenhouse effect, the oceans will boil away in 10^9 years and the Earth will be cooked into lifelessness like Venus today (Sagan & Mullen, 1972; Kasting, 1988). Life has already existed on Earth for three quarters of its allotted time. Perhaps the last organisms to live will be the thermophilic bacteria that started the whole thing. *Sic transit gloria mundi.*

In about 5×10^9 years the Sun will expand beyond the present orbit of the Earth and become a red giant (Penrose, 1989) like Aldebaran in the constellation Taurus (the Bull) or Betelgeuse in the constellation Orion (the Hunter). Thus, instead of ending in ice as was thought in the 19th century, the evidence from astronomy holds with Robert Frost that the Earth will end in fire.

There is one important fly in the ointment in the theory of the generation of energy in stars. Notice that the thermonuclear reactions given in section 8.1.2 produce neutrinos as well as nuclei and energy. The emission of β particles forms an energy spectrum and very few if any β particles have the known energy of the quantum transition. Neutrinos are particles that were proposed to be emitted simultaneously and to account for the missing energy. The time interval between the arrival of light from the 1987 supernova and the detection of a pulse of neutrinos by two laboratories allows a neutrino mass, in energy units, of no more than 14 electron volts (Woosley & Philips, 1988; Bahcall & Glashow, 1987). R. Davis, Jr has measured the flux of neutrinos from the Sun with a non-directional detector. He finds only about one-third the neutrino flux that current models of the Sun and current nuclear physics call for (Bahcall, 1978). Those who follow the scientific philosophy of Popper (1959, 1963, 1972) may regard this experiment as a falsification invalidating the entire theory that the energy of the Sun is generated by the set of reactions in section 8.1.2. According to the philosophy of Kuhn (1970), this is not a falsification but merely a puzzle to be solved in the current paradigm; astronomers and nuclear physicists concur (Newman & Rood, 1977). Hirata *et al.* (1989) have detected neutrinos from the Sun in a directional detector at Kamiokande II near Tokyo, Japan. Other solar neutrino detectors are being built in Canada, Italy and the Soviet Union (Wolfenstein & Beier, 1989). Physicists and astronomers will certainly know much more about energy generation in the Sun by the beginning of the twenty-first century.

A much more detailed discussion of the evolution of the solar system and the generation of energy in stars is given in Bowers & Deeming (1984) and in Audouze & Israël (1985).

8.1.4 The origin of the Earth

Planetesimals are believed to have been formed as a result of gravitational instability in the dust and gas of the solar nebula. Ohtsuki, Nakagawa & Nakazawa (1988) have calculated that the time for a body of 10^{18} grams to grow to 6×10^{27} grams, the present size of the Earth, is about 10^7 years. The abundance of certain elements such as ^{129}Xe, which is generated by the decay of ^{129}I with a half-life of 17 million years, serve as a clock and prove that the collapse to solid objects such as meteorites was accomplished in less than 50 million years in a solar system having a star of the mass of the Sun (Canuto *et al.*, 1983). Perhaps at one stage there were, near the Earth's orbit, about 100 objects of the size of the present Moon, 10 as big as Mercury and a few objects the size of Mars (Newsom & Taylor, 1989). Current models used by astronomers for the accretion of the Earth yield time estimates between 23 and 88 million years. The main source of heat in the early Earth was the energy of the heavy bombardment of impacting objects (Matsui & Abe, 1986). This energy, and the gravitational energy released by the segregation of Fe and Ni to form the Fe–Ni core, was probably enough to result in melting of the proto-Earth (Walker, 1976). The crust of the Earth formed in a time of about 10×10^6 years (Canuto *et al.*, 1983). The course of significant events may have occurred within a time of 50–100 million years, namely: (a) the collapse of the solar nebula to form the proto-sun; (b) the accretion of the Earth; (c) the dissociation of the nebular remnant; (d) the formation of the Earth's crust and the core of Ni–Fe; and (e) formation of the early atmosphere by bolide volatiles and outgassing.

The age of the Earth is usually given as 4.6×10^9 years; the Earth was still extremely active at that time. The oldest rocks on the Earth's surface have been dated to be 4.2×10^9 years old (Compton & Pidgeon, 1986), so the crust, and perhaps the oceans, must date from that time. Kasting (1988) believes that a runaway greenhouse atmosphere was present on the accreting Earth. According to his calculations the temperature of the surface was about 1500 K, which means that the surface was molten during this period. As the Earth approached its final mass, the sum of the accretional heating and the insolation dropped below the critical threshold for a runaway greenhouse effect. At that time the steam atmosphere began to condense and the rain-out created the oceans. In the case of Venus the insolation was above the critical value for a greenhouse atmosphere and rain did not fall (Zahnle *et al.*, 1988).

The early cratering from bolide impacts has been erased by the geological action of plate tectonics and by erosion. The cratering history of the Moon is far better preserved. Maher & Stevenson (1988) and Sleep *et al.* (1989) used that data, correcting for gravitational focusing and increased impact speed because of the larger mass of the Earth, to determine the heavy bombardment of the early Earth. This bombardment did not level off until about 3.5×10^9 years ago. The tsunamis generated by the impacts of all sizes of objects in the oceans must have had major world-wide effects. The shock heating of the atmosphere would have created enormous quantities of NO and NO_2. The torrential acid rains as the steam atmosphere rained out would have had serious consequences (MacDougall, 1988; Waldrop, 1988).

The early Earth was much more volcanically active than today. The volcanic section of the Fig Tree and Moodies group in South Africa is 8 to 10 km thick and may have had an original area of at least 30 by 100 km (Lowe *et al.*, 1989). Beds of spherules of the size of sand are found, in the Fig Tree formation and in the Barberton Greenstone belt, that were formed 3.4×10^9 years ago. This is the first tangible evidence of the Archean late heavy bombardment of the Earth. It is fair to say that enormous flood basalts, corresponding in magnitude to the maria on the Moon, were caused or were strongly influenced by the enormous energies of the bolide impacts (Lowe *et al.*, 1989). Flood basalt provinces occupying areas of 0.2×10^6 to 2×10^6 km^2 and possibly exceeding 1 km in thickness are dated even in the last 250 million years (Hallam, 1987). Some may have been initiated by core–mantle instability (Richards, Duncan & Courtillot, 1989) and may have been initiated by bolide impacts (Rampino & Stothers, 1988). There are now at least 80 asteroids with orbits that cross that of the Earth (Garwin, 1988). The early accreting Earth may have been impacted by an object slightly larger than Mars (Newsom & Taylor, 1989). If so, part of the impactor accreted to form the Moon in an orbit just beyond the Roche limit at which the tidal stress is sufficient to break up Moon-sized objects. The Roche limit is about 2.86 times the radius that the Earth had at that time. And so the Moon escaped breakup by tidal stress from the Earth and did not become a ring system like that around Saturn. Nevertheless, the tides and tidal currents were enormously greater than those of today because the Moon was so close to the Earth (Newsom & Taylor, 1989) and the day was approximately 15 hours (Walker *et al.*, 1983).

The heavy environmental insults that accompanied and were a part of the formation and evolution of the Earth tapered off about $3.8–3.4 \times 10^9$ years ago. Raup & Valentine (1983) suggest that there were multiple origin of life events but that the Earth was sterilized at intervals of 10^5 to 10^7 years by the climatic consequences of bolide impacts. The frequency and energy in specific impacts, according to present understanding (Sleep *et al.*, 1989) indicates that the upper time limit on the last sterilizing event is about 4.0×10^9 years ago. Since the sedimentary rocks at Isua (Schidlowski, 1988) show that life pullulated on Earth 3.8×10^9 years ago, the occasion when life originated is bracketed within these two times.

The bolide impacts and violent volcanisms continued to cause or to contribute to massive extinctions (Alvarez *et al.*, 1980; Alvarez, 1987; Garwin, 1988). Kyte, Zhou & Wasson (1988) report evidence of a collision by an asteroid of at least 0.5 kilometers in diameter in the south of the Pacific Ocean as recently as 2.3×10^6 years ago. This long history of torrential rainstorms and world-wide tsunamis left few places for Darwin's 'warm little ponds' or Sidney Fox's hot dry rocks where amino acids could self-organize (section 10.3.1).

Detailed studies of many factors relevant to the evolution of the atmosphere of Earth and comparisons with that of Venus are given by Kasting (1988) and by Zahnle *et al.* (1988). The emerging consensus is that both Earth and Venus had steam atmospheres and magma oceans (Newsom & Taylor, 1989). Venus lost its water because of its proximity to the Sun. Earth retained an ocean of liquid water because it was exactly the appropriate distance from the Sun (Stevenson, 1988).

8.1.5 The evolution of the atmosphere

The very strong solar wind, during a phase in the Sun's evolution similar to that of the star T-Tauri as it approached the main sequence of the Hertzsprung–Russell diagram, caused the loss of the primeval atmosphere (Fisher, 1976; Walker, 1976; Zahnle *et al.*, 1988). The evidence for this is that the abundances of the rare gases in the current atmosphere of the Earth are many orders of magnitude lower than in the solar nebula. The current atmosphere is thought to have been generated by extensive out-gassing when the Earth was very hot.

Melosh & Vickery (1989) proposed that atmospheric erosion may have been caused by bolide impacts. Their calculations suggest that Venus, Earth and Mars lost atmosphere in the heavy impact period of planet formation. The oceans may have been formed from juvenile water released in volcanic eruptions, by out-gassing and from water brought in by the collision of comets. Chyba (1990) calculated that he could account for between 0.2 and 0.7 ocean masses on Earth from the total accretion during the period of heavy bombardment. Similar processes are thought to have released considerable water on Venus (Matsui & Abe, 1986; Zahnle *et al.*, 1988) and on Mars (Greeley, 1987), in the same time period. A review of the evidence for water on Mars is given by Squyres (1989).

As the energy from bolide impacts and the gravitational energy from the segregation of the Fe–Ni core tapered off, the input of internal heat coming from radioactive decay of uranium and thorium isotopes and ^{40}K, plus the insolation from the 'faint young sun' was sufficient only to maintain a surface temperature in the range of 250–5 K (Sagan & Mullen, 1972). Had the temperature fallen to that level it would have resulted in total glaciation. Yet sedimentary rocks dating to that time are found in Greenland, South Africa and Australia – clear evidence that liquid water was in abundance on the early Earth. A detailed and extensive account of the current knowledge of the early Earth is given by Nisbet (1987).

It is accepted doctrine that the Earth escaped total glaciation due to a greenhouse effect (Sagan & Mullen, 1972; Hart, 1978; Newman & Rood, 1977; Holland, Lazar & McCaffrey, 1986; Kasting, 1988; Zahnle et al., 1988). After considering a number of candidate gases that might have caused the greenhouse effect, Sagan & Mullen (1972) suggested ammonia, which has an infrared absorption in the proper spectral window. This followed the suggestion of Oparin (1972) that the atmosphere of the early Earth was reducing and composed of methane, ammonia, water and hydrogen. The nomination of ammonia as a candidate gas is contrary to the work of Miller & Urey (1959), who showed that ammonia would be carried into the ocean by the torrential rains of the time and would have been largely decomposed. More recent work supports the proposal that ammonia would have been dissolved in the ocean or irreversibly converted to N_2 by photolysis in the intense ultraviolet. Owen, Cess & Ramanathan (1979) showed that the same greenhouse effect could be the result of a decreasing fraction of CO_2. CO_2 would have been gradually dissolved in the ocean to react with $CaSiO_3$ to form $CaCO_3$ plus SiO_2 (Miller & Urey, 1959). Kasting (1988), in his studies of the atmospheres of both Earth and Venus, found that the greenhouse effect can be attributed to the H_2O window at 8 to 12 mm.

Had the atmosphere of the early Earth been composed of a substantial fraction of methane, as proposed by Miller & Orgel (1974), it would have polymerized to alkanes in 10^6 to 10^7 years, thus producing an oil slick one to 10 meters thick (Lasaga, Holland & Dwyer, 1971). Yung & Pinto (1978) made the same analysis for Mars, assuming an early atmosphere in accordance with the model of Sagan (1977) except that ammonia was omitted. The model has CH_4, H_2 and N_2 in the proportions of 60:6:1 by volume. According to their calculations the methane would have polymerized in 10^7 years thus giving rise to a layer of alkanes about one meter thick. According to Abelson (1966) and Schopf (1972) there is no evidence of an ammonia–methane atmosphere on the early Earth. The life-time of the methane was extremely short due to photolysis by the intense ultraviolet of the atmosphere of the early Earth (Canuto *et al.*, 1983). There is no evidence of alkanes generated by other than biological processes on Earth and no evidence for alkanes on Mars.

The view that the early atmosphere was reducing was proposed because amino acids and other 'building blocks' are formed by the action of ultraviolet radiation, lightning and corona discharge in reducing atmospheres (Miller, 1953; Miller & Urey, 1959; Oparin, 1972; Gabel & Ponnamperuma, 1972; Sagan, 1977; Ponnamperuma, 1983). Walker (1976) concluded his analysis of the atmosphere of the early Earth with the remark that '... a highly reducing atmosphere, if it ever existed, would have a short life and a violent end'.

It is a *non sequitur* that the synthesis of amino acids in a particular mixture of gases proves that those gases were present in such quantities on the early Earth (Miller & Urey, 1959). The evidence for the constitution and evolution of the Earth and its atmosphere must come independently from astronomy and geophysical chemistry (Abelson, 1966). Scenarios for the origin of life must accommodate accordingly. The modern view is that the early atmosphere, in the era between 4×10^9 and 3.8×10^9 years, was neutral and composed of N_2, CO_2 and H_2O, with perhaps some NH_3 (Walker, 1976; Hart, 1978; Canuto *et al.*, 1983).

8.1.6 *The effect of the Earth's orbit on the evolution of the atmosphere*

The evolution of the Earth, and in particular its atmosphere, is very sensitive to the dynamics of the orbit. If the Earth–Sun distance were $\leqslant 0.95$ of its current value, a greenhouse effect would have made the climate similar to that of Venus (Hart, 1978; Kasting, 1988). According to Hart (1978), if the Earth–Sun distance were $\geqslant 1.01$ times its current value, runaway glaciation would have occurred about 2×10^9 years ago. This has been disputed by Henderson-Sellers & Henderson-Sellers (1988), who report a computer model that predicts that, once formed, convective heat circulation would prevent a near-global ocean from freezing even if the insolation were only 70 % of the present value. The effect of land masses, if the ocean were not near-global, would modify this model considerably. By the same token, an orbit with a little more eccentricity would have produced a runaway glaciation. It is believed that slight variations in the

Earth's orbit are the fundamental cause of the Pleistocene ice ages (Imbrie & Imbrie, 1980). If the inclination of the Earth's axis of rotation to the plane of the ecliptic had been much larger, like that of Uranus, the environment would, to say the least, have been quite different. If the period of revolution about the Earth's axis had been much longer, like that of Mercury and Venus, the oceans might have frozen because of the build-up of albedo (the proportion of incident light reflected) and run-away glaciation.

8.1.7 The number of 'planets suitable for life' in the Milky Way galaxy

The Sun is in one of the outer arms of the spiral Milky Way galaxy and interstellar dust obscures the center. Like many other galaxies, our galaxy has an active nucleus at the center that has a population of stars much more dense than in the outer arms. Inspection of photographs of similar spiral galaxies such as the famous one in Andromeda, which we see perpendicular to its plane of rotation, shows that the luminosity is enormously greater near the center. Other galaxies seen edge on show the same phenomenon. These 'active galactic nuclei' (AGN) generate enormous amounts of power. At least 10% of this luminosity is in hard X-rays (2–20 keV) (Pounds *et al.*, 1990). There is reasonable speculation that a black hole of 10^6 to 10^8 solar masses exists at the center of our galaxy (Rees, 1988). Roughly speaking, a black hole is a general relativity object so massive and with a gravity so strong that an 'event horizon' develops from which light cannot escape (Bowers & Deeming, 1984). Although their existence has not yet been fully confirmed, black holes are believed to be violent sources of X-rays from in-falling matter and would have the ability to disrupt and swallow stars by tidal action (Shapiro & Teukolsky, 1988). Galaxies have been found by the Infrared Astronomical Satellite that have infrared luminosities $> 10^{12}$ that of the Sun. It is believed that this enormous energy output is due to the accretion of matter into central black holes (Hernquist, 1989).

The processes that operated in the formation of the solar system are the same as those in any other condensing nebula and so planetary systems may be presumed to be very common. Planets 'suitable for the origin of life' are another matter, however. It is as well that the solar system is on one of the outer arms of the galaxy rather than nearer the extremely unpleasant AGN. Furthermore, only stars of the G type (see the upper horizontal axis of Figure 8.1) can have 'suitable planets'; the others are too hot and short-lived or too cool (Hart, 1979). Because of the very narrow range in distance of the habitable zone from the mother star there can be at most one 'suitable' planet per star and, of course, in many cases it happened that no planet accreted in exactly the correct orbit. In the case of double stars, the revolution of the two central stars about each other would produce a large varying gravitational dipole moment. This would greatly perturb the orbits of any planets that might be attached to the stars. Nearly circular orbits would be impossible unless the planets were very far from the center of mass of the system. Therefore that there are many double stars indicates that planets with a climate similar to that of the Earth may be extremely rare.

The detailed stochastic history of the planets may have played a decisive role in determining the 'suitability' of the Earth. For example, the absence of a large satellite of Venus, comparable to the Moon, suggests that the fickle goddess Fortuna contributed the essential event that distinguished the history of Venus and the Earth. It is plausible that the collision of a Mars-sized object with the Earth created the Moon (Newsom & Taylor, 1989) and contributed to the removal of the Earth's primeval atmosphere as a secondary result. The consequent removal of the greenhouse effect allowed the Earth to cool more rapidly than Venus and eventually made possible a climate where the oceans rained out and in which life could originate and survive (Kaula, 1990). Modern astrophysics (Peebles & Silk, 1990) shows that almost everything going on in the universe is incompatible with 'suitable planets'. Those who believe that there are at least 10^9 'suitable planets' in the Milky Way galaxy (Shklovskii & Sagan, 1966) should revise their estimates drastically downward.

8.1.8 Major extinctions and astronomical catastrophes

Alvarez *et al.* (1980) (see also Alvarez, 1987), proposed that the Cretaceous/Tertiary boundary extinctions were caused by the environmental insults from a bolide impact. They suggested that the Earth was enveloped in a blanket of dust that absorbed sunlight and stopped photosynthesis, thus starving most of the animal life found to be missing in the fossil record after that time. The shock heating and ionization of the atmosphere in the impact created enormous amounts of nitric acid of much higher acidity than known today (MacDougall, 1988; Waldrop, 1988). This acid rain would have reacted with limestone and generated quantities of CO_2. O'Keefe & Ahrens (1989) suggested that the impact occurred on a carbonate-rich shallow terrain and calculated the evolution of CO_2 from volatilization of $CaCO_3$. Both of these mechanisms would have released large quantities of CO_2 to the atmosphere. The time for CO_2 to approach equilibrium with the ocean is 10^3 to 10^4 years. Reaction equilibrium with crustal rock $CaSiO_3$ is of the order of 10^5 years. According to this paradigm, the extreme cold of the short period of darkness was followed by a greenhouse heating of the atmosphere of about 2–10 K over a period of between 10^4 and 10^5 years.

Fischer & Arthur (1977) proposed that major extinction events have a nearly periodic time structure. Raup & Seproski (1984) improved the data base and found a significant periodicity of 26×10^6 years in 12 major extinction events. They suggested an extraterrestrial cause. Stigler & Wagner (1987) criticized the method of analysis used by Raup & Seproski (1984) and concluded that their reported periodicity is an artifact of the method of analysis. The issue of whether periodicity exists continues to be disputed between Raup & Seproski (1988) and Stigler & Wagner (1988).

A great deal of attention is also being given to alternate models that consider the effect of volcanism (Hallam, 1987; Officer *et al.*, 1987). Rampino & Stothers (1988) proposed that other extinctions besides those of the Cretaceous/Tertiary boundary were caused by the environmental insults associated with flood basalt episodes triggered by heavy bolide impacts, with a mean period of 32 ± 1 million years. This period does not match the 26 million year period found by Raup & Seproski (1984, 1986, 1988).

Given that there is a periodicity in major extinctions, the next question is to identify the cause of that periodicity. Matter in the outer arms, in the form of stars, clouds of dust and gas is distributed near the plane of the galaxy. The Sun is believed to execute a harmonic motion across this plane thus encountering more dense matter as it passes through the plane than in the extremes of the harmonic motion. Rampino & Stothers (1984) and simultaneously Schwartz & James (1984) proposed that the solar system collides every half period with clouds of dust and gas of mass 10^3 to 10^4 times that of the Sun. This disturbs the Oort cloud (Oort, 1950) of about 10^{12} comets that is believed to be in distant orbit around the Sun (Weissman, 1990). A shower of comets would fill the sky and some of them would collide with the Earth. A total of 41 impact craters younger than 600 million years have been dated. Older ones have been destroyed by erosion and plate subduction. The period of this oscillation is believed, on astronomical grounds, to be about twice the periodicity for major extinction events reported by Raup & Seproski (1984, 1986, 1988) and by Rampino & Stothers (1984). This is consistent since the Sun crosses the galactic plane twice in each period of oscillation. However, Thaddeus & Chanan (1985) believe that the layer of galactic dust and gas near the Sun is too weak for a statistically significant period to be extracted from the nine extinctions analyzed by Rampino & Stothers (1984).

A second astronomical mechanism by which the Oort cloud of comets and other debris could be directed, periodically, toward the solar system is the Nemesis or Death star, a putative dark companion of the Sun, proposed simultaneously and independently by Davis, Hut & Muller (1984) and by Whitmire & Jackson (1984). Rampino & Stothers (1984) report that at present the Sun is close to the galactic plane but is passing through a locally poor region of gas and dust. After about 3 million years the Sun will move into a region where it will again be vulnerable to collision with intermediate and large-sized interstellar clouds of dust and gas. Davis *et al.* (1984) estimate that Nemesis is about 2.4 light years from the Sun and will present no danger until approximately 15000000 AD. Popular accounts of the Nemesis star are given by Muller (1989) and by Raup (1986). This putative object may be a brown dwarf or a substellar object that has a mass less than the limit for thermonuclear ignition of hydrogen. However, Nemesis may eventually be seen in infrared sky surveys if the mass is about 7% of the mass of the Sun, which is sufficient for thermonuclear ignition of hydrogen. It is expected to have an annual parallax motion of ± 1.4 arc seconds and a proper motion of 0.01 arc seconds. The eventual detection of Nemesis is essential for the establishment of this paradigm.

The issue of the Cretaceous/Tertiary extinctions is not settled and has been the topic of vigorous debate at a recent conference, 'Global Catastrophes in Earth History' (Garwin, 1988; Kerr, 1988a, b).

8.1.9 The significance of astronomical catastrophes for the origin of life and evolution

The significance of this astronomical background material for molecular biology is that the Earth was a very unpleasant place in the period of the late heavy bombardment between 4.0×10^9 and 3.8×10^9 years ago, which was when life appeared. No one doubts that it also became a very unpleasant place at a number of times in the Earth's history when major extinctions occurred. A number of major extinctions is very likely to happen before the final extinction of life that will be caused by the increased insolation about 10^9 years from now (section 8.1.3).

Thus, in considerations of the origin of life and evolution, biologists must take into account the origin and history of the solar system and that of the Earth. This means that the role played by major extinctions due to massive environmental disasters must be considered along with the classical Darwinian survival of the fittest. Many niches are left open after each major extinction and speciation from the survivors begins all over again after an evolutionary bottleneck. We are now ready to discuss the paradigm of the *primeval soup*.

8.2 Chemical evolution and the primeval soup

The paradigm of the *primeval soup*, upon which the modern effort in the origin of life is based, was established by the paper of Oparin (1924) and, independently, that of Haldane (1929). The mechanistic conception of life prevailed, supported by the fact that the heat developed in an organism was the same as that in a burning candle if the same amount of CO_2 was generated. Compounds such as urea had been synthesized *in vitro* by Friedrich Wohler in 1828 and it seemed that living matter would, in time, be produced in the same way. Since life was believed to be just complicated chemistry (Loeb, 1912), it was easy to believe that there was certainly enough time for natural processes of physical chemistry to cause life to arise on Earth in the early ocean. As Simpson (1964) put it later: 'Virtually all biochemists agree that life on Earth arose spontaneously from nonliving matter and that it would almost inevitably arise on sufficiently similar young planets elsewhere'. This notion had been expressed much earlier by Darwin in his now famous 'warm little pond' letter.

The first attempt to study the generation of biomolecules in the highly ionizing conditions of a beam of 40 MeV He^{2+} ions was that of Garrison *et al.* (1951). They found formaldehyde, but since they did not include nitrogen compounds they found no amino acids. The first successful attempt to simulate a putative primitive atmosphere of the Earth was that of Miller (1953). He found small quantities of Asp, Gly and Ala after circulating CH_4, NH_3, H_2O and H_2 in glass apparatus in which a corona discharge was supplied by a handy Tesla coil of the type used in chemical laboratories to find leaks in glass apparatus. A substantial quantity of tar or polymer was found in the system. That Gly can be synthesized by a corona discharge in an atmosphere of CO, NH_3 and H_2O had been reported by Loeb (1913), as Miller (1955) pointed out.

Miller & Urey (1959) proposed that organic molecules, which are called the 'building blocks of life,' were generated in the atmosphere, during the chemical evolution phase, by the action of ultraviolet, corona discharges and lightning and were rained out to the early ocean. Sagan (1965) and Canuto *et al.* (1983) showed that the ultraviolet energy flux on the early Earth was 100 times that at present. The gases considered to have been in the early atmosphere are transparent to ultraviolet. In order for the ultraviolet flux to be effective in the synthesis of amino acids, an ultraviolet photon accepter is needed. Sagan & Khare (1971) proposed H_2S, a component of the gases from volcanism. They found Ala, Gly, Cys, Glu and Asp in an apparatus similar to that used by Miller (1953), although, as in that case, most of the material was a tarry mixture. The synthesis of amino acids in shock tubes was discovered by Bar-Nun *et al.* (1970, 1971). Shocks are approximately 10^4 times more energy efficient than ultraviolet in prebiological synthesis (Sagan & Khare, 1971). Shock waves are produced in the atmosphere by lightning strokes and by micrometeorites and meteors (Hochstim, 1971). The period of heavy bombardment on the early Earth and the torrential rains that accompanied the formation of the ocean provided an enormous amount of these two forms of efficient energy. Purines and pyrimidines have been reported by Yuasa *et al.* (1984). A review of the status of chemical evolution has been given by Ponnamperuma (1983).

There is considerable experimental evidence that perhaps as many as 14 of the proteinous α-amino acids are synthesized in gases that presumably were in the atmosphere of the Earth during the time between 4.0×10^9 and 3.8×10^9 years ago. Yet the processes of chemical evolution do not produce proteinous α-amino acids in the proportion found in modern proteins. There is, for example, an embarrassingly large amount of Gly and Ala. It was regarded as plausible that these α-amino acids, along with the other compounds that were formed, were rained out into the ocean thus forming the 'hot dilute soup' of Oparin and Haldane. For followers of this paradigm, belief in the existence of the primeval soup is enough to believe that life would emerge spontaneously from it (Oparin, 1972).

8.2.1 *Microfossils: the earliest evidence of life*

Evidence of life on the early Earth comes from microfossils found in the Onverwacht Group, the Fig Tree Group in South Africa (Schopf, 1972) and in the Warrawoona Group in Australia (Schopf & Packer, 1987). They describe fossil micro-organisms seen in additional carbonaceous cherts found in the Warrawoona Group in Australia. The identification of the objects seen in the petrographic thin samples is dependent on the skill of the observer and their designation as fossils has been frequently disputed. Yet, by a number of tests for authenticity, they find it reasonable that the objects are fossil cyanobacteria that were photoautrophic and oxygen-producing.

8.2.2 *Evidence of life from the carbon isotope ratio*

It has been known for many years that conversion of inorganic CO_2 to living matter by biochemical pathways results in a fractionation that favors ^{12}C. 'From the viewpoint of biogeochemistry, it seems that sedimentary organic compounds must increasingly be viewed as products of a microbial community, not merely as residues of primary production' (Freeman *et al.*, 1990). Thus the carbon isotope ratio is a reliable and indestructible signature in determining whether carbonaceous material was derived from living organisms or from conversion by physical chemical pathways from primordial CO_2.

As we have seen in sections 8.1.4 and 8.1.5, the environment on the early Earth until about 3.4×10^9 years ago was often most unpleasant and of the kind that is believed to have contributed to the mass extinctions since the pre-Cambrian. Nevertheless, there is hard evidence that life pullulated by 3.8×10^9 years ago and by inference may have originated as early as 4.0×10^9 years ago (Schidlowski, 1988). The evidence cited by Schidlowski (1983, 1988) is based primarily on the decrease of ^{13}C in kerogen[1] found in

---

[1] Differences in the carbon isotope composition are expressed as the $\delta^{13}C$ values that give the per mil difference of the $^{13}C/^{12}C$ ratio of a sample relative to that of a conventional standard that defines the zero per mil on the δ-scale:

$$\delta^{13}C = \left[\frac{(^{13}C/^{12}C)_{sample}}{(^{13}C/^{12}C)_{standard}} - 1 \right] \times 1\,000\,(\%_0\,PDB)$$

The name of the standard (PDB) is a modified acronym of Peedee belemnite, referring to the carbonite skeleton of a fossil cephalopod (*Belemnitella americana*) from the Cretaceous Peedee Formation for which $^{12}C/^{13}C = 88.99$ (Schidlowski, 1988).

terrestrial sediments at Isua in Greenland that have been dated at 3.76×10^9 years ago. This is consistent with the correction of the calculations of Maher & Stevenson (1988) by Overbeck & Fogelman (1989), who find that the sterilization effects of the heavy bombardment were limited in such a way that life *could* have originated between 4.0×10^9 and 3.7×10^9 years ago.

The carbon isotope data is hard to dispute. One must conclude that life originated and established itself in a wide distribution on the Earth at some time in the 200×10^6 year period between 4.0×10^9 and 3.8×10^9 years ago. This time is remarkably short for both chemical evolution and the origin and establishment of life to occur, even in comparison with the time required for geological and evolutionary events. Writers on the origin of life have often appealed to time as a substitute for the selective role that enzymes play in life today (Wald, 1954). Yet the time available is one-tenth or less than has been thought (Wald, 1954; Calvin, 1961; Quastler, 1953, 1964; Shklovskii & Sagan, 1966).

Schlesinger & Miller (1983a, b) reported that Gly is produced in the Miller–Urey type of sparking experiment in an atmosphere of CO_2 if the concentration of H_2 is low or zero. There is only a very low yield of other amino acids. They found also that the so-called prebiotic molecules such as HCN and H_2CO, the chemical precursors to purines, pyrimidines and amino acids and to sugars, respectively, are produced in very low amounts in a neutral atmosphere. They agree that this situation is incompatible with the origin of life in the ocean.

Since the atmosphere of the early Earth is no longer believed to be reducing (section 8.1.5), Chyba *et al.* (1990) were led to consider the contribution of organic molecules which they presume to have been accumulated by the impact of carbonaceous asteroids and comets during the period of heavy bombardment by extraterrestrial objects. They suggest that $\sim 10^{20}$ kg of organics of cometary origin impacted on Earth during the period of heavy bombardment. Anders (1989) pointed out the obvious fact that the extremely high temperatures reached by incoming objects would pyrolyze organic molecules in those of larger size. Only very small cometary dust particles would survive accretion by the Earth. During the process of settling to the surface they would be subject to photolysis by ultraviolet radiation and any of the 'building blocks' would be destroyed. The fate of the carbonaceous asteroids and comets would not necessarily be incorporation in the ocean. Sleep *et al.* (1989) point out that the 'building blocks' would not survive in the very hot ocean and atmosphere during the heavy bombardment period which phased out in a brief interval before 3.8×10^9 years ago. Nevertheless, Chyba *et al.* (1990) have reviewed a number of scenarios in which they believe substantial carbonaceous material would have survived impact without becoming pyrolyzed. Their article is written in the subjunctive mood and few unqualified statements are made. This carbonaceous material arrives in the form of organic polymer or kerogen incorporated in a matrix of clay minerals, magnetite and nickel-rich iron sulfides (Wilkening, 1978). The scenario offers no means by which it can be shown that this carbonaceous material is 'prebiotic'. If the 'building blocks' of life in the primeval soup were accumulated in the manner of the scenario of Chyba *et al.* it is difficult to see why the surfaces of Mars and the Moon are not covered with a veneer of graphite.

8.3 Evidence against the existence of a primeval soup

8.3.1 Thermodynamic calculations of the concentration of amino acids in the early ocean

Hull (1960) was the first to calculate the equilibria of various compounds in the ocean. His calculations, based on thermodynamics, predicted vanishingly small concentrations of organic compounds. (See the comment on his paper by Bernal, 1960.) Sillén (1965) calculated the concentrations of Gly, Asp and Asn and found them to be very low, namely, less than about 10^{-6} M. These figures throw doubt on the assertion that the early ocean had the consistency of a thin soup from dissolved amino acids. He bluntly called the primeval soup a myth. Shapiro (1986) discusses the primeval soup paradigm and concludes rather forcefully that it has become a mythology.

Hulett (1969) considered the balance between synthesis and degradation of formaldehyde, hydrogen cyanide, amino acids and organic phosphates. He calculated a maximum concentration of 10^{-8} M for Gly. Rein, Nir & Stamadiadou (1971) emphasized the importance of ultraviolet radiation in destroying organic molecules. Dose (1975) calculated the concentration of amino acids to be about 10^{-7} M taking into account photochemical reactions. Pocklington (1971) reported that the dissolved amino acids in the North Atlantic are in the range of 6 to 47 μg per liter with an average of about 22 μg per liter. The mixing time of the contemporaneous oceans is 500 to 800 years and the primitive ocean could not have been much different. One must conclude that amino acids and intermediates that reached the ocean by rain-out would be circulated to a few tens of meters of the surface, there to encounter the degrading effect of the intense ultraviolet that existed at the time. (Hulett (1969) is in error in the case of the amount of energy supplied by shock waves from lightning and meteorites. His figure of 4×10^{-5} cal/cm^{-2} year was corrected to 0.10 cal/cm^2 year by Bar-Nun *et al.* (1971). Thus much more synthesis must have existed, but the degradation from ultraviolet remains.)

Some authors believe the early Earth was cold (Miller & Orgel, 1974); however, according to the scenario described above, the oceans were hot (Stevenson, 1988; Schwartzman & Volk, 1989). Organic molecules that composed the primeval soup would have passed through submarine hydrothermal systems at ocean ridge crests and pyrolized at these high temperatures and pressures (Nisbet, 1985, 1986a, b). Hydrothermal vents have been discovered in the modern ocean and it is likely that such vents existed on the early Earth.

The hypothesis that life originated near such vents has been examined by Miller & Bada (1988). They show that the retarding effect of pressure in the decomposition of amino acids is very much overcome by the effect of temperature. They conclude that hydrothermal vents contribute to the destruction rather than the synthesis of amino acids. The ocean currently passes through the vents in about 10^7 years. Since there were presumably more such vents associated with the extensive volcanism of the early Earth, this process was a limiting factor in the organic concentration of the early ocean, and not a site for the origin of life (Miller & Bada, 1988). This is disputed by Russell *et al.* (1988) who believe that the evidence is strong enough to continue to consider submarine hot springs as a site for the origin of life. See the reply from Miller & Bada (1989). See also the proposal of Wächtershäuser (1988a, b, 1990) in section 10.2.

8.3.2 The prebiotic organic polymers formed by natural causes from the primeval soup

The authors of all methods of simulating amino acid synthesis remark on the tarry, polymeric material usually left unanalyzed, which is by far the major portion of the results of such experiments. The formation of this tarry material must have preceded the emergence of the protobiont on the early Earth (Calvin, 1969). Furthermore, many scenarios call for the polymerization of the amino acids in the primeval soup to form, for example, coacervates or proteinoid microspheres from which the protobiont is alleged to have emerged (section 10.3). Most of such material was not part of the pathways to life since most of the chains of peptide bonds in the primordial soup would have been composed of both optical isomers, thus making folding of the protein sequence impossible. Much of the tarry material was insoluble and large amounts must have accumulated in the sedimentary rocks found today (Lyle, 1988; Arthur, Dean & Pratt, 1988; Freeman *et al.*, 1990).

If this material had existed, it would have survived as a non-biological kerogen in ancient sedimentary rocks and one would expect to find it in these rocks. One of the scenarios for the origin of life is that the amino acids were adsorbed on clay (Bernal, 1967; Cairns-Smith, 1965, 1971, 1982); accordingly, again, one expects to find the products of degeneration to kerogen in sedimentary rocks. There is, however, no geological evidence for kerogen generated by non-biological means as noted by Abelson (1966) and Schidlowski (1983, 1988). Brooks & Shaw (1973) also pointed out that no such non-biological nitrogenous cokes or graphite-like materials containing large amounts of nitrogen have been found – in fact, sediments found in Greenland, South Africa and Australia are extremely short of nitrogen.

Bar-Nun & Shaviv (1975) explain the absence of non-biological kerogen in sedimentary rocks by the postulated greenhouse heating of the surface by water vapor and acetylene, a scenario that I, for one, find hard to accept. The best evidence of the biological origin of the kerogen in the Greenland Isua rocks is from Schidlowski (1988), who reports a prolific data base of about 10000 measurements of the $\delta^{13}C$ in rocks of age 3.8×10^9 years. All carbon divides distinctly into two groups, marine carbonate limestone and biogenic kerogen. In the case of the marine carbonate, $\delta^{13}C$ values are close to zero, while the values in kerogen range from -10 to -50 per mil. This is extremely good evidence of the lack of large quantities of abiogenic kerogen and therefore that there never were the quantities of dissolved amino acids that could be called a dilute soup.

Chang *et al.* (1983) report isotope effects of the Miller–Urey corona discharge experiments in which the $\delta^{13}C$ values are reduced. Schidlowski (1988) disputes this and points out that in the bulk products, which are the rejected 'building blocks,' the carbon isotope effects are -10 per mil or less. He regards the tarry material that results from Miller–Urey corona discharge syntheses as being untypical and amenable to easy detection if present. Schidlowski (1988) does not address the question of the existence of the primeval soup. He confines his attention to identifying the kerogen in the Isua deposits as being of biotic origin.

8.3.3 Was the primeval soup a source of food for the protobiont?

The suggestion that the primordial soup provided a vast source of food for the first protobionts has been made by Haldane (1929), Cairns-Smith, (1971), Oparin (1972), Orgel (1973b) and Horowitz (1986). They neglected to notice that the organic substances in the primordial soup must have been racemic. In order for the paradigm to produce the protobiont, almost all of the amino acids must have been in polymer form. This is inconsistent with the notion that the protobiont would have been immersed in a sea of dissolved amino acids that could readily be used for food. Most of the polymerized amino acids would, as the Miller–Urey experiments show, be in an insoluble tarry mixture. Only those peptide bond chains composed solely of one optical isomer would have been available for food – an extremely small fraction (section 9.2.4).

Even if we suppose that the early protobiont had a means of selecting for food only the desired optical isomer, half of the insoluble tarry polymers would have remained to be preserved as kerogen in the ancient rocks. As Schidlowski (1988) has shown there is no evidence of such kerogen, in spite of intense search.

8.4 The conclusion that there is no evidence for a primeval soup

Although the Oparin–Haldane paradigm is now just a relic of the cosmology of the time when it was invented, it certainly deserved extensive research and much has been learned in investigating it. The same can be said for many other failed paradigms (Kuhn, 1957, 1970). Nevertheless, like the luminiferous ether, one has to conclude that there is no evidence that a 'hot dilute soup' ever existed. In spite of this fact adherents of this paradigm think it *ought* to have existed for philosophical or ideological reasons. For example, Chang *et al.* (1983) state: 'The record of biological evolution manifest in the chemistry of living organisms and preserved in the geological record probably provides the most compelling evidence for a period of chemical evolution early in Earth history'. However, to support this 'compelling evidence', they call for more direct evidence from, for example, extraterrestrial environments and for unambiguous evidence in the geological record on Earth. 'Clearly, the uncovering of unambiguous evidence in the geological record of an early period of chemical evolution would constitute a major confirmative discovery'. By the same token, finding unambiguous evidence (*sic*) that there is *no trace* of a major signature of the chemical evolution paradigm would be fatal to the paradigm. The creative skeptic demands positive evidence rather than faith, ideology or philosophy to support a belief.

It is an essential part of the chemical evolution paradigm that once the primeval soup was well established there was a gradual transition to biological evolution. Among the first signatures suggested to mark this transition were the presence or absence of isoprenoid compounds in kerogen found in very old sedimentary rocks. Unfortunately for this proposal, it was found that isoprenoid compounds can be formed by non-biological synthesis (McCarthy & Calvin, 1967).

Extraterrestrial environments have been examined in meteorites, lunar samples and the 'life on Mars' effort. Although traces of several racemic α-amino acids have been found in certain meteorites there is no evidence that they are of biological origin. Lack of evidence of organic matter or any other evidence of life on Mars proved to be *total* (Horowitz, 1986, 1990). The call of Chang *et al.* (1983) for unambiguous evidence to support the primeval soup paradigm in extraterrestrial environments and in the geological record has gone unanswered. Chang (1988) speculates that a period of plate tectonism in the history of a planet is a prerequisite for the origin and sustenance of life. He agrees that the lack of such a history on Mars argues against the origin of Martian life. In spite of that, and the total lack of evidence of life on Mars, he writes that: 'Because only further geological and biogeochemical exploration of the planet can place these qualitative speculations [sic] on firm ground, the search for evidence of extinct life on Mars continues to be of highest scientific priority'. From the point of view established in this book one must assign a probability zero for the degree of belief in the origination of life on Mars at any time in the past.

Cairns-Smith (1982) has a much more detailed discussion of the chemistry of chemical evolution and comes to the same conclusion, namely, that the primeval soup never existed. The discussion in the book by Day (1984) treats the origin of life in detail. He also comes to the same conclusion. Pflug (1984) analyzed the data from the microstructures found in the Isua metasedimentary sequences in Greenland, in Onverwacht, South Africa and in Warrawoona, Australia. The evidence shows a well-evolved biosphere before 3.8×10^9 years ago. He points out that all known bacteria, even the most primitive, are highly evolved organisms consisting of at least 10 000 different organic compounds. He concludes that such organisms could not have evolved in a time of 200–300 million years and therefore that there is no evidence for a prebiotic evolution.

Chemical evolution and its putative consequence, a high concentration of organic matter in the oceans of the prebiotic Earth, is an essential part of the belief that life arose spontaneously from this non-living matter and that it would almost inevitably arise on 'sufficiently similar young planets elsewhere' (Calvin, 1961; Simpson, 1964; Ponnamperuma, 1983). Woese (1979), who has been particularly outspoken, believes that 'the Oparin ocean thesis' is incorrect in its basic assumptions and therefore in its conclusions. Woese bases this opinion on the fact that dehydration does not proceed in an ocean and on the element of the proposed paradigm that requires the origin of life to have involved a long series of highly improbable events, i.e. miracles. He has further philosophical objections. Furthermore, one must echo the statement, made 21 years ago by Hulett (1969): 'The Oparin–Haldane model leading to the origin of life from simple components of a primitive reducing atmosphere may be incorrect. It is certainly incomplete.' Shapiro (1986) concludes that a major fault of the Oparin–Haldane paradigm is that there is no evidence that a primeval soup ever existed. Delbrück (1986) believes that the origin of life is a fundamental question in biology. However, he points out that there is no geological record of 'prebiotic evolution'. It is often said that one cannot prove a negative statement by experiment; nevertheless this lack of evidence means that the probability of the existence of a primeval soup must be very nearly zero. Accordingly, the degree of belief must also be very nearly zero (section 1.2.5).

It is regrettable that the creative skepticism which is applied to the Big Bang paradigm (Arp *et al.*, 1990) is not applied to the paradigm of the primeval soup, which is widely regarded by authors of papers on the origin of life as being so well established that no mention of justification is necessary (Joyce, 1989; Chang, 1988; Küppers, 1986, 1990; Waldrop, 1990; Chyba *et al.*, 1990). It is taught as scientific fact in college text books (Weiner, 1987), and in popular books on science (e.g. Beatty, O'Leary & Chaikin, 1981; Chaisson, 1981; Lewin, 1982; Friday & Ingram, 1985; Dulbecco, 1987). Those who, from time to time, have raised technical objections to the existence of the primeval soup or to the estimated number of 'sufficiently similar planets' are regarded as ideological deviationists, heretics and spoil-sports (Yockey, 1977c, 1981; Tipler, 1985a, b, 1988).

In response to the convincing evidence that no primeval soup ever existed on Earth or on Mars, the argument in support of the primeval soup scenario has moved to starting mass movements, *non sequiturs* and slogans (Dyson, 1979a, b). A typical slogan is: 'The absence of evidence is not evidence of absence! (Project Cyclops, 1973; Sagan & Newman, 1983)'. Happily for me, the local law enforcement authorities are satisfied that the absence of evidence that I am a bank robber is sufficient to justify the absence of a warrant for my arrest!

9

Did life emerge by chance from a primeval soup?

What! out of senseless Nothing to provoke
A conscious Something to resent the yoke
Of unpermitted Pleasure, under pain
Of Everlasting Penalties, if broke!

Omar Kháyyám, The Rubáiyát

9.1 Could life have originated in a primeval soup containing the 'building blocks' of life had such a soup existed?

That the early ocean on Earth contained a primeval soup with the 'building blocks of life' where life originated and that life would inevitably originate in a similar primeval soup on any 'suitable' planet elsewhere is an axiom of all origin of life scenarios (Simpson, 1964; Calvin, 1969). These scenarios fail because an obvious consequence is not found, to wit, there is no geological evidence independent of the existence of life that a primeval soup ever existed. It is necessary, nevertheless, to address the question as to whether life could have originated had there been such a soup.

9.1.1 How many genes are required by the protobiont in its genetic message?

John von Neumann, near the end of his very productive life, was interested in computers and their role in science and human affairs. In particular, he was interested in a robot that had a control message of sufficient information content that it could direct its own actions and reproduce itself. Since the processes by which living organisms direct their actions, and in addition reproduce themselves, are recorded in a genetic message that message must be algorithmically describable. This algorithm must be a logical sequence of steps and therefore must be a Turing machine (section 2.4.6) which can perform the same algorithm. Let us think of the protobiont as a kind of von Neumann self-replicating machine (von Neumann, 1966). The built-in set of instructions that directs its action has an information content and is analogous to a genetic message. Von Neumann (1966) thought that there was a minimum degree of complexity below which the process of self-replication is degenerative. Above this level, the process of self-replication was, as he put it, explosive. Thus the complexity of the built-in algorithm must be sufficient to allow the operation of the robot but it must also have an additional information content sufficient to allow it to self-replicate. We have seen in Chapter 5 and will see again in Chapter 11 that the built-in set of instructions that is required for the operation of the robot or organism will degenerate because of noise and eventually become insufficient for its function. The protobiont, like the self-replicating robot, must have a genome of sufficient complexity both to metabolize and to self-replicate. Von Neumann measured the complexity of his self-replicating machine by the number of its parts using as a model the computing machines of his time. He thought the number of parts must be in the millions. That is, in terms of the information theory of today, the information content must be in the millions of bits. Langton (1984) has shown that self-replicating machines which are simpler than that of von Neumann can be conceived. Nevertheless, these machines and the algorithms involved are still very complex. Let us consider how the information content of the set of instructions incorporated in the von Neumann machine leads us to an estimate of the minimum information content of the protobiont.

What is the smallest information content in the instructions required to direct the actions and replication of the protobiont? We may draw our estimates of its minimum information content from the genomes of the most primitive free-living organisms in order to ascertain the lowest threshold the various scenarios for the origin of life must attain. The smallest free-living organisms known today are mycoplasmas and spiroplasmas. Many spiroplasmas contain plasmids that have between 2000 and 50 000 base pairs (Bové, 1984) or between 4000 and 100 000 bits. Other estimates can be taken from the genome of viruses, for example, the genome of the virus ΦX 174 has 5375 nucleotides that transmit nine proteins (Fiddes, 1977). Another estimate of the minimum information content of the protobiont was given by Niesert (1987), who proposed that life could have started with a genome of three genes, or perhaps 300 to 400 amino acids, that is, about 2000 bits.

We must also consider the size of such ancient proteins as the bacterial ferredoxins, which have at least 56 amino acids and are believed to date nearly to the time of the origin of life, and the histones, which are also believed to be ancient and have at least 125 amino acids. The pea histone H3 and the chicken histone H3 differ at only three sites, showing almost no change in evolution since the common ancestor (Dayhoff, 1976). Therefore histones have 122 invariant sites. As we learned in Chapter 6, the information content of an invariant site is 4.139 bits, so the information content of the histones is approximately 4.139 bits × 122, or 505 bits required just for the invariant sites to determine the histone molecule. This is considerably larger than the information content we found for iso-1-cytochrome c in Chapter 6.

These estimates are, of course, highly speculative but they do establish criteria for the size of the set of instructions in the genome of the protobiont on which the various scenarios for the origin of life may be judged. The discussion above demonstrates clearly, however, that the minimum information content of the protobiont must be in the range of hundreds of thousands to several million bits. Scenarios on the origin of life must show how a complexity of that magnitude, which is characteristic of organisms, was generated (Yockey, 1977c, 1981). However, before considering these scenarios, we must first establish the formalism with which this judgement can be made.

9.1.2 The relation of 'order' and 'complexity' to the origin of life

The usual distinction authors make between living matter and non-living matter is to say that living matter has 'order'. This is because in considerations of life, these authors have confused the concepts of order and complexity. Not infrequently, biological molecules are said to be 'highly complex and ordered'. The Shannon entropy concept shows that these terms are contradictory (section 2.4.3). In other words, a highly ordered system has small entropy and cannot be complex, while a complex system has small order and high entropy. The meaning of the words 'order' and 'complexity' must come from the mathematics, not from intuition. We learned also, in section 2.4.1, that the complexity of a sequence is measured by its entropy. The entropy of a sequence measures its randomness so that when we speak of the amount of complexity in a sequence we are speaking of the amount of its randomness as well.

Authors who confuse 'order' and 'complexity' offer many examples of transitions from chaotic to orderly states to draw analogies to living systems. For example, it is often said that the crucial event in the origin of life is the saltation from a disordered to an ordered state (Dyson, 1982).

Accordingly, the thrust of many scenarios has been to show how to generate order out of chaos by analogy to some processes that are called order–disorder transitions in physics or physical chemistry. In his Tarner lectures, Dyson (1985) drew an analogy (somewhat apologetically) of what he called biological order to the emergence of order from disorder in ferromagnetism, the ferromagnetic transition that occurs abruptly at the Curie temperature in iron and iron nickel alloys. (Above the Curie temperature, such materials are disordered and not magnetic.)

Ferromagnetism is rather far removed from biology and a better example of the order–disorder transition can be chosen from a number of organic crystals that exhibit the phenomenon of *ferroelectricity*. Rochelle salt (KNa tartrate·$4H_2O$), KH_2PO_4 and triglycine sulfate are typical materials that exhibit ferroelectricity because of an order–disorder transition. The crystals are spontaneously electrically polarized in the ordered ferroelectric state. The order–disorder transition takes place abruptly at a Curie temperature above which the phenomenon disappears, as in the analogous ferromagnetic case. Nevertheless, neither the order–disorder transitions in ferromagnetism nor those in ferroelectricity are pertinent analogies to the origin of life because these transitions have nothing to do with the generation of a genome: a genome is complex, not orderly.

A theory in which events far from thermodynamical equilibrium would generate the 'functional order' essential to living organisms was proposed by Prigogine & Nicolis (1971) and Prigogine, Nicolis & Babloyantz (1972a, b). This theory also fails because it concerns the generation of order instead of complexity.

Unfortunately, these and other authors are looking through the wrong end of the telescope as we have seen in section 2.4.3. We must find a scenario that generates 'complexity,' not 'order,' if we are to understand the origin of life. The question to be addressed in this chapter is to calculate the probability that a genetic message of the complexity reflected by the proteins mentioned above and a genome composed in the range of 100 thousand to several million bits was generated by chance. See further comment on this point relevant to evolution in section 12.1.

9.2 The probability of generating protein sequences by chance

9.2.1 Previous attempts at calculating the probability

Oparin (1957), to his credit, never entertained the belief that protein sequences were formed by chance because, among other reasons, it shuts the door on scientific study of the origin of life. The notion that life emerged by chance from a prebiotic soup was, nevertheless, widely believed in the decade in which the papers by Oparin (1924) and Haldane (1929) were written. The opinion continues to persist (Monod, 1972; Doolittle, 1983) in spite of calculations that show an extremely low probability and therefore an extremely low degree of belief. Of course, to ascribe causality to an extremely rare event is the same as a belief in miracles.

Proponents of the origin of life by chance seldom actually calculate the probability and when they do so, they do it incorrectly. I shall show in this section how to calculate the probability of the origin of life by chance from a primeval soup (even if it never existed) given enough of it and enough time.

Scenarios on the origin of life can be divided into two categories: (a) first life started with proteins; (b) first life started with nucleic acids and in particular with tRNA (Eigen & Schuster, 1977; Eigen & Winkler-Oswatitsch, 1981; Joyce, 1989). These scenarios are further divided into categories that rely on chance and those that require the intervention of a medium from which the 'ordering' is accomplished by natural law.

Our calculations will be carried out in terms of proteins first. However, because of the mapping by the genetic code, these calculations can be carried out in terms of nucleotides in the same way. But this is awkward because if the primeval genetic code is redundant, there is more information in the nucleotide sequence than in the protein sequence.

In addition to a genome, a secondary necessary, but not sufficient, distinction of living organisms is that they must have some means of reproducing the genetic message. This involves the origin of the genetic code discussed in Chapter 7. Origin of life speculations may be put in two categories, namely DNA or RNA first, or proteins first. In the case of proteins first, the question of a genetic code between proteins is not mentioned by those whose propose it since that is a violation of the Central Dogma as originally stated by Crick, although it is not necessarily a violation of the Central Dogma as given in section 4.4. Information can be transmitted from protein to protein, provided the entropy conditions of the protein–protein genetic code are met and provided that a mechanism exists that plays the role of tRNA. Such mechanisms have been suggested by Root-Bernstein (1983). These conditions may (or may not) be met in scrapie, as Crick mentioned (section 4.4).

The authors of early attempts to calculate the information content of proteins thought it necessary to specify the location of each atom in the molecule. According to the sequence hypothesis, however, one needs to know only the sequence of amino acids, and that information is enough to control the folding of the protein (Anfinsen, 1973). For the purposes of calculating the probability of the generation of a genome by chance, we need to deal only with the genetic message.

The first attempt using the sequence hypothesis to calculate the probability of the emergence of life from the primeval soup by chance was made by Quastler (1964). On the basis of an estimate of the information content and an estimate of the number of opportunities, and using the provisions of Theorem 2.2 (section 2.3.2), he found 10^{-255}. The smallness of that number indicated to him that the first genome did not appear by chance. Quastler (1964) then pointed out correctly that the activity of proteins does not require that all amino acids be uniquely specified. Not having the data that was available later, he assumed, incorrectly, that only the five to 15 amino acids in the active pocket need be specified (see Chapter 6). By thus reducing the effective length of the protein under consideration, he was able to convince himself that the probability of the emergence of proteins from a primeval soup was such that, given enough of it and enough time, the events leading to life would occur. Later research on protein folding has shown that some replacements destabilize various intermediates in the folding pathway. That is, they may play no role in the activity of the folded protein but are necessary to achieving the folded state (Goldenberg *et al.*, 1989). Thus far more information is needed to specify a protein sequence than was proposed by Quastler (1964).

The next attempt was in two papers by Kaplan (1971, 1974). He made the usual mistake of ignoring Theorem 2.2, thus neglecting the fact that amino acids are not equally probable and the drastic effect this has on the number of sequences. He recognized that at most sites there are more than one functionally equivalent amino acid, but he grossly overestimated their number. A third attempt was made by Argyle (1977), who was unaware of the previous papers and made the same mistakes.

Sir Fred Hoyle (1980, 1982) has speculated on the probability of the generation of life by chance in a primordial soup. 'The trouble is that there are more than 2000 different independent enzymes necessary to life, and the chances of getting them all produced by a random trial is less that $10^{-40\,000}$. This minute probability of obtaining all the enzymes, only once each, could not be faced even if the entire universe consisted of organic soup' (Hoyle, 1982). Sir Fred offered the problem of the origin of life and his speculation on its probability as an objection to the Big Bang theory and a finite age of the universe which he regards as 'offensive to basic logical consistency'.

9.2.2 Correct methods of calculating the probability

The correct method of making this calculation (Yockey, 1977c) is the one I shall now bring up to date. In addition to solving this problem it is useful to demonstrate how one takes account of events that are not all of the same probability in contrast with what is usually the case in games of chance. All amino acids are not equally probable, so it is not correct simply to multiply the number of functionally equivalent amino acids at each site to get the total number of sequences. One must first calculate the *effective number* at each site according to the Shannon–McMillan theorem (Theorem 2.2). This theorem is not strictly correct for short sequences; however, it takes into consideration the probability of occurrence of each amino acid and gives us the best approximation of the *effective number* of amino acids at each site. One needs a model protein for which the functionally equivalent amino acids are known. I have chosen iso-1-cytochrome c, which was analyzed in Chapter 6, so the first step is to calculate the number of sequences in the high probability set. This is a worthwhile number to know in itself. An estimate of the probability of selecting one member of the iso-1-cytochrome c homologous family that differs by at least one amino acid at one or more sites in the high probability set is found by dividing the *effective* number of iso-1-cytochrome c sequences by the *effective* number of sequences of all amino acids at 110 sites in the high probability set. An estimate of the number of all sequences in the high probability set is, by the same token, the product of the *effective number* of all amino acids at 110 sites.

   In order to calculate the entropy of the functionally equivalent amino acids at each site, we must renormalize the probabilities for the probability of each amino acid at the site l. Let p_j be the probability of the jth amino acid. Let $\Sigma_j p_j$ be the sum of the probabilities of all amino acids at the site.

Then the renormalized probability p'_j is given by the following equation:

$$p'_j = p_j / \sum_j p_j \tag{9.1}$$

The entropy of the renormalized probability distribution of the functionally equivalent amino acids at site l considering only one isomer is

$$H^l_1 = - \sum_j p'_j \log_2 p'_j \tag{9.2}$$

In this example I shall take the probabilities as proportional to the number of codons as before. The calculation can easily be repeated with any set of probabilities that seems appropriate.

According to Theorem 2.2 the effective number of functionally equivalent amino acids at site l is

$$N^l_{\text{eff}} = 2^{H^l_1} \tag{9.3}$$

If all amino acids are of the same optical isomer, it is easy to see from equations (9.1) and (9.2) that for a given sequence

$$H^l_1 = \log_2 r - \left(\frac{x}{r}\right) 6 \log_2 6 - 8 \left(\frac{y}{r}\right) - \left(\frac{z}{r}\right) 3 \log_2 3 - \left(\frac{w}{r}\right) \tag{9.4}$$

where r is the sum of the number of codons of the functionally equivalent amino acids present at site l, x is the number of amino acids present with six codons, y is the number present with four codons, z is unity if Ile is present and zero otherwise and w is the number of amino acids present with two codons.

The biochemical questions of the polymerization of the amino acids in an aqueous medium to form a sequence that would fold up to make a protein with a biological activity are not discussed in the papers cited in section 9.2.1. Such subjects are skipped over with vague statements to the effect that 'the desired substances will arise spontaneously'. If the scenario is thought of being carried out with nucleotides the fact that 2',5' linked isomers predominate in non-biological reactions (Orgel, 1986) is usually ignored. The authors of ancient Greek plays used a *deus ex machina* (feminine *dea ex machina*, plural *dei ex machina*), a god or goddess as a machine, as a means to release humans from their otherwise inextricable predicaments. These authors have done the same thing to bypass these problems. For dramatic effect and to make it quite clear that these

Table 9.1. *Effective number of amino acids for iso-1-cytochrome c arranged according to functionally equivalent pairs, triplets and quadruplets*

Amino acid site	Functionally equivalent residues known, plain text / Predicted residues, italics	N_{eff}^l All residues	H_2^l One isomer	N_{eff}^l All residues	H_2^l Two isomers
23	(Ala–Gln) Gly *Pro* Ser *Thr* Asn Glu *Asp* Lys *Val Ile* Leu Cys Trp	13.513	3.756328386	25.510	4.672995053
36	(Ala–Gln) Gly *Pro* Ser *Thr* Val Glu Asn *Asp* Lys Ile Leu Cys Trp	13.513	3.756328386	25.510	4.672995053
48	(Ala–Gln) Gly *Pro* Ser *Thr* Asn Glu *Asp* Lys Ile Leu Val Cys Trp	13.513	3.756328386	25.510	4.672995053
52	(Ala–Gln) Gly *Pro* Ser *Thr* Asn Glu *Asp* Lys *Val Ile* Leu Cys Trp	13.513	3.756328386	25.510	4.672995053
65	(Ala–Gln) Gly *Pro* Ser *Thr* Asn Glu *Asp* Lys Val Ile Leu Cys Trp	13.513	3.756328386	25.510	4.672995053
66	(Ala–Gln) Gly *Pro* Ser *Thr* Asn Glu Asp Lys Val Ile Leu Cys Trp	13.513	3.756328386	25.510	4.672995053
68	(Ala–Gln) Gly *Pro* Ser *Thr* Asn Glu Asp Lys Val Ile Leu Cys Trp	13.513	3.756328386	25.510	4.672995053
70	(Ala–Gln) Gly *Pro* Ser *Thr* Asn Glu Asp Lys Val Ile Leu Cys Trp	13.513	3.756328386	25.510	4.672995053
97	(Ala–Gln) Gly *Pro* Ser *Thr* Asn Glu Asp Lys Val Ile Leu Cys Trp	13.513	3.756328386	25.510	4.672995053
100	(Ala–Gln) Gly *Pro* Ser *Thr* Asn Glu Lys *Asp* Val Ile Leu Cys Trp	13.513	3.756328386	25.510	4.672995053
108	(Ala–Gln) Gly *Pro* Ser *Thr* Asn Glu Asp Lys Val Ile Leu Cys Trp	13.513	3.756328386	25.510	4.672995053
111	(Ala–Gln) Gly *Pro* Ser *Thr* Asp Asn Glu Lys Val Ile Leu Cys Trp	13.513	3.756328386	25.510	4.672995053
112	(Ala–Gln) Gly *Pro* Ser *Thr* Asp Asn Glu Lys Val Ile Leu Cys Trp	13.513	3.756328386	25.510	4.672995053
6	(Ala–Tyr) Gly Pro Ser Gln Val Thr Asn Asp Lys Ile Leu Arg *His Cys Phe Trp*	16.977	4.085487054	32.420	5.018820387
11	(Ala–Tyr) Gly Pro Ser Thr Asn Glu *Gln Asp* Val *Ala* Arg His Ile Leu Lys Phe Cys *Trp*	16.977	4.085487054	32.420	5.018820387
12	(Ala–Tyr) Gly Pro Ser Thr Gln Glu Asn Asp Arg His *Val Ile* Leu Lys Phe Cys *Trp*	16.977	4.085487054	32.420	5.018820387
24	(Ala–Tyr) Gly Pro Ser Thr Asn Glu *Gln Asp* Lys Arg His *Val* Ile Leu Cys Phe *Trp*	16.977	4.085487054	32.420	5.018820387
28	(Ala–Tyr) Gly Pro Ser Thr *Gln Asn* Glu Asp Lys Arg His Val Ile Leu *Phe* Cys *Trp*	16.977	4.085487054	32.420	5.018820387
29	(Ala–Tyr) Gly Pro Ser Thr Asn *Gln* Glu Asp Lys Arg His *Val Ile* Leu Cys Phe *Trp*	16.977	4.085487054	32.420	5.018820387
41	(Ala–Tyr) Gly [*Pro*]* Ser Thr Gln Asn *Gly* Asp Lys Arg His *Val Ile* Leu Cys *Phe Trp*	15.986	3.998708270	30.427	4.927279699
69	(Ala–Tyr) Gly Pro Ser Thr Gln Asn Glu Asp Lys Arg His *Val Ile* Leu Cys Phe *Trp*	16.977	4.085487054	32.420	5.018820387

#	Sequence				
74	(Ala–Tyr) Gly Pro Ser Glu Gln Thr Phe Asn Asp Lys His Val Ile Arg Cys Trp	16.977	4.085487054	32.420	5.018820387
96	(Ala–Tyr) Gly Pro Ser Thr Asn Glu Gln Asp Lys Arg His Val Ile Leu Cys Phe Trp	16.977	4.085487054	32.420	5.018820387
31	(Ala–Lys) Gly Pro Ser Thr Asn Glu Asp Val Ile Leu Cys Trp	12.631	3.658900271	23.784	4.571943749
33	(Ala–Lys) Gly Pro Ser Thr Asn Glu Asp Val Ile Leu Cys Trp	12.631	3.658900271	23.784	4.571943749
55	(Ala–Lys) Gly Pro Glu Ser Thr Asp Asn Val Ile Leu Cys Trp	12.631	3.658900271	23.784	4.571943749
58	(Ala–Lys) Gly Pro Ser Thr Asn Glu Asp Val Ile Leu Cys Trp	12.631	3.658900271	23.784	4.571943749
95	(Ala–Lys) Gly Pro Ser Thr Asn Val Ile Glu Leu Cys Trp	12.631	3.658900271	23.784	4.571943749
104	(Ala–Lys) Gly Pro Ser Thr Asn Asp Val Ile Glu Leu Cys Trp	12.631	3.658900271	23.784	4.571943749
109	(Ala–Lys) Gly Pro Ser Thr Asn Asp Val Ile Glu Leu Cys Trp	12.631	3.658900271	23.784	4.571943749
110	(Ala–Lys) Gly Pro Ser Thr Asn Glu Asp Val Ile Leu Cys Trp	12.631	3.658900271	23.784	4.571943749
9	(Ala–Arg) Gly Pro Ser Gln Glu Val Thr Asn Asp Lys Ile Leu Cys Trp	14.342	3.842216900	27.249	4.768142826
13	(Ala–Arg) Gly Pro Ser Thr Gln Asn Glu Asp Lys Val Ile Leu Cys Trp	14.342	3.842216900	27.249	4.768142826
15	(Ala–Arg) Gly Pro Ser Thr Asn Glu Gln Asp Lys Val Ile Leu Cys Trp	14.342	3.842216900	27.249	4.768142826
30	(Ala–Arg) Gly Pro Ser Thr Gln Glu Asn Asp Val Lys Ile Leu Cys Trp	14.342	3.842216900	27.249	4.768142826
61	(Ala–Arg) Gly Pro Ser Thr Gln Asn Glu Asp Lys Val Ile Leu Cys Trp	14.342	3.842216900	27.249	4.768142826
62	(Ala–Arg) Gly Pro Ser Thr Gln Asn Glu Asp Lys Val Ile Leu Cys Trp	14.342	3.842216900	27.249	4.768142826
4	(Ala–Met) Gly Pro Ser Gln Glu Val Thr Asp Asn Lys Ile Leu Arg Cys Phe Trp	15.843	3.985798567	30.182	4.915623128
7	(Ala–Met) Gly Pro Ser Gln Glu Val Thr Asn Ap Lys Arg Ile Leu Cys Phe Trp	15.843	3.985798567	30.182	4.915623128
10	(Ala–Met) Gly Pro Ser Gln Glu Thr Asn Leu Ile Lys Arg Val Phe Cys Trp	15.843	3.985798567	30.182	4.915623128
20	(Ala–Met) Gly Pro Ser Thr Gln Asn Glu Asp Lys Arg Val Ile Leu Cys Phe Trp	15.843	3.985798567	30.182	4.915623128
63	(Ala–Met) Gly Pro Ser Thr Gln Asn Glu Asp Lys Arg Val Ile Leu Cys Trp	15.843	3.985798567	30.182	4.915623128
47	(Ala–His) Gly Pro Ser Thr Gln Asn Glu Asp Lys Arg Val Ile Leu Cys Phe Trp	15.213	3.927279699	30.182	4.915623128
107	(Ala–His) Gly Pro Ser Thr Gln Asn Glu Asp Lys Arg Val Ile Leu Cys Trp	15.213	3.927279699	30.182	4.915623128
27	(Ala–Ser) Gly Pro Thr Val Ile Leu	7.804	2.964299088	14.201	3.850013374
89	(Ala–Ser) Gly Pro Thr Val Ile Leu	7.804	2.964299088	14.201	3.850013374
32	(Ala–Glu) Gly Pro Ser Thr Asp Val Ile Leu Trp (Val required to complete Hamming chain.)	10.039	3.27567157	18.734	4.227567157
51	(Ala–Glu) Gly Pro Ser Thr Asp Val Ile Leu Trp	10.039	3.27567157	18.734	4.227567157
37	(Ala–Gly) Pro Thr Val Ile Leu	6.868	2.779889185	12.484	3.641958150
59	(Ala–Gly) Pro Thr Val Ile Leu	6.868	2.779889185	12.484	3.641958150
39	(Ala–Asn) Gly Pro Ser Thr Glu Asp Val Ile Leu Trp (Thr required to complete Hamming chain.)	10.893	3.445307006	20.394	4.350006911
64	(Ala–Asn) Gly Pro Ser Thr Glu Asp Val Ile Leu Trp	10.893	3.445307006	20.394	4.350006911
14	Gly	1.000	0.000000000	1.000	0.000000000

Table 9.1. (cont.)

Amino acid site	Functionally equivalent residues known, plain text; Predicted residues, italics	N_{eff}^i All residues	H_2^i One isomer	N_{eff}^i All residues	H_2^i Two isomers
16	(Thr–Arg) *Gly Pro Ser Asn Asp Lys Ala Gln Glu Cys Trp*	11.534	3.527806882	21.559	4.43024 5907
18	Phe	1.000	0.000000000	2.000	1.000000000
19	(Ile–Lys) *Gly Pro Ser Thr Asn Glu Asp Val Leu Cys Trp*	11.636	3.540545101	21.786	4.44537006
21	(Arg–Lys) Gln	2.586	1.370950594	5.173	2.370950594
22	(Cys–Ala) *Gly Pro Ser Thr Val Ile Leu Tyr (UCU or UCC Ser complete Hamming chain.)*	9.447	3.239903821	20.839	4.381180103
25	Cys	1.000	0.00000000	2.000	1.000000000
26	His	1.000	0.0000000	2.000	1.000000000
34	(His–Ser) *Gln Asn Glu Asp Lys Arg (Only AGU and AGC codons used for Ser.)*	7.192	2.846439345	14.384	3.84643 9345
38	Pro	1.000	0.000000000	2.000	1.000000000
40	Leu	1.000	0.000000000	2.000	1.000000000
42	Gly	1.000	0.000000000	1.000	0.000000000
44	(Ile–Tyr) *Phe Gly Pro Ser Thr Gln Asn Glu Asp Lys Arg His* Val Leu Cys Trp	15.986	3.998708270	30.427	4.927279699
45	(Asn–Val) *Gly Ser Pro Thr Trp*	6.336	2.663465190	11.341	3.503465190
46	Arg	1.000	0.000000000	2.000	1.000000000
49	Gly	1.000	0.000000000	1.000	0.000000000
50	(Gln–Thr) *Gly Pro Ser Asn Glu Asp Lys*	8.220	3.039148672	14.890	3.896291529
53	(Gly–Ser) *(Only AGU and AGC codons used for Ser because of Hamming chain.)*	1.900	0.918295341	2.381	1.251629167
56	(Tyr–Lys) *Gln Arg His Trp*	5.002	2.339572262	10.123	3.339572262
57	(Thr–Ser)	1.960	0.970950595	3.920	1.970950594
71	(His–Val) *Gly Pro Ser Thr Gln Asn Glu Asp Lys Arg Cys Trp*	12.424	3.186642370	18.150	4.181897960
72	(Phe–Met) *Leu Gly Ser Thr Vale Ile Cys Trp*	8.645	3.111835407	15.896	3.990623286
76	Leu	1.000	0.000000000	2.000	1.000000000
77	(Glu–Leu) *Thr Lys Gly Pro Ser Asn Val Cys Trp*	9.781	3.290006068	18.150	4.181897960
78	(Asn–Asp) Glu	3.000	1.584962501	6.000	2.584962501
79	(Pro–Ile) *Gly Val Ser Thr*	5.870	2.553269690	10.507	3.393269690
80	(Arg–Ser) *Gln Glu Asp Lys Asn (Only AGU and AGC codons used for Ser.)*	6.420	2.641604168	12.481	3.641604168
81	(Lys–Glu) *Asn*	3.000	1.584962501	6.000	2.584962501

No.	Description				
83	(Ile–Met) Val Leu (Cys Trp) (Leu needed in the Hamming chain; Cys and Trp are not.)	3.459	1.788450457	6.909	2.788450457
84	(Pro–Leu) Gly Ser Thr Val Cys Trp	7.244	2.856791471	13.249	3.727759213
85	(Gly–Lys) Ser Asn Glu Asp (Only AGU and AGC codons used for Ser.)	5.742	2.521640636	9.421	3.235926351
86	(Thr–Asn) Gly Pro Ser	4.745	2.246439345	8.262	3.046439345
87	Lys	1.000	0.000000000	2.000	1.000000000
88	Met	1.000	0.000000000	2.000	1.000000000
90	Phe	1.000	0.000000000	2.000	1.000000000
92	Gly	1.000	0.000000000	1.000	0.000000000
93	(Phe–Met) Gly Ser Thr Val Lys Ile Leu Cys Trp	9.514	3.250013374	17.578	4.135727659
94	(Gln–Pro) Ser Asn Glu Lys Asp (Gln to complete Hamming chain between Lys and Pro)	6.261	2.646439345	12.522	3.646439345
98	(Gln–Asp) Glu (Asx identified as Asn.)	4.000	2.000000000	8.000	3.00000000
99	(Arg–Leu) Gly Pro Ser Thr Gln Asn Glu Asp Lys Val Cys Trp	12.367	3.628433000	23.317	4.543326617
101	(His–Asp) Gln Asn Glu Lys	6.000	2.584962501	12.000	3.584962501
106	(Leu–Met)	1.507	0.5916727786	3.014	1.591672779
113	Glu	1.000	0.0000000000	2.000	1.000000000
8	(Ala–His–Phe) Gly Pro Ser Thr Gln Glu Asp Asn Lys Arg Val Ile Leu Cys Trp	16.092	4.008253193	28.685	4.842216900
17	(Ala–His–Phe) Gly |Pro|* Ser Thr Gln Glu Asp Asn Lys Arg Val Ile Leu Cys Trp	15.908	3.916290974	17.566	4.134710786
35	(Gly–Leu–Tyr) Ser Thr Lys Gln Asn Glu Cys Trp	9.553	3.255922907	17.566	4.134710786
54	(Tyr–Phe–Pro) Gly Ser Asp Asn Glu (Ser required to complete Hamming chain)	7.237	2.855388542	12.895	3.688721876
Total			310.1508314656		408.08378722

* Pro is predicted by the prescription but found by Hampsey, Das & Sherman (1986, 1988) not to be functionally equivalent.
Source: Hampsey, Das & Sherman (1986, 1988).

important questions are bypassed by the authors advocating these scenarios, I shall assign the task of selecting amino acids and of polymerizing them to form proteins or DNA to the three Fates acting as *dei ex machina*.

Let us see how the Fates could go about selecting an iso-1-cytochrome c sequence from a primeval soup. Lachesis, the caster of lots, casts her 110 icosahedral dice, suitably weighted to reflect the probabilities of each amino acid. Clotho, who spins the thread of life, polymerizes them. Atropos watches the progress of the spinning of the thread of iso-1-cytochrome c and cuts it when Lachesis assigns an amino acid to a site that is not among those that are functionally equivalent.

In the case that all amino acids (except Gly, which is symmetrical) are of the same optical isometry, what is the probability that Lachesis and Clotho will complete a chain of 110 amino acids in the iso-1-cytochrome c homologous family without having it cut by Atropos? The number of iso-1-cytochrome c sequences in the high probability set obtained from Table 9.1 is 2.316×10^{93}. The total number of sequences in the high probability set in a sequence of 110 sites if $H = 4.139$ is 1.15×10^{137}. Therefore an estimate of the probability that Clotho will complete her chain and find an iso-1-cytochrome c molecule in one trial is 2.00×10^{-44}. Using the Poisson distribution it is easy to calculate that the Fates must carry out 1.5×10^{44} trials to have a probability of 0.95 of finding one molecule of iso-1-cytochrome c.

9.2.3 The effect of chirality on the probability

The situation becomes much worse the more realistic one makes the scenario. I pointed out (Yockey, 1977c) that all references on chemical evolution that lead to the primeval soup report that many non-proteinous amino acids and analogues are formed along with the proteinous amino acids. And, of course, all asymmetric molecules made by non-biological means are racemic. In the case of the formation of nucleotide sequences, D-mononucleotides in the *anti* conformation and L-mononucleotides in the *syn* conformation may form very close structural homologues when bound in a complementary fashion. Thus the synthesis of one enantiomer is strongly inhibited by the other and prevents chain elongation (Joyce, Visser *et al.*, 1984; Joyce, Schwartz *et al.*, 1987; Joyce, 1989).

Furthermore, the activity is destroyed by the incorporation of one analogue or one wrong optical isomer because that prevents the folding of the protein chain. Since Gly is symmetric, there are 39 proteinous amino acids that can be incorporated by the *dei ex machina* into the protein chain. To take this into account let us renormalize the probability distribution of functionally equivalent amino acids at each site counting each asymmetric amino acid as a separate element. Thus the probability of each optical isomer is $1/2$ the probability assumed before. For example, the probability for each Leu codon is $3/61$, not $6/61$. Gly codons retain the probability of $4/61$. It is easy to show that the entropy of the selection at each site, H_2^l, in this second case where chirality is included is

$$H_2^l = \log_2 r - \left(\frac{x}{r}\right) 6 \log_2 3 - 4\left(\frac{y}{r}\right) - 8\left(\frac{g}{r}\right) - \left(\frac{z}{r}\right) 3 \log_2 \left(\frac{3}{2}\right) + \frac{v}{r} \quad (9.5)$$

where r, x and y have meanings as in equation (9.4), g is unity if Gly is present and zero otherwise and v is the number of amino acids with one codon. The values of H_2^l and N_{eff}^l are given in Table 9.1. The entropy calculated in the case where the isomers of all amino acids are present is 5.074 and, from equation (9.3), the effective number of all amino acids including both isomers is 33.675. The estimate of the total number of all sequences in the high probability set is 1.005×10^{168}. When Atropos recognizes that the amino acids selected by Lachesis must be all of the same symmetry, the estimate of the probability of selecting one iso-1-cytochrome c sequence where all the amino acids are of one optical isomer is $2.316 \times 10^{93}/1.01 \times 10^{168}$, which is equal to 2.3×10^{-75}.

We must give the scenario credit for the number of amino acids in the primeval soup. Let us suppose that the Fates use all the amino acids in the primeval soup at each trial. Let us accept the concentration of the amino acids in the primeval soup from those who believe in it. Shklovskii & Sagan (1966) and Eigen (1971) estimate the primeval soup to have contained $\sim 10^{44}$ amino acid molecules. Bar-Nun & Shaviv (1975) estimate $\sim 5.4 \times 10^{41}$. In order to favor the scenario we choose the larger number. Using the Poisson probability distribution, it is easy to calculate that Lachesis must throw her icosahedral dice selecting one from all $\sim 10^{44}$ amino acids in the primeval soup once each second for 10^{23} years to have a probability of 0.95 that her nimble-fingered sister Clotho will complete one iso-1-cytochrome c molecule. If this were the correct scenario, the fatal sisters would be just beginning since the universe is only about 1.5×10^{10} years old.

9.2.4 Did the genome evolve from shorter sequences?

To say that modern protein families evolved from smaller and more primitive ones is a legitimate part of the primeval soup scenario. Let us calculate the longest chain belonging to a protein family that may be formed by chance. There are 39 amino acids since the protobiont regards the optical isomers as being distinct molecules. The effective number, calculated by equation (9.3), is 33.675. The geometrical average of the number of members of the iso-1-cytochrome c homologous family is the 110th root of 2.316×10^{93}, i.e. 7.0594. Now let Atropos again be aware, and wield her shears accordingly, that once an isomer is incorporated in the chain, all isomers of the amino acids must be of the same handedness. She knows, of course, that Gly is symmetric. An estimate of the probability that Lachesis will select one member of the functionally equivalent amino acids at each of N sites is

$$(7.0594/33.675)^N \tag{9.6}$$

The product of the number of trials and the probability of success at each trial is equal to the Poisson parameter λ. Let N be the length of the longest chain that is a member of the homologous protein family that the goddess chooses, with a probability 0.95; then $\lambda = 3$. Making one choice each second, an estimate of the probability that Lachesis will select from a primitive soup of 10^{44} racemic molecules a chain of length N that is part of a member of the cytochrome c family in time t is found from

$$(7.0594/33.675)^N \, (10^{44}/N) \times t = 3 \tag{9.7}$$

We now limit the work of the goddesses to the 200×10^6 years that are allowed by our knowledge of geology. Let $t = 6.3 \times 10^{15}$, the number of seconds in 200×10^6 years. The integer value of N that satisfies equation (9.5) is 85. The time allowed for the work of the goddesses is generous by at least a factor of 10, since Maher & Stevenson (1988) point out that the Earth may have been sterilized by bolide impacts at intervals of 10^5 to 10^7 years. It is clear that in this scenario, only sequences far too short to meet the criteria of section 9.1.1 could be created by chance in the time allowed.

Now let us make this calculation in a more realistic fashion by assuming that there is a mutation frequency β per nucleotide per year. Then using all 10^{44} amino acids we may write the following equation:

$$(7.0594/33.675)^N \, (10^{44}/N) \, 3N\beta t = 3 \tag{9.8}$$

If β equals 3.5×10^{-8} per year per nucleotide, a value reported for redundant (silent) mutations in mitochondria by Miyata *et al.* (1982), and t equals 2×10^8 years, then the integer value of N that satisfies equation (9.7) is 69. The effect of the non-biological amino acids and the analogues is not taken into account, so that assumptions from which these calculations begin are very favorable to the scenario, nevertheless, the values of N are still far too small to meet the criteria for a genome discussed in section 9.1.1. Let us remind ourselves that we have calculated the probability of the generation of only a single molecule of iso-1-cytochrome c. Of course, very many copies of each molecule must be generated to form the protobiont.

9.3 Life did not originate from the 'building blocks' by chance

In this chapter I am using probability as a measure of degree of belief. It is clear that the belief that a molecule of iso-1-cytochrome c or any other protein could appear by chance is based on faith. And so we see that even if we believe that the 'building blocks' are available, they do not spontaneously make proteins, at least not by chance. The origin of life by chance in a primeval soup is impossible in probability in the same way that a perpetual motion machine is impossible in probability. The extremely small probabilities calculated in this chapter are not discouraging to true believers (Hoffer, 1951) or to people who live in a universe of infinite extension that has no beginning or end in time. In such a universe all things not *streng verboten* will happen. In fact we live in a small, young universe generated by an enormous hydrogen bomb explosion some time between 10×10^9 and 20×10^9 years ago. A practical person must conclude that life didn't happen by chance (de Duve, 1991).

Exercises

1. Substitute values that you believe to be reasonable for the numbers that appear in the equations in this chapter. For example, Kaplan (1974) guessed that the average number of functionally equivalent amino acids at a given site is 13. Do you think that number is reasonable? How many iso-1-cytochrome c sites have 13 or more functionally equivalent amino acids (Table 6.4)? How sensitive is the result to the number of amino acids assumed? Can you find a value of N by this means that corresponds to the length of the genome of a virus?
2. Do the argument and the calculations in this chapter agree with the philosophy of chance and necessity of Jacques Monod (1972)? (see section 5.1.3).

3. Why is the geometric rather than the arithmetic average the correct one to use to find the average number of functionally equivalent amino acids in a cytochrome c sequence?

4. Plot the value of N as a function of t in equations (9.7) and (9.8). Choose other values of β and t that you think appropriate.

5. Doolittle (1983) criticized the calculation given in Section 9.2.4 on several grounds, especially that the value of the mutation frequency β per year is much too small. Calculate the length N in equation (9.6) using $\beta = 35.0 \times 10^{-8}$. Calculate the length of the chain, allowing values of time and of β that you consider appropriate.

6. The number of amino acids with one codon does not appear explicitly in equation (9.34). The number of amino acids with two codons does not appear explicitly in equation (9.5). Work through the derivation of these equations and find the reason for this.

7. Sir Fred Hoyle (1980, 1982) speculated that the probability that 2000 different enzymes would appear at random in a primeval soup is $10^{-40\,000}$. What figure would you get from the considerations in this chapter? Was Sir Fred optimistic or pessimistic?

8. What is the effective number of amino acids calculated from equation (9.3) in the case of one isomer? Answer: 17.621.

10

Self-organization origin of life scenarios

If you can look into the seeds of time
And tell me which will grow and which will not,
Speak then to me, who neither beg nor fear
Your favors nor your hate.

Banquo, Macbeth, Act I, Scene III

10.1 The primeval soup and non-stochastic scenarios

Today, most authors who follow the primeval soup paradigm do not believe that life arose in a hot dilute soup by chance. They believe that first there must have been a means of concentration of the 'building blocks' and then a self-organization into informational biomolecules. The scenarios to be discussed in this chapter propose to solve four problems, namely: (a) the question of concentration of the 'building blocks' from an extremely dilute primeval soup, (b) the question of how the peptide bond was formed in an aqueous medium, (c) the question of the formation of the 'building blocks' into a genetic message and (d) the problem of chirality. The purpose of this chapter is to examine the evidence for these scenarios from the mathematical formalism that was developed in Part I and establish a degree of belief in each case.

10.1.1 Concentration of amino acids and the origin of life from sea foam

Bernal (1960, 1967) and Banin & Navrot (1975) have suggested that one means of concentration of organic matter from a dilute soup is in the form of sea foam. Perhaps short chains of the compounds were generated in the soup, floated to the surface and then were driven by wind and waves to accumulate on the shores of the ancient seas. By this means, dilute amino acids and other biomolecules could have become very concentrated. According to these authors there should have been no great difficulty in polymerizations (and presumably the generation of a protobiont). The credibility of this scenario is supported by the authority of Hesiod

(*Theogony*, *c.* 700 BC) who tells us that Aphrodite (her name means foam-born), arose from the sea foam on the island of Cythera from where she was then wafted to Cyprus. However, it must be noted that her emergence was aided by the god Cronus who castrated his father Uranus (the god, not the planet) and threw the genitals into the sea near Cythera where they served as an *élan vital* for the emergence of Aphrodite. The goddess, being immortal, was able to escape the effect of the intense ultraviolet that bombarded the Earth at the time and that would have exposed mortal organisms to a lethal dose in less than 0.3 seconds (Sagan, 1973). Aphrodite (Homer, *c.* 850 BC; Euripides, *c.* 440 BC) is still widely worshipped today, especially near colleges, universities and army camps. Nevertheless, as far as the emergence of mortal beings is concerned, one must be a bit more skeptical about this mode of origination since we assume only the established processes of physics and chemistry and no *élan vital*. Let us, therefore, consider other means of concentration.

10.1.2 *Adsorption of the 'building blocks' of life on clay particles*

Bernal (1951, 1967) and Cairns-Smith (1965, 1982) are the principal investigators who propose that the 'building blocks' of life were concentrated by adsorption on clay. The very thin crystallites that compose clays are about 10 nanometers in thickness. Their adsorption ability is due to the very large surface available (Katchalsky, 1973). Bernal (1951, 1967) suggested that protobionts were formed on clay and played the role of viruses, the hostess being the Earth. He recognized that a primary difficulty with the primeval soup paradigm is the extreme dilution of the system.

There is considerable geological evidence that clays were present in large quantities at the time of the origin of life. Clay is a well-known industrial catalyst. Clays are used in the petroleum industry as catalysts in cracking and reforming reactions. It is therefore not surprising that clay can serve as a catalyst for the synthesis of some biochemical compounds. Experimental work on the proposal that clay particles concentrate and polymerize amino acids was done by Paecht-Horowitz, Berger & Katchalsky (1970) and by Katchalsky (1973). They found that copolymerization proceeds quickly between alanyl adenylate and prolyladenylate in the presence of mont-morillonite. The peptide bond is formed even though the process is carried out in an aqueous medium. Although chains of as many as 50 amino acids were found, there were clear and distinct patterns in the distribution. Katchalsky (1973) reported that Ala and Ser or Ala and Asp copolymerize strongly but, while Ala and Asp polymerize, each in the presence of the other, there is essentially no copolymerization between the two. Lahav, White & Chang (1978) studied the effect of thermal cycling on the polymerization of Gly on several days. The synthesis of diGly and short oligopeptides was consistently found in the presence of clays. This work shows a great deal of intersymbol influence not typical of proteins.

Cairns-Smith (1965, 1982) suggested that clay particles themselves were the substance of the first organisms. He suggested that the information content reflected by the imperfections in the mineral crystallites of clay were the first genes and that kaolinite was a good candidate for the clay in which these genes appeared. Cairns-Smith (1965, 1982) gives a very good description of crystal imperfections, which are submicroscopic steps, dislocations, interstitials, vacancies, impurities and other faults that are a part of the natural growth of all crystals. The gamma rays, alpha particles and perhaps even neutrons from the extinct natural radioactivity in the ancient clays also created crystal imperfections. For example, the electronic structure of transparent non-conducting crystals or glasses may be affected by exposure to ultraviolet or X-rays. Electrons that are ionized by that radiation are trapped in defects in crystals or glasses called *color centers*. When heated, the materials give off light as the temperature passes through certain regions. The plot of temperature against light output is known a *thermoluminescent glow curve*. The crystallite structure of graphite and other materials used as neutron energy moderators in nuclear reactors suffers *radiation damage*, the well-known Wigner effect. These imperfections are associated with stored energy that may be released upon annealing. As we shall see, the entropy associated with crystal imperfections is the Maxwell–Boltzmann–Gibbs entropy of thermodynamics, not the Shannon entropy or the Kolmogorov–Chaitin algorithmic entropy of information theory. This distinction is important because the entropy of the genome is the Shannon entropy or the Kolmogorov–Chaitin algorithmic entropy.

One of Cairns-Smith's (1982) interesting suggestions is that molecular life grew from crystal life by *epitaxy* of the 'building blocks' on the surface of the crystallites that compose clay. *Epitaxy* is the growth of a crystal of one substance on the surface of the lattice of another. The best-recognized example of epitaxy is the growth of GaAs crystals on substrates of Si crystals. This process is flexible in that an exact match of the lattice dimensions is not necessary. McPherson & Schlichta (1988) have used epitaxial nucleation to grow protein crystals on mineral surfaces. They point out that an alternative growth process reflects the submicroscopic steps, faults, dislocations and other imperfections that are a part of the natural growth of mineral crystals. This reaction may be termed heterogeneous nucleation. Once the nucleation has started, the growth of the protein crystal is on the lattice of the protein crystal itself, not on the clay surface. It should be emphasized that, in contrast with the work of Paecht-Horowitz *et al.* (1970) and of Katchalsky (1973), no formation of proteins from a solution of amino acids is described in the work of McPherson & Schlichta (1988).

Cairns-Smith (1965, 1982) proposes that the information content of these clay genes was obtained from the complex patterns of imperfections in the surface of the crystallites. He gives an interesting discussion of the question of chirality and arrives at the conclusion that there is, at present, no explanation of the emergence of chirality from the racemic mixture that is decreed by thermodynamics. In this Ponnamperuma, Shimoyama & Friebele (1982) concur.

Cairns-Smith (1982) gives the following equation for the information content, I:

$$I = \log_2 w \qquad (10.1)$$

where w is the number of *possible* ways that 'the replicable could be disposed'. He makes an inappropriate application of equation (10.1) to *E. coli.* This equation does not recognize that some 'ways that the replicable can be disposed' are not as probable as others, and therefore it is not an acceptable measure of information content (Theorem 2.2).

The clay scenario does not solve the problem of the generation of a *linear* informational biomolecule, whether that might be DNA or RNA. The growth of biomolecules on the surface of the clay particles is an epitaxy that generates a three-dimensional object. Cairns-Smith (1982) does not make a clear distinction between 'order' and 'complexity' or between information theory entropy and Maxwell–Boltzmann–Gibbs entropy (sections 2.2.2 and 12.1.1). His discussion loses force through its lack of rigor and quantification.

The imperfections in crystals are all associated with energy levels. Upon heating, the stored energy anneals out, as the activation energy of each kind of imperfection becomes comparable with kT, where k is the Boltzmann constant. This annealing process tends to remove imperfections and to return the crystal to a more ordered and more nearly perfect condition. The entropy of the crystal is the Maxwell–Boltzmann–Gibbs entropy, which can be measured by physical methods in the annealing process. Discussion of any information content that is presumed to be reflected in the complex patterns of crystal imperfections on the surface must involve the spectrum of the energy levels associated with each imperfection and the Maxwell–Boltzmann–Gibbs entropy. To pass from the entropy of crystal imperfections to the Shannon entropy of a biomolecule, one must have a mapping of the probability space that describes these imperfections onto the probability space that describes a genetic message.

When this is discussed in words there seems to be no difficulty in believing that the defect structures of crystals are the nearest analogues of the covalent structures of molecules in molecular life. However, the theorems discussed in sections 2.1.2, 2.3.2 and 2.3.3 tell us otherwise. The existence of this transition requires a probability space based on an extension of an alphabet of four letters with a Shannon entropy equal to the Maxwell–Botzmann–Gibbs entropy on the crystal surface. As I have emphasized, no probability space which reflects the mathematics of sequences and which is isomorphic with the probability space of the crystal surface exists. Consequently, these two probability spaces do not have the same entropy and, since there is no mapping, or code, which relates them, they have nothing to do with each other.

The theorems pertaining to isomorphism discussed in sections 2.1.1, 2.2.1 and 2.4.6 apply even to such simple systems as Bernoulli sequences generated by the tossing of a coin. The transfer of information from clay surfaces to organic macromolecules that is presumed to be a pseudo-DNA/RNA/protein system is mathematically *impossible*, not just unlikely, if the entropies of the two probability spaces are not equal. To say that crystal life is a modified perfection while molecular life is a tamed chaos is merely a play on words.

It is true that the probability spaces of DNA and of proteins do not have the same entropy, but this is not a fundamental difficulty because there is a code between the alphabets of these spaces. A code exists whereby information goes, unidirectionally, from the DNA sequences to the protein sequences. No such code exists that makes possible the transcription of messages from the two-dimensional clay particles to a one-dimensional informational biomolecule. The clay scenario is one of the attempts to use the 'order' that is characteristic of a crystal as an analogue of the 'order' that is supposed to characterize informational biomolecules. This has been discussed in sections 2.4.3 and 9.1.2. The question of an equivalence between Maxwell–Boltzmann–Gibbs entropy and Shannon entropy, with regard to evolution, is discussed in sections 12.1.1 and 12.1.2.

The best that can be imagined from any of the clay scenarios is a racemic biomolecule with a great deal of intersymbol influence and consequently a low entropy. A feature of this and other origin of life scenarios is the presumption that a simple system with a low entropy, once established, will grow into a more complex system as entropy increases. (This is related to Fox's 'instructions in the amino acids themselves'.) It is true that evolution *requires* an increasing algorithmic Kolmogorov–Chaitin entropy (section 12.1.2). However, the crystal patterns that Cairns-Smith (1982) shows exhibit far too much regularity to reflect the information theory entropy of even the simplest enzymes. This contradicts his statement that the theoretical information density in crystals is comparable to that in DNA (Cairns-Smith, 1965, 1982). The progress of an ensemble of these highly ordered sequences to highly organized sequences must be by Markov chain processes.

We have seen in section 9.2.3 that the sequences of biological specificity, namely proteins, are quickly lost in an enormous number of sequences of all amino acids in the high probability set. The progression of the sequences derived from clay to proteins is essentially the same process conceived in the origin of life by chance discussed in Chapter 9. Therefore, for this reason also, the clay scenario provides no pathway from the crystal imperfections in clay particles to informational biomolecules.

Cairns-Smith (1965) first considered protein as the 'takeover' molecule but in 1982 suggested RNA. Strigunkova, Lavrentiev & Otroshchenko (1986) have produced oligonucleotides from adenosine monophosphate adsorbed on kaolinite by radiation with ultraviolet. Some fragments had the 2'–5' bond and others had the 3'–5' bond. The yield of such compounds was reported to be about 10^{-2} wt%. The discussion in section 2.2.1 is independent of the specific informational biomolecules involved.

The clay scenario solves the problem of the formation of a peptide bond in an aqueous medium. Clay materials do adsorb and concentrate biological monomers from aqueous solution. There is some selection on a purely chemical basis since basic molecules are adsorbed more than neutral or acid ones. Biological and non-biological molecules are adsorbed equally and there is no evidence of asymmetrical adsorption. Accordingly, Ponnamperuma, Shimoyama & Friebele (1982) stated that experimental research bears out only partially the role of clays in chemical evolution as proposed by Bernal (1951, 1967) and Cairns-Smith (1966, 1982).

Cairns-Smith (1982) argues against the existence of a primeval soup, a conclusion in which I concur (section 8.3). However, clearly in a final stage of his scenario, molecular life must take over from clay life. It is hard to understand how this could have happened in the absence of a supply of the 'building blocks' from a primeval soup. Consequently, the adsorption of organic chemicals on clay, which has been shown to occur in the laboratory (Ponnamperuma *et al.*, 1982), should have been the source of an abiological kerogen that one would expect to find in the 3.8×10^9 year old Isua sedimentary rocks. No abiogenic kerogen was found by Schidlowski (1988) in these or other rocks.

10.2 Pyrite formation and the origin of life

10.2.1 Wächtershäuser's theory of carbon fixation in the formation of pyrite

Wächtershäuser (1988b, c, 1990) has put forward substantially the same objections to the primeval soup paradigm as those I have discussed in Chapter 8. He has made a very ingenious proposal for the fixation of carbon and the generation of membranes of lipids and anionic peptides. Central to his theory is the idea that early life was autotrophic (in contrast to the heterotrophy of the primeval soup paradigm and the adsorption of amino acids on clay). The fixation of carbon proceeds directly on the surfaces of minerals such as FeS and produces iron pyrite FeS_2. This is proposed as the first source of energy for life. This process is described (Wächtershäuser, 1990) by the following exergonic reactions, where $\Delta G°$ is the free energy for pH zero and $\Delta G°'$ is the free energy for pH 7:

$$4CO_2 + 7H_2 \rightarrow (CH_2\!-\!COOH)_2 + 4H_2O \quad \Delta G° = -151 \text{ kJ/mol}$$

$$4HCO_3^- + 2H^+ + 7H_2 \rightarrow (CH_2\!-\!COO^-)_2 + 8H_2O$$
$$\Delta G°' = -160 \text{ kJ/mol}$$

The free energy is much larger in the presence of H_2S and FeS:

$$4CO_2 + 7H_2S + 7FeS \rightarrow (CH_2\!-\!COOH)_2 + 4H_2O + 7FeS_2$$
$$\Delta G° = -420 \text{ kJ/mol}$$

and

$$4HCO_3^- + 2H^+ + 7H_2S + 7FeS \rightarrow (CH_2\!-\!COO^-)_2 + 7FeS_2 + 8H_2O$$
$$\Delta G°' = -429 \text{ kJ/mol}$$

These reactions provide the energy for the fixation of carbon and also a binding surface for the organic constituents that are formed. Any process by which the constituents lose their surface bonding is irreversible and the material lost by dissolution in the ocean. The propensity for thermal degradation in a surface-bonded state is less than in solution so that the process may proceed at the higher temperatures typical of submarine hot springs (see section 8.3.1).

Several forms of iron sulfides have been found in extremely primitive sulfur bacteria which live in anaerobic conditions (Farina *et al.*, 1990; Mann *et al.*, 1990; Williams, 1990). This seems to be a corroboration of Wächtershäuser's proposal (Popper, 1990; Russell, Hall & Gize, 1990).

Among the advantages of his proposal are that it generates the organic molecules without the concomitant large amount of tarry material that should have been absorbed in sedimentary rocks and become kerogen. It generates the compounds in place thus avoiding the need for a method of concentration. It is consistent with energy requirements and thermodynamics. This is a serious defect of the primeval soup paradigm, which requires that amino acids polymerize in aqueous solution whereas peptides actually hydrolyze. Rather than requiring ready made 'building blocks' it assumes only the CO_2, H_2S and FeS found in submarine hot springs. The reaction products remain in place and there is no requirement for concentration, a serious difficulty with the primeval soup paradigm. The reactions shown above are far from thermodynamical equilibrium and are irreversible as must be the case in living systems.

The proposal is incomplete in that it generates a two-dimensional monomolecular layer while all informational molecules are one-dimensional. It shares with other theories the idea that life is just complicated chemistry and does not take into account the need for the sequence hypothesis. The generation of complexity and of a genetic code is not adequately addressed. Neverthelesss, it is free of many of the objections to the primeval soup paradigm and therefore it is a step forward. Although Wächterhäuser did not mention the papers by Schidlowski (1983, 1988) and Schidlowski *et al.* (1983) those papers are among the strongest support for his theory.

10.3 Self-organization of amino acids to form proteinoids

10.3.1 The proteinoid scenario for the origin of life

Sidney Fox and his disciples have championed the 'proteinoid' scenario for the origin of life, a latter-day form of Oparin's coacervate paradigm. They used the well-known reaction of heating amino acids to 150–180 °C (Miller & Urey, 1959; Katchalsky, 1951). These 'proteinoid microspheres are prepared by heating a mixture of amino acids that must contain a large proportion of Asp and Glu together with other proteinous amino acids in various proportions. This mixture is heated for 2–5 hours at 180 °C. At the end of this time a 1 % solution of NaCl is poured slowly over the mixture and boiled for 30 seconds. When the solution is cooled one may see under the microscope numerous spheres of material'. This reaction requires a large excess of Asp and Glu, whereas it is well known that Gly and Ala are by far the most abundant amino acids under the presumed prebiotic conditions. Fox and his followers believe that these microspheres are a model of first-life based on protein-first. They believe that the evolutionary role of the proteinoids is that of the first informational molecule. They purport to see the beginning of cell walls, growth and budding typical of living cells (Fox & Dose, 1972). According to Shapiro (1986), Fox feels that he has largely solved all the origin of life problems. He argues that the proteinoid scenario must be accepted because there is no other alternative. This is the classical *fallacy of the false alternative*. I shall show, in this section, how well his solution stands up to criticism, both experimental and theoretical.

This scenario proposes that the amino acids in the primeval soup were concentrated by evaporation in pools and lagoons and heated by the action of hot volcanic rocks. After formation, the proteinoids were washed to the ocean by rivers fed by rain storms carrying with them enormous numbers of proteinoid microspheres. These proteinoid microspheres evolved, by unspecified gradualist and uniformitarian processes, the proclivities of living organisms, including the appearance of the current genetic code.

Three themes appear consistently in Fox's voluminous and perhaps repetitive publications: (a) 'the instructions for protobiotic proteins have been explained as arising from the amino acids themselves'; (b) the amino acid sequences in proteinoids are non-random (Nakashima *et al.*, 1977) and are ordered, as are proteins; (c) proteinoids are weak catalysts and therefore are progenitors of proteins. These ideas are not original to Fox;

they were expressed by Wald (1954) and were no doubt current for some years before. Fox (1973, 1975) has his own *non-mathematical* definition of 'information', namely, '...we use as a definition of information the capacity of a molecule or system to interact selectively with other molecules or systems'. This is a property shared widely with all manner of chemicals. With such a definition there is no way to comment on his belief that proteinoids are the first 'informational molecules'. Fox's notion of randomness is more primitive than that discussed in Yockey (1977c) and in section 2.4.2, namely, 'A random process is one in which only chance factors determine the particular outcome of any single trial of the experimental operation'. This 'definition' is circular. He also states (Fox, 1984a) that: 'A generally satisfactory definition of "order" is elusive'.

10.3.2 Objections to the proteinoid scenario

A number of authors have published objections to the proteinoid scenario on chemical grounds, among them, Miller & Urey (1959), Bernal (1967), Miller & Orgel (1974), Jukes & Margulis (1980), and Day (1984). Most of the objections to the proteinoid scenario come directly from the work of Fox and his followers. Dose (1976) points out the following: proteinoids yield 50% to 80% amino acid upon hydrolysis, but contain 40% to 60% fewer peptide links than typical proteins; such crosslinks disturb most of the sequencing methods typically used in protein chemistry; the optically active amino acids racemize during the heating process; and no helicity was detected (Fox, 1973). Fox (1975) reports that proteinoids differ from proteins in that most of the constituent amino acids are racemized by the heat treatment necessary for their formation. The low yield of amino acids after prolonged hydrolysis indicates extensive crosslinkages (Folsome, 1976). Andini *et al.* (1975) and Temussi *et al.* (1976) showed by nuclear magnetic resonance and paramagnetic resonance that thermal poly-peptides contain only β-peptide linkages. They state that their results cast serious doubt on the role that proteinoids may have had on prebiotic polypeptide synthesis because only α-peptide linkages are characteristic of proteins. Fox (1976b, 1984b) is aware of the results of Andini *et al.* (1975) and Temussi *et al.* (1976). He defends his scenario with the following comment: 'With catalysis of peptide formation by lysine-rich proteinoid, it could well be that only modern proteinous structures and bonds would result. This is being investigated. The criticism is rooted in a popular anachronism demanding that a primitive form be fully like a modern form;

the critic is ignoring stepwiseness in evolution'. English translation: Keep hope alive! In spite of these comments, Dose (1976) supports proteinoids as possible candidates for ancestors of the first proteins. Katchalsky (1973) and Miller & Bada (1988), on the other hand, do not regard Fox's microspheres as being on the evolutionary pathway to the first living organism.

Side reactions lead to largely unknown substances (tar) that are linked to the polymerized amino acids (Dose, 1976). Amino acids decompose at the temperature needed, 150–180 °C, so that the heating period must be artificially short. Furthermore, since Gly and Ala are by far the most abundant amino acids formed by chemical evolution, it is doubtful that the appropriate amounts of Glu and Asp, required by the proteinoid synthesis, existed in the lagoons as postulated. 'Prebiotic' experiments show conclusively that mixtures of pure α-amino acids, similar to the experiments of Fox, did not exist on the primeval Earth. The proposed polymerization would certainly be stopped by a variety of molecules that have been shown to be formed in simulation experiments (Horowitz & Hubbard, 1974). A glaring weakness of the proteinoid scenario is that it requires a highly concentrated and intimate mixture of a specific selection of amino acids as noted by Ponnamperuma (1983). Fox invokes the evaporation of ocean water as a concentrating process. Evaporation also concentrates salts that in the primeval ocean, as well as in the contemporary ocean, would have been many orders of magnitude more concentrated than that of the amino acids. Therefore, the effect of any concentration of amino acids by evaporation is limited (Folsome, 1976). Beds of salt and salt domes exist in numerous locations throughout the world. The concentration of amino acids in the primeval ocean is discussed in section 8.3.1. Those who believe in the proteinoid scenario should search for evidence that amino acids and proteinoids in the ocean had been concentrated at the time such salt beds were laid down.

The explanation of protein folding, its relation to the amino acid sequence and to specificity is one of the fundamental unsolved problems in molecular biology. In the proper folding of the polypeptide chains the formation of helices is prevented by the random presence of only one D isomer in the sequence. Furthermore, studies of the effect of amino acid replacements have shown that some replacements destabilize various intermediate states of the protein folding process (Goldenberg *et al.*, 1989). Thus, among the functions of functionally acceptable amino acids at given sites is the role they play in the protein folding process. The fact that proteinoids are a heterogeneous mixture of sequences of racemic amino acids that cannot fold to produce an active protein is a serious difficulty in the hypothesis that proteinoids were precursors of true proteins. There is a chasm between the proteinoid spheres of Fox and the simplest protein. Fox proposes no transition from racemic amino acids to those of one optical isomer (sections 9.2.4 and 8.3.3). In contemporary polypeptides, the only sequences of amino acids that contain the D isomer are anti-biotics.

The description of the resemblance of thermal proteinoids to proteins has fluctuated in time since they were first prepared. In 1975 proteinoids were 'artificial proteins' (Fox, 1975). Yet in two publications in 1976, Fox states (Fox, 1976a) that proteinoids are not proteins, reiterating the assertion (Fox, 1976b) in his comment on the work of Temussi *et al.* (1976), (see Temussi's reply, 1976). In 1980 they were thermal proteins (Fox & Nakashima, 1980). They became 'protein-like polymers' in 1981 (Hart-mann, Brand & Dose, 1981). By 1984 they had again become 'thermal proteins' (Fox, 1984a), but in another publication that year (1984b), Fox states, 'Proteinoids are in the main much like proteins, but *the very name indicates that they are not proteins* (my italics)'. However, by 1986, Fox writes (Fox, 1986), 'Biomacromolecular information emerged at the stage of nonrandom thermal reactions of sets of amino acids to yield informed protein'.

Fox (1973) claims catalytic activity for proteinoids while admitting that such activity is many orders of magnitude smaller than that of enzymes. Special support for the proteinoid scenario cannot be drawn from this feature because it is shared with many other substances, for example, clays (section 10.1.2).

From the point of view of information theory the proteinoid scenario is hoist by its own petard by Fox's insistence on the non-randomness of the proteinoid sequences and 'the instructions in the amino acids themselves'. These 'instructions' are presumably the conditional probabilities of one amino acid following another. An algorithm may be written containing the 'instructions' in so far as they are known. It is with this algorithm that the first genetic messages are to be written and those messages cannot contain more information than that algorithm (Chaitin, 1987a). The stronger the statistical structure is claimed to be, the less is the information of the so-called messages that it is claimed are generated and the fewer there are of them (Theorem 2.2). These 'instructions' remain a barrier to the generation of a genome until they are removed as they are by the present day tRNA-based mechanism (Yockey, 1981). Orgel (1968) pointed out that if the 'instructions' are strong this strictly limits the number of polypeptides.

In their answer to Yockey (1981), Fox & Matsuno (1983) object to the treating of amino acids as playing cards. Since a function of tRNA is to eliminate the intersymbol influence between amino acids, the analogy of amino acids to playing cards is actually a good one. As Jukes & Margulis (1980) point out in commenting on the proteinoid scenario, it was learned very early that there are no intersymbol restrictions on the possible sequences of amino acids. This was one of the observations that showed the impossibility of overlapping genetic codes. Fox and his followers, in the papers cited, simply do not meet the objections to self-organization presented in Yockey (1981) and in this chapter. Fox's arguments are a sequence of *non sequiturs*, the wish being the substance of things hoped for. One must conclude that Fox's proteinoids are not the seeds of time of which Banquo spoke.

10.4 Self-organization and the hypercycles of Eigen

Eigen's first paper was given as the Robbins Lecture at Pomona College, California in the spring of 1970 and was published in 1971. His contribution together with that of his colleagues to the theory of self-organization was developed further by publications in 1978, 1981 and 1982. A discussion of the theory of self-organization of the Göttingen school is given by Küppers (1983). It was during this time that Kolmogorov (1958, 1959, 1965, 1968), Martin-Löf (1966) and Chaitin (1966, 1969, 1970, 1974, 1975a, b, 1977, 1982, 1987a) published their work on computational complexity and randomness which culminated in algorithmic information theory. Although I had called attention to the application of algorithmic information theory to molecular biology during this time (Yockey, 1977c, 1981), as often happens, these separate demes in the groves of academe were not on speaking terms. Accordingly, although Eigen and his colleagues use some theorems from information theory, their work is not based on the axioms of information theory.

One's premises or axioms must be stated clearly in the beginning in forming an argument. It is not acceptable to improvise or to adopt *ad hoc* notions that are not derivable from those axioms as one pursues the discussion. One must not make a play on words by first using a word in one sense and later in the argument use it in another. In science, words are merely names for mathematical ideas. I pointed out in the Prologue the perils of reasoning with words, exemplified somewhat even by Aristotle, the originator of formal logic. I have emphasized that a number of theorems cited in this book are counter-intuitive. For example, one may mention Shannon's channel capacity theorem (section 5.2), the Shannon–McMillan theorem (Theorem 2.2) and the Perron–Frobenius theorem (section 1.3.3). Accordingly, words and intuition are poor guides here and one must stick to the mathematics.

The axioms of information theory and coding theory are based on fundamental mathematical considerations as discussed in Part I. Information theory and coding theory are experimentally justified. They have made important contributions to communication technology and have given rise to many applications that would not be possible otherwise. The various fields in theoretical physics and mathematics are governed by axioms such as Euclid's axioms in geometry, the laws of thermodynamics, Maxwell's equations, Schrödinger's wave equation and so forth. These laws cover their respective fields uniquely. The theorems of information theory and coding theory are also unique and follow from the basic principles of mathematics. However, the work of Eigen and his colleagues is a sequence of empirical assumptions and improvisations guided by intuition. Therefore the derivation of their equations is, unfortunately, *ad hoc* and incorrect.

Eigen (1971) asks the question: 'Can we use information theory to solve our problem of self-instruction?'. He recognizes the requirement for the generation of specificity in the macromolecules and bases his paradigm on a self-organizing nucleic-acids-first scenario. He cites equation (2.24) as a measure of 'information content'. Yet Eigen (1971) states that: 'Such sequences cannot yet contain any appreciable amount of information'. Along with many other authors (e.g. Johnson, 1970) he makes a play on words by using 'information' in the sense of knowledge, meaning and specificity. Eigen & Schuster (1977) use 'information' in the correct sense according to equation (2.24) but in the same section of their paper they use 'information' in the semantic sense. This is done in spite of the fact that in his second paragraph Shannon (Shannon, 1948; Shannon & Weaver, 1949) warned against confusing the semantic aspects of a message with the aspects of the communication system that apply to any message. This point is discussed in sufficient detail in section 2.1.1 where Shannon's second paragraph is quoted. To remedy what he sees as an inadequacy in 'classical information theory' he calls for a purely empirical 'value parameter' that is characteristic of 'valued information'. He states that this 'valued information' is reflected by increased 'order'. This is looking at the problem through the wrong end of the telescope. The relationship between 'order' and 'complexity' is discussed in section 9.1.2. These words are merely names for mathematical functions and they take their precise meaning *from* the mathematical formalism.

The work of Eigen and his colleagues is based on the primeval soup paradigm (section 8.4), but he rejects the origin of life by chance scenario, although he says that life must have started by self-organization from random events in a molecular chaos without functional organization (Eigen, 1971). He defines 'random' as follows: 'The term 'random', of course, refers to the non-existence of functional organization and not to the absence of physical (i.e. atomic, molecular or even supramolecular) structures'. Of course, this is not the definition of 'random' that is used in this book (section 2.4.1).

Eigen (1971) and Eigen & Schuster (1977, 1978a, b) address the question of the effect of errors and they arrive at the following equation:

$$v_{\max} = \frac{\log_e s_m}{(1 - \bar{q}_m)} \tag{10.2}$$

where s_m is the 'superiority parameter' for the sequence and $v_{\max}$ is the maximum length of a sequence in which the expectation of error is below a geometrical average quality factor $\bar{q}_m$. In section 6.3.2 I discussed the fact that amino acids at each site are either fully functionally equivalent or non-functional. The existence of one non-functional amino acid in the sequence makes the sequence non-functional. Except for the invariant sites there is more than one functionally equivalent amino acid at each site. The method of calculating that number, taking into account the fact that amino acids are not all equally probable, is in section 9.2.2 and the results are given in Table 9.1 for iso-1-cytochrome c. Thus there is a very large number of functionally equivalent 'master sequences' in the high probability set. The landscape of specificity resembles the buttes of Monument Valley in Arizona, not the rolling hills of Appalachia or the ridges, gendarmes and peaks of the Alps. (The abstract landscape representation resembling the rolling hills of Appalachia or the ridges, gendarmes and peaks of the Alps may be satisfactory for the study of evolution, however.) Except for proteins where nearly all sites are invariant, such as the histones, there is no *unique* 'master copy', 'master sequence' or 'consensus sequence' in homologous proteins to be measured by a 'superiority parameter' as some have thought (Eigen, 1971; Eigen & Schuster, 1977, 1978a, b; Eigen, Winkler-Oswatitsch & Dress, 1988).

Eigen & Schuster (1977) put forward equation (10.2) as defining a threshold at which an 'error catastrophe' will occur: 'If this limit (i.e. ψ_{max}) is surpassed, the order, i.e. the equivalent of information, will fade away during successive reproductions'. Here a play is made on the words 'order', 'organization', 'information' and 'function'. The means by which a message will 'fade away' is given by equation (5.15) in section 5.3. There is no need for such improvisations as 'value parameter' or 'superiority parameter' since the communication system treats all messages the same whether they have 'meaning' or not. It is always assumed that the source has chosen a message that has some importance. A message generated by random noise will 'fade away' in the same way and at the same rate as a message which carries 'valued information' (section 5.3). The proper treatment of the question of 'error catastrophe' is taken up in Chapter 11.

Almost all organisms that have lived at one time are now dead and those that are now alive will die. It is both Malthusian doctrine and canonical Darwinism that far more creatures of all kinds are produced than can possibly survive to reproduce. It follows from the effect of genetic noise on the genome that there will be a spectrum of information content (Yockey, 1958). The selection criterion of the genomes of organisms now alive is that the portion of the population that has the largest information content and therefore the most vitality reproduces itself exponentially.

All Eigen's derivations are carried out without reference to the Shannon–McMillan theorem (Theorem 2.2) and therefore he grossly overestimates the number of sequences that need to be considered. For example, to illustrate his point with regard to randomness he refers to a dice game, icosahedral dice for proteins and tetrahedral dice for nucleic acids. Eigen (1971) supposes that: 'We have already seen that a protein with 100 amino acid residues has about 10^{130} different choices of sequences and we would have to throw the die a corresponding number of times in order to arrive at a specified one'. There is a bit of carelessness here in using the Poisson approximation. As I discussed in section 9.2.4, assuming there are n events with a probability p, if $np = \lambda$ and $\lambda = 1$, then one has a probability of $1/2$ of 'arriving at a specified one'. If one wishes a probability of 0.95 of 'arriving at a specified one' then $\lambda = 3$. Eigen (1971) and Eigen & Schuster (1982) calculate the number of natural peptide sequences as 20^v and the number of polynucleotide sequences as 4^v where v is the number of sites in the sequence under consideration. This does not take into account the chirality of amino acids nor the fact that they are not all equally probable. I showed in Chapter 9 that, using the Shannon–McMillan theorem (Theorem 2.2) the entropy in that case is given by equation (9.5). The entropy of each protein site calculated in the case where the isomers of all amino acids are present is increased from 4.139 to 5.074 and, from equation (9.3), the effective number of all amino acids including both isomers is increased from 17.621 (Example 9, Chapter 9) to 33.675. The estimate of the total number of all sequences in the high probability set for 110 sites is: $N_2^l = 1.005 \times 10^{168}$. The calculations of Eigen and Schuster are wrong on three counts. First, neither all nucleotides nor all amino acids are equally probable. Therefore the number of sequences in the high probability set (Theorem 2.2) is much smaller than the total possible. Second, it is the number of sequences of the second extension triplets (i.e. codons) that is of interest, not the number of sequences of the letters of the UCAT alphabet. Third, the codons are mapped onto the amino acids by the conditional probability matrix elements $p(j|i)$ of the genetic code (equation (5.1)). The probability space of the amino acids, which is related to the protein sequences, and the probability space of the codons, which is related to the DNA sequences, each has its own maximum Shannon entropy independently. The protein sequences have a maximum entropy when all p_j are equal to each other and the DNA sequences have a maximum entropy when all p_i are equal to each other. The Shannon entropy of the protein sequences and of the DNA sequences cannot have a maximum simultaneously, as one can understand from inspection of equation (5.1).

Eigen (1971) states 'At the "beginning", whatever the precise meaning of this may be – there must have been *molecular chaos* without any functional organization among the immense variety of chemical species. Thus the self-organization of matter we associate with "the origin of life" must have started with random events'. The hypercycle could have started to be effective only when the concentration in the primeval soup had attained a sufficiently high level. Eigen & Schuster (1978b) state that no less than 10^8 *identical copies* must be present to start the hypercycle. 'The hypercyclic link would then become effective only after concentrations have risen to sufficiently high level...There is only one solution to this problem: the hypercycle must have a precursor, present in high natural abundance, from which it originates gradually by a mechanism of mutation and selection'.

Eigen (1971) attempts to squeeze mathematical meaning out of words. He considers that: '...The question of this form, when applied to the interplay of nucleic acids and proteins as presently encountered in the living cell, leads *ad absurdum*, because "function" cannot occur in an organized manner unless "information" is present and this "information" only acquires its meaning *via* the "function" for which it is coding . . . What is required in order to solve such a problem of interplay between cause and effect is a *theory of selforganization* which can be applied to molecular systems, or more precisely, to special molecular systems under special environmental conditions'. This is intuitively correct if one assumes that 'function' and 'information' are implemented by an algorithm with a high information content as that term is understood in this book. However, Eigen's statement that the self-organization of matter associated with 'the origin of life' must have started with random events (Eigen, 1971) means that these random events were genetic noise. I emphasized in section 2.4.3 that highly organized sequences that contain a 'set of instructions' have a large entropy and therefore are embedded in the high entropy end of the scale. The highly organized sequences are lost among the ensemble of genetic noise sequences and cannot initiate the self-organization process. For example, I showed in section 9.2.2 that the number of iso-1-cytochrome c sequences in the high probability set obtained from Table 9.1 is 2.316×10^{93}. The total number of sequences in the high probability set in a sequence of 110 sites if $H = 4.139$ is 1.15×10^{137}. Therefore the iso-1-cytochrome c sequence is only a fraction 2×10^{-45} of the number of sequences in the high probability set. The situation is much worse if one includes the effect of chirality, as discussed in section 9.2.2.

Thompson & McBride (1974) solved Eigen's equations (Eigen, 1971) exactly and found that one 'master sequence' will emerge. Anderson (1983) expressed the opinion that the hypercycles are likely to be so specific that only one outcome is permitted. This, of course, means a very low information content. Eigen (1971) states that: 'The information resulting from evolution is a 'valued' information and the number of bits will not tell us too much about its functional significance'. This is true, if one is charitable with what one now sees as Eigen's terminological inexactitudes. However, I have shown (Yockey, 1977b) that the information content of cytochrome c is between 298 and 374 bits. In Chapter 6 that calculation is brought up to date with the result that the genetic message must contain between 233 and 374 bits to record the instructions to construct one of the molecules of iso-1-cytochrome c in the high probability set. In Section 9.1.1, regarding the protobiont as a von Neumann machine, I showed that a generous lower limit of the information content of the genetic message in its genome to be several million bits. If the genome is not at least that large it cannot record the 'valued information' of which Eigen writes. On the contrary, Eigen & Schuster (1977) stated that their $v_{max} \leqslant 100$ nucleotides for the first reproductive units.

One of the more important objections to the hypercycle paradigm is that there is no evidence that a primeval soup dense enough for the hypercycle to start (section 8.4) ever existed.

The hypercycle of Eigen (1971) is primarily biochemical and I leave further criticism to others (Maynard Smith, 1979; Bresch, Niesert & Harnasch, 1980; Niesert, Harnasch & Bresch, 1981; Anderson, 1983). See also the reply to Bresch, Niesert & Harnasch (1980) by Eigen, Gardener & Schuster (1980).

10.5 The word game model of self-organization and evolution

10.5.1 The information content in the 'instructions in the words themselves'

The idea of modelling evolution with a word game is due to Maynard Smith (1972, 1979). To play the game one starts from a word and must find another word in a minimum number of steps without going through a non-sense sequence, i.e. a sequence that has no function. For example, *word* may be converted to *gene* by the following steps: *wore*, *gore*, *gone*, *gene*. The player uses his knowledge of the language to determine the change he wishes to make.

Eigen & Schuster (1977, 1978a) use a variant of the word game in modelling self-organization. By means of a computer program they start with a non-sense sequence and by changing letters approach and eventually find a target sequence. Dawkins (1986), in a popular book, purports to show by a similar computer program on a 64 kilobyte computer that chance, by cumulative selection, can develop meaningful sequences. Unlike proteins, letters and words have an intersymbol influence similar to that which Fox claims for his proteinoid sequences. Intersymbol influences are reflected in the digram, trigram and *n*-gram frequencies and the spelling, grammar and rules of composition. Paraphrasing Fox ('the instructions in the amino acids themselves') these regularities are to be regarded as 'instructions in the letters and words themselves'. It is in this way that Fox, in many of his papers, Eigen & Schuster (1977, 1978) and Dawkins (1986) propose to illustrate, in analogy, the self-organization of a genome.

In Jonathan Swift's *Gulliver's Travels*, written in 1726, the projectors, as they were called, at the Grand Academy of Lagado put 'the instructions in the letters and words themselves' to the useful purpose of writing learned works without the trouble of mastering any of these difficult subjects. One of the savants at that great institution of research and higher learning had constructed a simple frame that recorded all the words in the language of that country in all their several moods, tenses, and declensions. Students turned cranks, rearranging words in random order, so that sentences might appear. The frame was eagerly watched and when fragments of meaningful sequences appeared, graduate students wrote them down. Thus a collection of fragments of meaningful sentences was accumulated to be collated later into works on philosophy, poetry, theology, law and other learned matters of concern to the savants of the Grand Academy.

Some success was indeed obtained. It was known in Lagado that Mars has two satellites of periods 21 h 30 min and 10 h. The machine may have been the source of this knowledge. The correct identification of just two satellites and the nearly correct periods could not have happened by chance and is evidence that the machine produced useful results. The satellites of Mars, Deimos (Terror) and Phobos (Fear), were rediscovered in 1877 by Asaph Hall using the more prosaic and laborious means of telescope observation. Modern values for the periods are 30 h 18 min and 7 h 39 min. Supporters of the word game scenario may excuse the small lack of accuracy of this machine due to the fact that it employed only word frequencies and not the digram, trigram and n-gram frequencies of the language. Of course, it did not have the immense speed of modern computers.

The idea was ahead of its time, but now the large memories, high speed and better programming of present day computers could put this project to use for mankind. If the information is contained in the letters and words themselves, then properly programmed, the large, fast, supercomputers of today could write books on philosophy, poetry, politics, law, theology, physics, chemistry, the mathematical foundations of molecular biology and even on the origin of life. The lost works of Aristotle, Archimedes, Sophocles, Aeschylus, Euripides, Aristophanes, and other ancient philosophers, poets and mathematicians could be recovered in the original Greek or translated correctly into English.

By modern standards the machine at the Grand Academy of Lagado was crude and slow, but nevertheless, the basic principle was in accordance with the self-organization ideas of Fox for the origin of proteins from proteinoids. It should be possible to measure the statistics of the interrelations of amino acids in proteinoids and to use this knowledge to write an algorithm that would reveal, without much ado, the way in which modern proteins arose from the proteinoids. By the same token, this procedure can be applied to the hypercycle principle of self-organization. The speed and very large memories of modern supercomputers make it possible to recapitulate by forward extrapolation the origin of life. To be sure, a large number of non-sense sequences will be generated but the algorithm could be written to instruct the computer to print out only the 'ordered,' and therefore significant, sequences for examination by graduate students. For example, a large library of protein sequences is now available. For test purposes, the computer could be instructed to search for and print these sequences when the non-random generating mechanism of Fox or the hypercycle principle of self-organization produced them. I made this proposal in Yockey (1981) but Fox & Matsuno (1983), in their comment on that paper, did not take advantage of it to prove their most important point.

However, word games are not an appropriate model for self-organization or evolution for the following reasons:

(1) There is no intersymbol influence among amino acids polymerized in proteins, whereas there is strong digram, trigram and *n*-gram intersymbol influence between letters. Words have intersymbol influences reflected in the frequencies of use and the spelling, grammar and rules of composition. It is easy to write English sentences that are grammatically correct but make no sense whatever. This reduces the information content of word sequences and drastically reduces the number of sequences in the high probability set.

(2) There is no way to get around the complexity of proteins, as I discussed in Chapter 6. The computational complexity of the problem is grossly underestimated. This was illustrated in Chapter 2 by the deceptively simple traveling salesman problem: given a set of *n* cities that the salesman must visit and the distance between all the cities, find the shortest tour that visits each city once and returns to the starting point. The number of computational steps in the algorithm that accomplishes this computation defines the complexity of the problem. In spite of intensive study, the most effective algorithms for finding the best tour require a number of steps that is exponential in *n* (Lawler *et al.*, 1985; Packel & Traub,

1987). Computational complexity that is exponential in the problem size, *n*, is called *intractable*. Expansion to the order of 100 or more sites results in the word game problem and the construction of evolutionary trees becoming *intractable* in the same way as does the traveling salesman problem (sections 2.4.3, 2.4.5 and 12.2.3).

(3) Eigen & Schuster (1987a, b) use the concepts associated with equation (10.2). As I have pointed out above, this equation and the concepts associated with it are incorrect because the Shannon–McMillan theorem (Theorem 2.2), Shannon's noisy channel theorems and the homology properties of proteins are ignored.

10.5.2 *The information content of the paradigms for self-organization*

Paradigms for the origin of life are essentially algorithms. They must contain as much or more information than the ensemble of genetic messages that they purport to generate (Chaitin, 1987a). In the order–disorder phenomena discussed in section 9.1.2, the atoms or molecules assume self-ordered positions like soldiers on parade because the information content required is very small and indeed 'is in the atoms or molecule themselves'. The information content of iso-1-cytochrome c was calculated in Chapter 6 and is far larger than the information content in the paradigms discussed in Chapters 9 and 10. Indeed, the information content of these paradigms is very much too small to select even a very small informational molecule. From the point of view of information theory this reflects at least an incompleteness in all the paradigms for the origin of life discussed in Chapters 9 and 10. This point is illustrated in the discussion of the information content of π, e and other recursively enumerable transcendental numbers (section 2.4.1).

Teleology has disappeared from science and appears only in religion, since the late nineteenth century. However, both Eigen & Schuster (1978a) and Dawkins (1986) have incorporated a teleological 'target phrase' in their computer program which they claim proves that they generated meaningful sentences by chance. Chaitin's theorem LB (Chaitin 1987a) proves that they put in their algorithms exactly what they wished to find. The projectors of the Grand Academy of Lagado used 'the instructions in the letters and words themselves' to accomplish the useful purpose of writing learned works without the trouble of mastering any of these difficult subjects, but since they knew little of such subjects they did not use a 'target phrase'.

10.6 Non-equilibrium thermodynamics and the origin of life

Non-equilibrium thermodynamics has been applied to biology by Prigogine & Nicolis (1971). This is not so much wrong as inappropriate when application to the genetic message and to the origin of life is attempted. The form in which it has been applied confuses information theory and coding theory with the management of energy and Maxwell–Boltzmann–Gibbs entropy (Prigogine, Nicolis & Babloyantz 1972a, b).

As I have pointed out (Yockey, 1974, 1977c) (see sections 2.2.2 and 12.1) the Shannon entropy of information theory has nothing to do with the Maxwell–Boltzmann–Gibbs entropy. Shannon entropy does not enter in the formalism of Prigogine *et al.*, and thus they attempt to force non-equilibrium thermodynamics to play the role that should be assigned to information theory and coding theory in describing complexity and order. This again results in a confusion between order and complexity. We are informed, for example, that: 'Biological systems are highly ordered and complex' (Prigogine *et al.*, 1972a). What is meant, of course, is that biological systems are *highly organized.*

In recent studies of chaotic physical or chemical systems it has often been remarked that irreversible processes such as turbulence may be a source of order (Coveney, 1988). Some realization is emerging that the Kolmogorov–Chaitin entropy must be applied in understanding complicated stochastic processes. This has been attributed to the fact that the prediction of chaotic processes requires an equally long and complicated sequence of symbols to be described (Chernikov, Sagdeev & Zaslavsky, 1988). If so, this is in the context of chaotic dynamic systems and has nothing to do with the origin of life because the Kolmogorov–Chaitin entropy will not be equal to the Shannon entropy of the genome. These chaotic dynamic processes will, accordingly, not be isomorphic with the genome.

However, the subject matter of non-equilibrium thermodynamics, namely, energy and the Maxwell–Boltzmann–Gibbs entropy, is appropriately applied to the use and management of energy in organisms which, it is clear, are far from equilibrium. In addition, non-equilibrium thermodynamics is appropriate to the mechanisms of proofreading the genetic message that are discussed in section 11.3.2.

10.7 RNA catalysis and the origin of life

As the role of nucleic acids and proteins in the genome and in the function of an organism developed, nucleic acids were assigned the role of recording, duplicating and transferring information and proteins were assigned the role of function. This was formalized in the Central Dogma. However, this assignment of roles provided no pathway from the primeval soup to a protobiont because the duplicating and transferring of the genetic message to protein required a complex system of syntheses that are themselves carried out by proteins formed by the direction of genetic messages. Crick (1968) suggested that: 'Possibly the first "enzyme" was an RNA molecule with replicase properties'. This was proved prophetic by the discovery of *ribozymes* in which RNA plays both the role of information transfer and the catalysis role of enzymes (Kruger *et al.*, 1982; Zaug & Cech, 1986; Cech, 1986a, b). This discovery was thought by some to have provided the pathway from the primeval soup to the protobiont (Gilbert, 1986). That is, it was proposed that a self-replicating system that did not depend on the enzyme action of proteins but rather on a pre-enzyme action of nucleic acids preceded the present system (Cech, 1986a, b; Gilbert, 1986; Darnell & Doolittle, 1986; Westheimer, 1986). Crick's proposal is not as promising as seemed at first, since RNA is not as versatile a catalyst as protein enzymes.

Taken at best, however, this proposal simply moved the search for an origin of life scenario one step nearer the primeval soup, where it encounters the need for a plausible mechanism for the prebiotic formation of ribose, a component of RNA. Ribose is only one of a number of sugars and never the primary product (Orgel, 1986). There is only one plausible synthesis of ribose that may be considered in the prebiotic milieu, namely, the polymerization of formaldehyde. Furthermore, the condensation of adenine or guanine with ribose leads to mixtures of the optical isomers that are very complex. From the point of view of the biochemist, this problem is as difficult as the one the discovery of ribozymes was supposed to solve (Orgel, 1986). Furthermore, non-biological reactions in the prebiotic milieu lead to 2',5' isomers (Orgel, 1986). This does not lead in the direction of the origin of life because the reactions that are typical of biochemical products are almost always the 3',5' linked isomers, whether or not they are catalyzed by protein-enzymes.

Furthermore, RNA is itself a complex molecule and even those who believe in a primeval soup do not regard it plausible that RNA existed in large quantities in the primeval soup. HCN is produced in the spark discharge experiments and has frequently been proposed as a percursor of nucleotides. Cytosine is produced in about 5% concentration from a 1.0 M aqueous solution of potassium cyanate and 0.1 M cynanoacetylene. This can hardly be called a prebiotic synthesis since it requires a well-educated chemist acting as a *deus ex machina*. HCN oligomerization has often been suggested as the prebiotic source of purines and pyrimidines but none of the five bases that occur in nucleic acids are found in the oligomerization products in dilute solution (Schwartz & Bakker, 1989). Joyce (1989) has reviewed the difficulties of generating RNA in the primeval soup. Much of this has been discussed in section 8.4.

No one doubts that the molecules in the primeval soup were racemic. Further research in the oligomerization of nucleotides showed that the process proceeds if the monomers are of one optical isomer. For example, poly (D-C)-directed oligomerization of the D enantiomorph of guanosine 5'-phospho-2-methylimidazole (2-MeImpG) is a very efficient reaction. The 3',5'-linked oligo(G)s are the major products and range from the dimer to the 30-mer. The incorporation of the opposite optical isomer, L-2-MeImpG, is much less rapid and when incorporated acts as a chain terminator (Joyce *et al.*, 1984). This discovery of *enantomeric cross-inhibition* means that the scenario that RNA is the elixir of life is at least incomplete. Some workers in the field now believe that life did not start with RNA (Joyce, 1989).

Joyce *et al.* (1987) have proposed that the RNA world was preceded by a protobiont with a simpler genetic system in which certain simple achiral analogues of nucleotides were involved. This suggestion proposes to move the problem a step nearer to the protobiont but still encounters the primary questions of the generation of the genetic message and of the genetic code between the alphabet of the genetic sequence that stores and replicates the genetic messages and the alphabet of the protein sequences that implement function.

10.8 Does life pullulate in the universe?

10.8.1 Does life exist in millions of other sites in the universe?

'... There is even increasing recognition of a new science of extraterrestrial life, sometimes called exobiology – a curious development in view of the fact that this "science" has yet to demonstrate that its subject matter exists'.

This sarcastic remark of Simpson (1964) has not lost its sting in 26 years, especially in view of the fact that no evidence of organic compounds or life was found by the 'life on Mars' program (Horowitz, 1990). Calvin (1961) stated: 'We can assert with some degree of scientific confidence that cellular life as we know it on the surface of the Earth does exist in some millions of other sites in the universe. We thus remove life from the limited place it occupied, as a rather special and unique event on one of the minor planets, to a state of matter widely distributed throughout the universe.' As Simpson (1964) put it: 'Virtually all biochemists agree that life on earth arose spontaneously from nonliving matter and that it would almost inevitably arise on sufficiently similar young planets elsewhere'. This belief is simply a matter of faith (Hebrews 11:1) in strict reductionism and is based entirely on ideology.

10.8.2 Did life come from outer space?

The suggestion has been made frequently that life came to the Earth from outer space (Crick & Orgel, 1973; Hoyle & Wickramasinghe, 1977a, b, c, 1978, 1988; Hoyle, 1980; Crick, 1981). Hoyle and Wickramasinghe base their belief in the prevalence of life in outer space on their quarrel with the Big Bang theory and its requirement for a short life of the universe. Sir Fred Hoyle is concerned with the information content of the universe. He is correct on this point; as I have pointed out the crucial issue in the origin of life is to determine the means by which information content has been generated (section 9.1.1; Yockey, 1977c; 1981). It is not entirely clear that what Hoyle means by information content is what it has been defined to be in this book. He believes that the probability of the generation of life by chance is $10^{-40\,000}$. Although this figure is incorrect (section 9.2.2), he concludes correctly that life could not have been generated by chance in a small, young universe of age between 10×10^9 and 20×10^9 years. He believes that life is a cosmic phenomenon. 'The only possibility for accumulating the needed information seemed to lie in an opening of the cosmic time-axis into the past, just as my old starting position in 1948' (Hoyle, 1980). That is to say, any event not *streng verboten* will happen in an ageless universe of infinite extent. D. T. Wickramasinghe and Allen (Wickramasinghe & Allen, 1986) found a band in the 3–4 mm region of the infrared spectrum of comet Halley which they ascribe to a tarry substance on the surface of the comet. Hoyle & Wickramasinghe (1987, 1988a, b) believe that this structure is evidence of bacteria in the comet.

This conclusion is strongly disputed by Chyba & Sagan (1987) and by Greenberg & Zhao (1988). They point out that the spectrum indicates only that the emission is from a substance containing CH_2 and CH_3 groups. The same comment applies to the assertion that the infrared spectrum in the same band indicates that interstellar dust gives evidence that life is a cosmic phenomenon widely distributed in space. 'We suggested that the origin of life on Earth involved essentially the assembly of complex chemical components which had already been provided in the last stages of the accumulation of the Earth and which exist widely and in comparatively great quantities within the interstellar medium' (Hoyle & Wickramasinghe, 1977c). Hoyle *et al.* (1977) propose that dust in interstellar clouds is composed of radiation resistant organic polymers or prebiotic polymers such as cellulose or starch. It is well known that cosmic dust is a major constituent in the universe (Ney, 1977). Hoyle & Wickramasinghe (1978) believe that this cosmic dust plays the role of the primeval soup on Earth, from which life is presumed to have evolved by chemical evolution. This scenario does have the virtue of increasing enormously the amount of material and, if the cosmic time axis is opened to the infinite past, also providing sufficient time for any event to happen. However, Hoyle and Wickramasinghe arrived at their conclusion by a series of *non sequiturs*, the wish being the substance of things hoped for.

An extensive search has been made for evidence of biogenic carbonaceous material in samples returned from the Moon. Most of this work provides no evidence that the 200 ppm of carbon in these samples is of biogenic origin (Ponnamperuma, 1982). The hard evidence against a biogenic origin is the fact that the $\delta^{13}C$ values do not correspond to biogenic carbon (Ponnamperuma, 1972, 1982, 1983). Other sources of carbon and amino acids are found in the Murchison meteorite and other chondrites. Although a large assortment of amino acids and other organic compounds is found, the selection is quite different from that found on Earth (Cronin, Pizzarello & Cruikshank, 1988). Zhao & Bada (1989) have found α-amino-isobutryric acid and racemic isovaline in K/T boundary sediments at Stevns Klint in Denmark. The indicator that isovaline is of extraterrestrial origin is that it is racemic. The only terrestrial source of isovaline is a rare occurrence in some fungal peptides in which it is always found in the form of the D enantiomer. The racemization of isovaline is extremely slow and the D enantiomer would survive its 65-million-year age. More detail and further references are given by Zhao & Bada (1989). Chyba *et al.* (1990) dispute the extraterrestrial origin of these compounds on the grounds of their calculations of the survivability of organic compounds from the object that caused the K/T boundary event. They suggest that these compounds were synthesized in the shock and post-impact quench. It is clear that from the present evidence extraterrestrial amino acids do not have the signature, based on isotopic ratio or on composition, that would support the thesis that life came from outer space.

10.9 Evaluation of the chemical evolution paradigm

10.9.1 Paradigms for the generation of information

Because they don't meet the criteria in section 9.1.1 for the minimum generation of information for the origin of life, the paradigms discussed in Chapters 9 and 10 are impossible in probability in the same way that perpetual motion machines are impossible. In the nineteenth and early twentieth centuries some practical mechanical engineers thought they could avoid that theoretical limitation and build perpetual motion machines by very clever mechanical design. However, the second law of thermodynamics is independent of any clever design. It is even independent of whether atoms exist or not. Some prominent chemists in the late nineteenth century regarded the proposed existence of atoms to be 'just a theory'. Having thermodynamics, they could manage very well without atoms. (It was Einstein's papers of 1905 and 1906 that put the final nail in the coffin of the continuous matter theory.) By the same token the authors of the paradigms discussed in Chapters 9 and 10 believe that they have already solved the problem of the origin of life or that with one or two clever experiments they will prove even to skeptics that, in spite of abstract mathematical theories, they have found the elixir of life. Delbrück (1986) believes that the theories proposed to explain the origin of life, although they may seem quite plausible and intelligent, in fact tell very little about the origin of life.

10.9.2 Is chemical evolution a latter day alchemism?

The alchemists believed that by finding the proper elixir they could prolong life indefinitely and make gold from dross metals. From the knowledge available at the time this was a very reasonable conjecture. It had been believed since antiquity that there were four elements: earth, fire, air and water. All substances must be combinations of these four elements. Also it was well known even in ancient times that a strong, useful and beautiful alloy, namely bronze, could be made by melting together two dross metals, copper and tin. Homer, in the Iliad, tells us that, although iron was known, bronze was the alloy of choice for weapons in the Trojan War. A similar alloy, brass, could be made by melting together copper and zinc. If only the proper elixir and treatment could be found, surely gold could be made from dross metals such as lead.

As pointed out by Cairns-Smith (1982), the alchemists believed that one has only to find some useful way to retain the malleability of lead and change the color from gray and dull to yellow and shiny and *voila*! gold. Their inability to do this immediately did not shake their faith. They could make very favorable progress reports to the aristocracy that supported their work. The fact that iron and copper combined with yellow sulfur make pyrites, which are yellow and shiny, encouraged them – and their patrons – in their search. More than one person has been taken in by iron pyrite or fool's gold.

Today we know that gold is not an alloy or a chemical compound and no amount of faithful work along the lines of paradigms by metallurgists will result in a method of making gold. We now know that gold is an element and was made by the r-process in exploding stars (Burbidge *et al.*, 1957; Cowen *et al.*, 1987), a notion *totally inconceivable* to the alchemists. Indeed, we now make materials more valuable than gold, namely, plutonium and tritium, from dross metals in nuclear reactors.

Perhaps we are now in much the same situation. Chemical evolution was a logical paradigm under the circumstances in which it was first put forward. However, let us now evaluate carefully what we have learned. In Chapter 7, I looked backward in time to find out how far in the past one could see, albeit through a glass darkly. I showed that one is justified in believing that life originated on Earth (and did not come from outer space, section 10.8) several to many times in the period between 4.0×10^9 and 3.8×10^9 years ago. Life originated using a doublet genetic code that was the first extension of the quaternary A, U, G, C, alphabet. It went through a bottleneck at the second extension that led to the modern triplet code. The genetic apparatus today can read doublet and quadruplet codons in addition to triplet codons. Several endosymbiotic events took place that resulted in the incorporation of the mitochondria in the eukaryotic cells. Looking forward in time, from 4.0×10^9 years ago, one finds that there is evidence that the 'building blocks of life' (in addition to those stones that must be rejected by the builder) may have been formed by natural processes that presumably existed on the Earth in this period. But there is no evidence for the existence of a primeval soup much different in its concentrations of amino acids from those in the ocean of today. In any event, all the various scenarios that have been put forward to explain how life emerged from this hypothetical primeval soup have serious drawbacks (Yockey, 1977c, 1981).

Oxymorons such a *soldier-scientist, military intelligence, pre-planning, student-athlete, pro-active, creation-science, criminal justice system, democratic socialism* and *Holy Roman Empire* pullulate in the English language. It is to be expected that those who proposed a scenario will coin names that represent it favorably. For example, Chang *et al.* (1983) state: '"Chemical evolution" should not be confused with Darwinian evolution with its requirements for reproduction, mutation and natural selection. These did not occur before the development of the first living organism, and so chemical evolution and Darwinian evolution are quite different processes'. Nevertheless, this does not prevent them from continuing to say: 'However, the term chemical evolution has an appeal because of the continuity it implies for the transformation of lifeless into living matter'. This is the classical *post hoc ergo propter hoc* fallacy. The crowing of the rooster does not cause the Sun to rise. The fact that life exists on Earth does not prove that there must have been a prebiotic soup. Oxymorons such as *chemical evolution, prebiotic soup, proteinoid, self-organization etc.,* prejudice the mind to the belief that there is a necessary connection with the origin of life. Even the term 'organic chemistry', a subject that simply deals with the compounds of carbon, tends to prejudice the mind to adopt a meaning implying biotic origin.

That path to 'The broad goal [which] is to arrive at an intellectually satisfying theory of how living forms could have emerged step by step from inanimate matter on the primitive Earth' (Dickerson 1978) *is not in sight,* contrary to Dickerson's optimistic statement. The considerable work on the origin of life in the last decades has resulted in many facts, most of them inconvenient. What is needed is creative skepticism, fettered by facts, however inconvenient, rather than endless unfettered speculation. There is no justification today for believing that life is about to be created *in flagrante delicto* in the laboratory as was proposed by Jacques Loeb (1912).

Although it was justified when the work started, chemical evolution, a latter-day alchemism, is still-born and no amount of work on that paradigm, however religiously it may be carried out, will tell us how life originated. However, this will not shake the faith of the true believers. A true believer confronted with evidence contrary to his doctrine regards this as merely a test of his faith. His greatest fear is heresy and treason to his infallible doctrines. He views the world behind a fact-proof screen that protects his infallible doctrine from heretics and unbelievers (Jukes, 1987a; Hoffer, 1951).

Science often finds itself attempting an explanation that is far beyond its capability at the time. Many examples of this are discussed by Kuhn (1957, 1970). We know that life originated on Earth and we know the approximate timespan in which this occurred. But we are frustrated in finding a means by which it happened. It is worth reading Sir Arthur Eddington's (1926) chapter entitled 'The source of stellar energy'. By the application of the basic laws of thermodynamics and of quauntum theory as known at that time, he correctly put his finger on the main source of stellar energy, namely, the burning of hydrogen. He was frustrated in finding a mechanism by which that might occur but, to his great credit, he did not resort to unfounded speculation and factoids (they only simulate facts). We are in the same situation as Eddington with regard to the origin of life. The currently accepted scenarios are untenable and the solution to the problem will not be found by continuing to flagellate these scenarios.

10.10 Bohr's axiomatic approach to the origin of life; algorithmic information theory, randomness, indeterminacy and incompleteness in molecular biology

There are two approaches to the origin of life problem. We may take the suggestion which Bohr made in 1933 that life is to be accepted as an axiom and proceed from there, or we may continue the search for the origin of life from non-living matter. Let us first consider Bohr's suggestion. Chaitin (1988) has pointed out that if we try very hard to solve a problem without success the reason may not be that we are not smart enough or haven't tried long enough; the reason may be that there is no solution. There may be no pattern or law that leads inexorably to the origin of life from non-living matter. Indeed, there are many problems in mathematics that are known to have no solution. Readers who have solved Socrates' problem of doubling the square, posed in the Prologue, are invited to go from two dimensions to three dimensions and find a method of doubling the cube. The second law of thermodynamics tells the engineer that perpetual motion machines are impossible, and that furthermore he can't even build an engine that doesn't waste most of the heat energy by transferring it to a lower temperature where it becomes less available to be converted to work. In physics we have the Heisenberg uncertainty principle, wave-particle duality and the general indeterminacies of quantum mechanics. In relativity we have the fact that neither matter nor energy can travel faster than the speed of light. These are elementary axioms of modern physics which have no explanation. Since axioms are elementary statements that can't be proven, we take them for our basis of reasoning in mathematics. Even so, we find limits to our reasoning powers, typically in Gödel's incompleteness theorem (Gödel, 1931) and in Turing's halting problem, sections 2.4.2, 2.4.5, 2.4.6, 2.5 and (Chaitin, 1982, 1987a, b, 1988). An excellent and extensive popular discussion of the Turing machine can be found in Penrose (1989).

The reason that a scientific explanation for the origin of life has not been found may be that the problem is intractable or indeterminate and beyond human reasoning powers (Chaitin, 1982, 1987a, b, 1988; sections 2.4.3 and 2.4.5). Algorithmic information theory, Gödel's indeterminacy theorem and Turing's halting problem have shown that incompleteness, indeterminacy and randomness are natural and widespread in science (Chaitin, 1982) and not pathological as was believed at first. If, as seems to be the case, genetic information is a program for constructing organisms (Chaitin, 1977) then the genetic information system is isomorphic with communication systems, Turing machines and mathematical logic. It must obey the same laws that these systems obey. Since there is randomness, indeterminacy and incompleteness in arithmetic (Chaitin, 1987a, b, 1988), how can molecular biology escape? To believe otherwise is vitalism. This conclusion may require some getting used to for the majority of molecular biologists, who for the most part believe in the deterministic view of the origin of life (Simpson, 1964; Calvin, 1969). 'We must further conclude that the evolution of life, if it is based on a derivable physical principle, must be considered an *inevitable* process despite its indeterminant course. The models treated ... and the experiments discussed ... indicate that it is not only inevitable in principle but also sufficiently probable within a realistic span of time. It requires appropriate environmental conditions (which are not fulfilled everywhere) and their maintenance. These conditions have existed on Earth and must still exist on many planets in the universe' (Eigen, 1971). On the contrary, the incompleteness theorems discussed in this book show that life is consistent with the laws of physics and chemistry but not derivable from them. We must, therefore, as Bohr

(1933) suggested, take life as an axiom in the form of the sequence hypothesis. As Simpson (1964) suggested long ago, the time of the molecular biologist is better spent on understanding life as it is, just as theoretical physicists spend their time understanding the limitations and the consequences of quantum mechanics and general relativity, having accepted the inexplicable axioms of these subjects.

I have no scenario to explain the origin of life. My purpose is to demonstrate that for a successful explanation to emerge it is important to eliminate *factoids* and clear away the encumbrances of failed scenarios and paradigms. It is as well to remember what Charles Darwin (1872) wrote about the origin of life: 'It is no valid objection [to the theory of natural selection] that science as yet throws no light on the far higher problem of the essence or the origin of life. Who can explain the essence of the attraction of gravity? No one now objects to following out the results consequent on this unknown element of attraction; notwithstanding that Leibnitz formerly accused Newton of introducing "occult qualities and miracles into philosophy"'.

10.11 Suggestions for the enlargement of origin of life research

The adoption of the existence of life as an axiom together with the notions of randomness and incompleteness in molecular biology may be philosophically repugnant to some. Accordingly, experiments of a much larger scope both in their size and in the time during which they are carried out may be attractive. Chemical evolution experiments have all been done in the short timespan of the graduate student's career and with apparatus that can occupy a corner of a laboratory room. Prebiotic electric discharge experiments are still being carried out by the use of a Tesla coil that can be held in one's hand in the same manner as in 1953 (Miller, 1953; Stribling & Miller, 1987). Corona discharge in putative prebiotic atmospheres is often stated to simulate lightning. Such small equipment simulates corona discharge, not the enormous transient energy in lightning, which occurs for nanoseconds, not days.

It is strange that, in this era of megascience, no one has used the large simulation facilities for experiments in chemical evolution. Large-flash X-ray machines can simulate a lightning discharge in real time, including ionization, ultraviolet, heat and shock environment. Such machines can generate 20 megavolts and a current of 800 kiloamperes in a pulse width of 40 nanoseconds. Exposure either to X-rays or the electron beam may be obtained. The temperature in the ambient gas may reach 10 000 K and is quenched in several nanoseconds. This is comparable to the environment in a lightning flash (Essene & Fisher, 1986). Whatever chemistry occurs in a lightning flash is simulated very closely, combining simultaneously all the elements in a lightning discharge mentioned above.

To my knowledge biologists have proposed a simulation of the prebiotic soup on only a tiny scale or a short timespan. On the other hand, physicists often operate on a much larger scale and for much longer times. A huge tank has been placed in a salt mine to detect the putative decay of the proton, at this date without success. The tank holds 8000 metric tons of highly purified water. The detection of decay particles is done with 2048 photoelectric tubes (Goldhaber, 1983). Other projects of this size have been built and monitored for many years to detect neutrinos from the Sun and supernova explosions. A solar neutrino observatory has been proposed to be built 2073 meters underground in a mine at Sudbury, Ontario. The detector is essentially 1000 metric tons of D_2O. Other large detectors and observatories of the solar neutrinos are being built in Italy, Japan and the Soviet Union (Wolfenstein & Beier, 1989). The problem of the origin of life is as important to biologists as the energy production in the sun is to physicists and astronomers. Putting chemical evolution to the test to find out the behavior of a prebiotic soup justifies several projects of this size or larger. Various simulations of proposed prebiotic atmosphere and energy sources could be tried and the consequences, if any, monitored with appropriate detectors and analysis. These projects must be conducted over the normal years of the career of a senior investigator, just as the physics and astronomical facilities are. Such experiments would produce interesting chemistry whether or not they shed light on the origin of life.

Exercises

1. Repeat the calculation of the entropy of a dice game in Exercise 2, Chapter 2 but this time choose an icosahedral die so that you are generating the sequences discussed by Eigen (1971). Choose a probability for each face of the die that reflects the probability that you think appropriate for the occurrence of each amino acid. Calculate the ratio of the number of sequences in the high probability set to the set of all possible sequences.
2. Read that portion of Plato's dialogue *Meno* where Socrates invites the slave-boy to solve the problem in geometry of doubling the square. In Socrates's theory of knowledge the slave-boy already knew how to double the square. It was only necessary to bring this out by asking questions.

Plato knew nothing of information theory. But from what you have
learned did Socrates actually transfer the algorithm for doubling the
square to the slave-boy by the specific questions he asked?

3. Eigen & Schuster (1977, 1978a) start with a nonsense sequence and, by
   means of a computer program that interchanges letters, approach and
   eventually find the target sequence 'Take advantage of mistake'. Dawkins
   (1986), in a popular book, entertains a similar computer program on a 64
   kilobyte computer purporting to show that chance, by accumulative
   selection, can develop meaningful sequences. Compare the self-
   organization procedure of Eigen & Schuster (1977, 1978a) and of
   Dawkins (1986) with Socrates' questioning of the slave-boy with regard
   to the algorithm for doubling the square. Can this same self-organization
   procedure be applied to doubling the cube?

11

Error theories of aging

None answered this; but after Silence spake
A Vessel of a more ungainly Make:
" They sneer at me for leaning all awry;
What? did the Hand of the Potter shake?"

Omar Kháyyám, The Rubáiyát

11.1 The need for a measure of aging

In this chapter I shall review current theories of aging and then apply what we have learned about mutual information to develop a theory of aging.

The notion of 'vitality' runs through the early papers on aging, which try to define and quantify it intuitively. However, 'vitality' turned out to be hard to define, let alone quantify. Mr. Justice Potter Stewart (U.S.A. Supreme Court) said that he couldn't define pornography, but he knew it when he saw it, so presumably on the same principle living animals have *vitality*, while dead or moribund ones do not, and thus we know the ones that have *vitality* when we see them.

A theory suggested by Brody (1923) and formalized by Failla (1958) defined vitality as the inverse of the mortality rate: vitality was thought to be a physiological property that decays at a rate proportional to the amount present and therefore exponentially. However, this is not a satisfactory definition because it defines vitality in terms of what organisms that have it are not, i.e. not dead, rather than what it is. Nevertheless, in each of a number of theories of aging (Mildvan & Strehler, 1960; Finch & Schneider, 1985), vitality was defined by the notion of loss of various functions or death rate. These definitions of vitality are expressed by the Gompertz function, an empirical relationship of mortality rate as a function of age in humans (Gompertz, 1825), which is as follows:

$$\lambda_X = \lambda_0 \, e^{\alpha X} \tag{11.1}$$

where λ_X is the age-specific mortality rate, sometimes called the force of mortality, λ_0 is the force of mortality at age zero, α is the slope parameter and X is the age. The two parameters of equation (11.1), λ_0 and α, are determined empirically from survivorship curves and are not otherwise properties of the individual organisms. Using the Gompertz function is not satisfactory for the purpose of understanding the process of aging because empirical theories just describe data and do not explain them. One must therefore seek a more fundamental way of approaching the problem.

In progressing from an empirical description of a phenomenon to a theoretical understanding of it based on first principles, it is essential in any scientific endeavor to identify the significant measurable variables. The information content of the genome and of proteins is clearly a measurable quantity as has been demonstrated in this book: I present in this chapter an argument to support designating the information content of the genome as a measure of what we are used to calling vitality. Thus I shall give the word 'vitality' a fundamental mathematical meaning as I have done already for *order, orderly, complexity, similarity, information* and other concepts. I shall then pursue the effect of genetic noise in reducing the information content of the genome and thereby the vitality of the organism (Yockey, 1956, 1974; Yockey *et al.*, 1958).

11.2 Error theories of aging

11.2.1 *'Inborn errors of metabolism' theories of aging*

Aging to some extent occurs in all organisms. Theories of aging are almost as numerous as the practitioners of the subject. The idea that aging is due to the accumulation of defects in the genome has received much consideration in the past. Theories based in one way or another on the accumulation of defects in the genome are fundamental in the sense that they involve the basic phenomena of molecular biology and do not depend on the particular physiology of a specific organism. To that extent they are not applicable to explaining the death of such semelparous animals as the Pacific salmon immediately after mating (Finch & Schneider, 1985). Thus we see that aging is clearly a complicated phenomenon involving many factors to one degree or another. However, if error theories of aging that involve the accumulation of defects in the genome can be put on a firm theoretical foundation, then the role of the accumulation of defects in the genome can be separated from other factors. I do not presume to present this as the only cause of aging. I am abundantly aware of the cross-linking of collagen when I get up from my desk and move about the room.

As the molecular basis of genetics became better understood, 'inborn errors of metabolism' drew consideration as the basis for theories of aging. A number of authors proposed that damage to the informational molecules in the genome caused the 'inborn errors'. Szilard (1959) proposed that the chromosomes in the somatic cells suffer 'hits' from random events until the surviving fraction reaches a critical value below which death occurs. Failla (1960) proposed somatic mutations caused by spontaneous chemical action or the 'hits' resulting from X-rays cause a progressive deterioration, aging and eventual death. Medvedev's theory of aging (1966) is based on molecular changes in the structure of the genes. Specifically, he considered the accumulation of molecular changes in the DNA that were transferred to the proteins causing their inactivation and eventually the death of the organism.

Holliday & Kirkwood (1981) included the possibility of errors in DNA in addition to the feedback errors in the synthetases as suggested by Orgel (1963, 1970, 1973a). They called their theory a *general theory of aging*. Holliday & Kirkwood (1983) distinguish the *protein error theory* (or *error catastrophe theory*) from the somatic mutation theory.

A selection of significant papers from the enormously large literature on protein theories of aging was edited and published by Holliday (1986). Detailed discussion of accuracy in molecular processes is available in the book edited by Kirkwood *et al.* (1986). A review of the whole field of the biology of aging was edited by Finch & Schneider (1985). The three works just mentioned contain many further references in this large field in biology and medicine.

11.2.2 Orgel's 'error catastrophe' theory of aging

Orgel (1963, 1970, 1973a), however, must be awarded the credit for the most frequently cited papers on the idea that aging is the reflection of the progressive accumulation of proteins with errors that affect their specificity. He divides the types of errors into two categories. The first is the transcription error resulting from the replacement of a 'correct' amino acid by an 'incorrect' one. Second, he proposes that certain of these transcription errors occur in the synthetases so that an increasing error frequency is fed back into the protein synthesis process, thus leading to reduced specificity of an information handling enzyme so that eventually the error frequency must reach a value at which the processes necessary for the viability of the cell are not sufficient. To reflect the accumulation and feedback of the errors, Orgel proposed the following equation, where p is the error frequency, t is the generation time and α is a constant of proportionality:

$$\frac{dp}{dt} = \alpha p \tag{11.2}$$

At time $t = 0$ the error frequency is p_0 and equation (11.2) may be integrated as follows:

$$p = p_0 e^{\alpha t} \tag{11.3}$$

According to equation (11.3), the population of erroneous proteins increases exponentially, resulting in what Orgel (1963) called an 'error catastrophe' that causes the death of the organism. This speculation, as he called it (consistent with the definitions of theory and speculation as used in this book), has been the inspiration of many experiments intended to find supporting evidence. He suggested that a more sophisticated consideration would include the variance of the error frequency in the population of cells. This is similar to what I had also proposed (Yockey, 1958).

Orgel's (1963) paper would appear to predict that all life forms will die out sooner or later because of the 'error catastrophe'. However, in 1970 he modified his 1963 paper to suggest that under some circumstances the error rate does not continue to increase exponentially, but rather that a small stable error frequency can be maintained so that an 'error catastrophe' is not inevitable. To determine whether an organism can exist with a small stable error frequency, Orgel made the following calculation: let c_n be the error frequency in the nth generation, R, the residual error frequency and b the proportionality between errors in the synthetic apparatus and errors in freshly synthesized protein. Then

$$c_{n+1} = R + bc_n \tag{11.4}$$

If $c_0 = 0$, then

$$c_n = R(1 + b + b^2 + \ldots + b^{n-1}) \tag{11.5}$$

If the sequence in equation (11.5) extends to indefinitely large n and if $b < 1$, it may be summed to $R/(1-b)$, which is the steady state error frequency.

Orgel (1963) pointed out that the protein synthesizing enzymes might have an error frequency of $\approx 10^{-4}$ in distinguishing between chemically similar amino acids such as Val and Ile. This is approximately the value found experimentally by Loftfield & Vanderjagt (1972) and Hopfield *et al.* (1976). Orgel regarded this error frequency as measured in generation time; however, according to the evolutionary clock theory, one should measure t in sidereal time rather than in generation time as was believed in 1963. Therefore, at first $bt \ll 1$ so that the exponential is approximately a linear function of bt. This means a very long time must elapse before bt becomes of the order of unity and therefore the calculation shows there is not an enormous increase in the error frequency except perhaps over very long times. This shows why the feedback of errors may be very small or not exist at all, thereby contradicting Orgel's original conclusion.

In 1963 it may have seemed reasonable that the mutation between active 'correct' synthetases and 'incorrect' ones was frequent. However, as I discussed in Chapter 6, it is generally true that placing a non-functionally-equivalent amino acid in the protein chain renders the protein inactive, rather than creating a new active but 'incorrect' protein. I discuss in section 12.3 how the mutation of an active synthetase to another active synthetase with a different specificity is similar to the evolution of *de novo* functions in protein evolution. The discussion there shows that the protein space is many orders of magnitude more sparse than has been realized. Therefore, the overwhelming probability is that a mutation renders the synthetase totally inactive rather than that a *de novo* function should appear (Yockey, 1974). The feedback of errors may be very small or not exist at all because of the sparseness of the protein space. Pütz *et al.* (1991) have provided experimental evidence which shows that the active 'correct' synthetases become inactive but do not become 'incorrect'. They find that for tRNA synthetase systems studied by their single mutations in determinant nucleotides specificity decreased from 10 to as much as 10^4 or 10^5 fold. It was only by means of four carefully selected mutations that they were able to change tRNA[Phe] to a substrate for yeast aspartyl-tRNA synthetase.

Although Orgel does not use the term 'feedback' (1963, 1970, 1973), other authors have used the term (Kurland, 1986; Rosenberger & Kirkwood, 1986) to describe the same phenomenon. Attempts to measure the feedback of errors have been made but no rigorous proof of this effect has been found (Kirkwood *et al.*, 1986, section 2.3.2). However, these authors also say that whether feedback errors can take place is not in serious dispute (Kirkwood *et al.*, 1986, section 2.5).

Thus we see that a correct analysis of a feedback channel that *increases* the error frequency remains to be made. However, the role of a feedback *correction* channel in the *correction* of errors in a message has been given some attention by communication engineers (Blahut, 1987). For example, Shannon (1956) proved that the ordinary capacity of a memoryless channel (like that of the protein synthesis channel) with a correction channel is equal to that of the same channel without feedback. Reprints of other significant papers on the effect of a feedback correction channel with a strong communication engineering bias can be found in Slepian (1973). No application of the large literature on control processes and stability theory has been made to justify the conjectures of Orgel with regard to equations (11.3) and (11.5).

We learned in Chapter 6 that a more sophisticated view of what constitutes errors than that of which Orgel was aware in 1963 is that only those replacements of a functional amino acid that are not functionally equivalent constitute errors in the protein sequence (Yockey, 1958). Therefore, although Orgel's papers (1963, 1970, 1973a) stimulated a great deal of research, his speculation must now be replaced because it is *ad hoc* and because, as we learned in section 5.2.1, the effect of genetic noise on the genome is not measured by error frequency but by the conditional entropy of the error frequency.

11.3 Protein error and the chemical thermodynamics of the protein synthesis process

11.3.1 Causes of error in the protein synthesis process

The protein synthesis process is astonishingly accurate. The Potter's hand seldom shakes. The existence of organisms, such as the coelacanth, that have preserved their morphology for 4×10^8 years is adequate evidence (Fricke & Plante, 1988; Fricke, 1988). It is remarkable that in less time other organisms have differentiated, evolved and become extinct. The simultaneous existence of creativity, stability and decadence is an astonishing property of the storage, processing and transfer of information in the genetic system. The occurrence of amino acid replacement in the protein synthesis process is related both to evolution and to aging: when replacement is creative, it results in evolution; when replacement is in error, it leads to aging. A brief discussion of causes of error in the protein synthesis process follows in order to put certain facts before us. More detailed material and further references can be found in standard texts such as Watson *et al.* (1987) and in a book edited by Kirkwood, Rosenberger &

Galas (1986). A recent review of errors in reading the standard genetic code has been given by Parker (1989).

11.3.2 Free energy and the accuracy of protein synthesis

Various authors, starting with Pauling (1957), have used the basic concepts of thermodynamics as a guide in attempts to understand the chemical dynamics of the charging of tRNA. However, biochemical reactions occur far from equilibrium and consequently far from the formalisms of equilibrium thermodynamics; even the formalisms of non-equilibrium thermodynamics are not strictly applicable (Ehrenberg, Kurland & Blomberg, 1986; Coveney, 1988). For example, the thermodynamical potentials, such as the Helmholtz and Gibbs free energies, do not exist except for the special case of equilibrium (Coveney, 1988).

Even though the formalisms of non-equilibrium thermodynamics are not strictly applicable, a model based on Michaelis–Menten enzyme kinetic mechanisms has been developed, primarily by Galas & Branscom (1978), Clayton *et al.* (1979) and Goodman *et al.* (1980). One proceeds by noting that in simple chemical reactions the equilibrium constant between two states is equal to the Boltzmann factor

$$e^{\Delta G/RT} \tag{11.6}$$

where ΔG is the free energy difference between the two states, R is the gas constant and T is the temperature. Expression (11.6) can be applied to calculating the probability of mispairing tRNA and also to calculating other errors in the protein synthesis process that depend on the equilibrium ratios of the populations of the states that reflect error to the populations of the correct states.

The average free energy, ΔG, of a hydrogen bond is approximately 3 kcal/mol. This is important because the purine and pyrimidine bases that make up DNA are held together by hydrogen bonds. At room temperature, about 300 K, the value of expression (11.1) is about 10^{-4}. The AT pair in DNA is held by two hydrogen bonds, so the probability that adenine will mispair to cytosine is about $10^{-4} \times 10^{-4} = 10^{-8}$. The GC pair is held together by three hydrogen bonds. However, the accuracy does not approach 10^{-12} at the stage of first pairing because the base selected has a probability of about 10^{-4} of being in the imino or enol tautomeric form that leads to a mispairing. Thus the use of the Boltzmann constant in expression (11.6) leads us to an estimate of the error probability that is, to a good approximation, that found in experiments.

Watson & Crick (1953, a, b) pointed out that purine–pyrimidine mispairing might explain spontaneous errors or mutations in the replication of DNA. The stability of the AU and GC pairs was measured by Grosjean, de Henau & Crothers (1978). They found the two pairs nearly equal in stability. They proposed that the stacking interactions require distortion in the hydrogen bonds, which, they suggested, is more costly in energy for the GC pairs than for the AU pairs.

Chemical mutagens cause errors in DNA replication. UV and X-rays cause damage to DNA. Many of these errors are removed by repair enzymes. The overall error rate in DNA ranges from 10^{-7} to 10^{-12}. For a review see Walker, Marsh & Dodson (1985).

11.3.3 The probability of the mischarging of tRNA

Van der Waals forces of repulsion and attraction are certainly involved in the charging of tRNA. They depend on the size and shape of the side chain in the amino acid and are weaker and less specific than hydrogen bonds. Pauling (1957) suggested that the genetic data processing system might have difficulty in discriminating between similar proteinous amino acids. On the basis of the free energy difference, considering only a single stage in the process without branching, he found, from the Boltzmann factor, an error frequency of about 0.05 for Val replacing Ile and the same for Gly replacing Ala. He also estimated the error frequency of Ala replacing Ser and Phe replacing Tyr to be 0.01.

Loftfield (1963) measured the frequency with which Val, Ile and Leu replace each other by means of ^{14}C labeling experiments in oviduct synthesis of ovalbumin and found a frequency of about 1/3000. Loftfield & Vanderjagt (1972) improved their techniques and, using the known sequence of rabbit and human hemoglobin, found that at a specific site, the frequency of a Val substitution for Ile is in the range of 2–6×10^{-4}. They noted the fact that the transition of G to A in the Val codons results in the three codons for Ile and AUG for Met. In Table 6.2 one sees that the BH distance of these closely related amino acids is 0.10 and that the Hamming distances of the codons are 1,2.

Although Pauling's (1957) values for the replacement frequencies of amino acids were much too large, he had posed the question and established the methods that led to the study of the accuracy of the genetic data processing system.

Woese (1990) and others have speculated that 'at sufficiently early stages in evolution the fundamental information-transferring processes . . . must have been error-ridden'. We are now in a position to calculate approximately the error frequency that can be tolerated by the protein generating system. Let the probability of a mischarging or other misreading event be α. Not all misreading events are errors. In the study of iso-1-cytochrome c we learned that the geometrical average number of functionally equivalent amino acids is 7.0594 per site. The geometrical average number of proteinous amino acids is 17.62062. The approximate probability that if a misreading event occurs it will be among the functionally equivalent amino acids is 7.0594/17.62062. The probability that the misreading will transpose from a functionally equivalent amino acid to a non-functional amino acid is roughly (17.62062–7.0594)/17.62022. Let us suppose that in a string of N amino acid sites a fraction 0.5 have no error, that is, have no non-functionally equivalent amino acids. Then we have the following, admittedly rough, approximate equation which should be sufficient for the present purpose if $\alpha \ll 1$:

$$\left(1 - \alpha \frac{7.0594}{17.62062} \times \frac{17.62062 - 7.0594}{17.16062}\right)^N = 0.5 \tag{11.7}$$

If we let $N = 100$, corresponding to the length of a small protein such as iso-1-cytochrome c, then $\alpha = 0.029$. If $N = 1000$, or 10 small protein lengths, $\alpha = 0.0029$. If $N = 10,000$ or 100 proteins (which may have been the minimum information content for the protobiont), then $\alpha = 0.00029$, very nearly the value found for modern protein synthesis by Loftfield & Vanderjagt (1972).

We must recall in this connection the efficient and sophisticated protein synthesis process of the overlapping genes. One finds in this calculation, as well as by other means, that as far back as one can see, although through a glass darkly, toward the origin of life, the simplest organisms must have been extremely complex and the protein synthesis system was not 'error-ridden'. The notion that the genetic logic system of the protobiont, progenote or genote was 'error ridden' is based entirely on ideology and is without facts to support it. It does not help the situation to introduce neologisms such as 'progenote' or 'genote' (Woese, 1990).

11.3.4 The proofreading process

The next important contribution to the study of protein error was made by Hopfield (1974) and independently by Ninio (1975) who both pointed out that tRNA charging takes place in more than one step and that non-cognate amino acids could be eliminated along branches in the succeeding steps of the process. Hopfield called this *proofreading* and pointed out that several steps with the same or comparable precision lead to the product of two or more Boltzmann factors and to estimates of the order of $0.01 \times 0.05 = 5 \times 10^{-4}$. Later Hopfield *et al.* (1976) and Yamane & Hopfield (1977) measured the frequency of replacement of Ile by Val using the radioactive tracers ^{3}H and ^{14}C. They found that kinetic proofreading reduces error by a factor of $1/180$, which leads to an overall error rate of the order of 10^{-4} which is consistent with the work of Loftfield & Vanderjagt (1972).

It is now accepted that the high accuracy of DNA replication is accomplished through some form of proofreading or double sieving action of the DNA polymerases in which the non-cognate substrates are discarded at a rate greater than that of the cognate ones (Goodman & Branscom, 1986). This is because incorrect substrates have a larger activation energy and associate with smaller probability: if they are activated they dissociate faster than the correct substrates and so have a reduced probability of entering the activated intermediate states of the replication process. They are thus eliminated more frequently than the cognate substrates in the branches of the secondary stages of the replication process. Grosjean *et al.* (1978) measured the life-times of the correct pairings to be between 65 msec and 150 msec with a few exceptions. The life-times of the incorrect pairings were usually between 1 msec and 10 msec.

Some error containing proteins will be formed in the cell. A protease scavenging system was evolved to remove such molecules to prevent their accumulation. This is the last stage of error protection in protein formation. Many proteins have a rather short half-life and the inactive ones seem to have even shorter ones (Capecchi *et al.*, 1974; Goldberg & Dice, 1974; Goldberg & St. John, 1976).

Amino acids that are sufficiently different in their properties do not require the proofreading mechanism. The activation of Cys is favored by a factor of 5×10^6 or more for the rejection of Ser and Ala (Ferscht & Dingwall, 1979). Ferscht, Schindler & Tsui (1980) found that tyrosyl-tRNA synthetase is adequately protected from charging with Phe because the activation of Tyr is favored over that of Phe by a factor of 10^5. (See comment on Phe–Tyr in sections 6.2.2 and 6.3.3 and see also Tables 6.2–6.4 and 6.6.)

The protein proofreading system is illiterate. It operates simply as a data processing system. Unlike the nineteenth-century telegrapher, it does not understand the message and it is unable to request a retransmission or to make corrections from the meaning or specificity of the context. In the case of a data recording and processing system with which the genetic system is isomorphic, there is no redundancy in the language or the meaning of the message that may be used for correction. The protein *proofreading* system is like an illiterate proofreader who replaces only those letters that are damaged and do not resemble the standard. The protein *formation* system is illiterate in the sense that all error corrections are made without regard to the specificity of the protein. In the case of natural languages, errors may be detected and in some cases corrected by the redundancy and meaning of the text. However, in the case of a data recording and processing system, with which the genetic system is isomorphic, there is no meaning in the sense we understand it that is analogous to messages in natural languages.

There have been considerable efforts to search for defective proteins uncorrected by the proofreading processes. These proteins will have amino acid replacements scattered along the protein chain in a random fashion. Those replacements, which occur at sites that are invariant or have few functionally acceptable amino acids, are more likely to be errors and to result in inactive proteins than replacements at those sites that have many functionally equivalent amino acids. This is a consequence of what we learned in Chapter 6 about the information content of sites in the protein chain. However, when one realizes that there are 2.316×10^{93} iso-1-cytochrome c sequences and 1.137×10^{137} sequences of 110 amino acids in the high probability group it is seen that the inactive proteins are lost among the 1.137×10^{137} sequences that are not cytochrome c (section 9.2.2). Therefore tests for specific errors in protein may well be very difficult and so the fact that they have not been found is not a criticism of protein-error theories of aging.

Proofreading must not be confused with editing (section 7.3.2).

11.3.5 Energy dissipation due to error in the genetic logic operation, and its relation to the Central Dogma

The rejection of non-cognate substrates by proofreading is accompanied by a dissipative loss of energy and an increase of the Maxwell–Boltzmann–Gibbs entropy (Hopfield, 1980). Energy use by the cell is priced in the sense that, should such losses become too large, the energy use by the cell for other needs suffers. Therefore, the energy use of the cell places a limit on the genetic noise that can be corrected. Nevertheless, some genetic noise can be tolerated and, consequently, perfect accuracy is neither necessary nor desirable because the ability to evolve depends on some flexibility in the genetic message carried by the DNA. This situation is much the same as the complexity problems in computer data processing systems (section 2.4.3) where the data is also subject to error and the complexity needed to correct error is priced in terms of computer memory and computer time and, therefore, in money.

Although there is no analogue for energy in communication theory, nevertheless, the actual equipment used by the communication engineer dissipates energy and the biological communication system does likewise. The question arises, in each case, of the minimum energy dissipation required by the basic physics of the operation. Both systems are subject to thermal noise because of the indeterminacy expressed in the laws of quantum mechanics. As I discussed in section 3.1.3, Kullback (1968) defined the mutual information $I(2|1)$ in statistical inference as the average information per observation for discrimination in favor of observation H_2 against H_1 as

$$I(2|1) = \int f_2(x) \log_e \left(\frac{f_2(x)}{f_1(x)} \right) d\lambda(x) \tag{11.8}$$

where $\lambda(x)$ is a probability measure. From this equation Kullback (1959) rigorously derived the correct equation for the channel capacity of a continuous signal:

$$I(\Delta\omega, \Delta t) = (\Delta\omega \times \Delta t) \log_e \left(1 + \frac{S}{P} \right) \tag{11.9}$$

where S is the signal power (energy per unit time), P is the thermal noise power, $\Delta\omega$ is the frequency band width of a continuous signal and Δt is the duration. Equation (11.8) had also been obtained by Shannon (1948) but without the same rigor. P is given by

$$P = kT\Delta\omega \tag{11.10}$$

where k is the Boltzmann constant and T is the absolute temperature. In the case with which we are concerned, the ratio S/P is small compared to unity. Accordingly, the logarithmic expression, $\log_e(1 + x)$, in equation (11.9) may be replaced approximately by x. If one makes that substitution and some rearrangement one finds the following equation for the minimum energy per bit:

$$\frac{S\Delta t}{I(\Delta\omega, \Delta t)} = kT\log_e 2 = 0.693\,kT \tag{11.11}$$

The ratio of the signal energy $S\Delta t$ to the number of bits in the signal, if the alphabet is binary, is $0.693\,kT$. In the case of interest to molecular biology, the discrimination is between four observations and the logarithm of four must appear. The energy per bit is $1.386\,kT$ or about 0.035 electron volts. This dissipation of energy occurs at each step in the genetic logic operation including proofreading. Since the ratio of signal power to thermal noise power is very small, the expression on the right of equation (11.11) is the minimum energy per bit. This result had also been obtained by Shannon (1948) and later by von Neumann (1966).

The question of the minimum dissipation of energy in computation and communication has received considerable attention in the theory of computers because the engineering problem of heat removal limits the compactness and therefore the speed of computing machinery. The basic principle in the calculations leading to equation (11.11) is that noise discards information and consequently the communication process is irreversible. The discarding of information by any means makes the genetic logic operation irreversible because the transition function lacks a single-valued inverse.

In the case of the genetic logic system, it is clear that information is discarded by the redundant genetic codes between mRNA and protein as well as by genetic noise. Therefore the process dissipates energy, generates Maxwell–Boltzmann–Gibbs entropy and is irreversible. This is addressed specifically in the derivation of equation (5.13), which gives the amount of information that must be discarded as required by the redundance of the standard genetic code. There is no *élan vital* in biology so the genetic logic system must obey the same fundamental laws as other logic operations. This supplements the discussion of the Central Dogma given in Chapter 4.

There is considerable discussion in the literature that leads to the result that an ideal computer can operate without a limit on energy dissipation if it operates slowly enough and if the computations leading to the output are not erased. Neither of these conditions applies to the genetic logic system. Once the tRNA has accomplished its coding operation, except in the case of Met and Trp, the information with regard to the exact codon is erased from the memory of the genetic logic system. The considerations given in this section of the question of energy dissipation in genetic logic operations are sufficient for the purposes of this book but the reader may wish to consult some significant papers in the field, namely, Landauer (1961, 1988), Bennett (1973, 1988), Bennett & Landauer (1985) and Zurek (1984, 1989).

11.4 The mathematical formulation of the genetic noise theory of aging

I made the suggestion that the quantity one is searching for, which plays the role of vitality, is the information content of the genome (Yockey, 1956). In that paper, written well before the genetic code was worked out, the central idea was clearly stated: that the genetic system is an information recording and transmitting system. This idea was developed further in later papers (Yockey, 1958, 1974, 1977b). In these papers I also showed that the Shannon conditional entropy is the correct measure of the effect of error or noise on the genetic message, not the error probabilities *per se* (sections 2.2, 4.4).

We are now in a position to put together what we have learned in previous chapters in this book to study the effect of noise on the genetic message or genome. Noise is described by the probability transition matrix **P** with matrix elements $p(j|i)$. The noise is calculated from these matrix elements by equation (5.14). The first cause of noise is that due to random read-off errors in DNA that mischarge tRNA. Mischarged tRNA molecules cause genetic noise by a random misreading of the established codons. This will produce a general background genetic noise.

The second cause of genetic noise is a Markov process in which the DNA code words are altered by single base interchanges. The altered codons, which are states in the Markov process, transmit correctly according to the genetic code until a change is made in the codon. The random walk progresses from one codon to another and this is reflected in a change of one amino acid to another. As the process proceeds it is described by the square probability transition matrix, **P**. Because it is square, **P** may be raised to a power, t, corresponding to each step (sections 1.3.1, 1.3.2). $\mathbf{P}(t)$ is a λ-matrix the elements of which are polynomials of degree t (section 1.3.3). Beginning with the first amino acid probability vector, $\mathbf{p}(0)$, the subsequent distribution, $\mathbf{p}(t)$, after t steps is found by this equation:

$$\mathbf{p}(t) = \mathbf{p}(0)\,\mathbf{P}(t) \tag{11.12}$$

As I pointed out in section 1.3.4, this formalism takes account of any number of back mutations in a natural fashion. The noise entropy at each step in the random walk is calculated by the matrix elements and vector components appropriate to that step. The effect of genetic noise may be calculated by substitution of these values in equations (5.13) and (5.14).

Suppose we start with an ensemble of organisms all of which use the same selection of sense codons and, of course, with the same information content. Some will take paths in the Markov chains that exhaust their redundance fairly quickly. Others will have a number of back mutations and the random walk out of the sphere of functionally equivalent amino acids will take more time. One notices immediately that the proper treatment of back mutations is critical in obtaining a correct calculation. A spectrum in the information content of the ensemble of organisms is produced naturally by the Markov process without *ad hoc* assumptions.

To illustrate this further, let us look again at the error correcting code in Table 4.1. The source was instructed to use the sense code words although other code words would be decoded correctly. The sense code words are those with the largest Hamming distance from other destination code words. If the noise level in the channel is sufficiently small, the source can get away with choosing any of the code words in the column headed by the destination code letter in Table 5.1. Let us assume that we have a number of sources each of which at first uses the sense code words. As time goes by, some sources, one by one, forget the sense code words. An ensemble of sources each with its set of code words is generated. Those that continue to use the sense code words will be able to survive the largest noise level. Those that use only code words one Hamming distance from error will send erroneous messages at any level of noise if the messages are long enough. This change from the original sense code words to sense code words less protected from noise is isomorphic with mutations to less protected functionally equivalent amino acids. Amino acids that are nearest to the center of the sphere of functionally equivalent amino acids, at each site, will take the longest time to diffuse out of the sphere. As a result, they will be able to continue to synthesize active proteins in the presence of considerable noise.

The idea of a probability distribution in information content in the genome was introduced in Yockey (1958), where I stated that we can associate a spectrum of information content in the algorithmic sense describing the ensemble of both telegraphers and proteins. Those that make full use of the redundance provided by the error correcting code will have the largest probability of continuing longest in the presence of noise generated in the random walk described above. Those that use it least will have the largest probability of erroneous transmissions.

Returning to the aging process of the organism, we may expect that the genetic noise will increase with time as the random walk proceeds. Those organisms that are less protected from genetic noise by redundancy will succumb first. Survival curves in those cases (e.g. the histones) where almost all sites are invariant, and that consequently have little or no redundancy, will be exponential since the probability of error is constant in time. Organisms will have survival curves with a 'shoulder' if the cognate codon is two or more Hamming distances from a non-cognate codon and is near the center of the sphere defining functionally equivalent amino acids. That is, as the age is small there is nearly complete survival. As the random walk proceeds, redundancy is exhausted and the survival curves turn sharply downward reflecting the flux of events out of the protection of redundancy, error correction mechanisms and proofreading. Survival curves with a shoulder are formed in a very natural fashion according to this theory and they are the ones most frequently reported. This is reflected in equations (5.22) and (5.23) in which the effect of white genetic noise is calculated. This process occurs in the germ line as well, and that contributes to the formation of a spectrum in information content in the ensemble of organisms.

There is no place in this theory for a small stable error frequency as postulated by Orgel (1970) or the error threshold of Eigen & Schuster (1977, 1978a, b). Is it necessary to prove that a stable error frequency exists in order to show that protein error theories do not require all life to die? Almost all organisms that ever existed have, in fact, died and all organisms now living will also die. How does it happen that living organisms still exist in the absence of stable error frequency or an error threshold? How then, is the accuracy of protein synthesis maintained in some organisms while others die? The answer is that the portion of the population that has the largest information content and therefore the most vitality reproduces itself exponentially. It is both Malthusian doctrine and canonical Darwinism that far more creatures of all kinds are produced than can possibly survive to reproduce.

12

Information theory and molecular evolution

I am in point of fact, a particularly haughty and exclusive person, of pre-Adamite ancestral descent. You will understand this when I tell you that I can trace my ancestry back to a protoplasmal primordial atomic globule.

Pooh-Bah, *The Mikado*, Act 1, W. S. Gilbert (1887)

12.1 Is molecular evolution contrary to the second law of thermodynamics?

12.1.1 Thermodynamics and the creation of 'order' by evolution

Thermodynamics and the theory of evolution by natural selection are among the great scientific theories of the nineteenth century. Yet almost immediately after the introduction of these two great theories they were pitted against each other. The opinion that evolution was contrary to the second law of thermodynamics was pushed by scientists who did not accept evolution. In the twentieth century this alleged conflict has been a favorite theme of the Biblical creationists and of the creation-science advocates (Wilder-Smith, 1981; Gish, 1989).

An isolated system is one which *does not* exchange heat or work with another system. The second law of thermodynamics states that the Maxwell–Boltzmann–Gibbs entropy of an isolated system does not decrease; it is, in general, not a conservation law since entropy is conserved only in reversible processes. Entropy is regarded as a measure of 'order', sometimes without bothering to define 'order'. Thermodynamics, as it was viewed in the nineteenth century, predicted a 'heat death' of the universe in which all bodies in the universe will be at the same temperature in a state of complete 'randomness' or 'disorder'. (Think of it as the death of heat exchange, not a death by heat.) Evolution, on the other hand, so it was thought, predicted an increase in 'order' in the sense that more highly evolved organisms were considered to be more 'orderly'. Thus the reason for the conflict between the evolutionists and the creationists is that the second law was perceived to predict increasing 'disorder' while the theory of evolution was perceived to predict increasing 'order'. The notion that a paradox exists between biological evolution and the second law of thermodynamics persists today and is perceived to require resolution (Waldrop, 1990).

As I shall show, a proper understanding of both evolution and the second law of thermodynamics demonstrates that they are not in conflict and that the supposed paradox is merely a play on words.

I pointed out in section 8.1.3 that our current knowledge presents a quite different picture from the one seen in the nineteenth century. Today we observe an expanding universe powered by the 'Big Bang' that is quite different from the 'heat death' theory of nineteenth-century physics and cosmology. An expanding universe never reaches equilibrium or a constant temperature and so astronomers and cosmologists see the situation in quite a different way today than in the nineteenth century.

Scientists who defend evolution have been caught, much as Br'er Rabbit by the Tar Baby, in a word-trap implying that evolution leads to organisms that are increasingly more 'orderly'. As I have mentioned in this book several times, it is perilous for the scientist to use words in their everyday meaning to express scientific concepts. One should remember that in science these words are merely names for mathematical functions and have no other meaning in themselves.

Thwaites & Awbrey (1981), in an attempt to defend evolution, are caught in this 'order–disorder' word-trap but get half-way out when they say: 'We should use a phrase like *high information content* in place of *ordered* [their italics]'. This starts us in the correct direction; however, only by attaching a mathematical definition to the words do they have utility and free scientists from word-traps.

Kitcher (1982) believes that there is a dilemma and that the way out is to declare that life is an open system receiving work and heat and that in such systems entropy *decreases* over time. This is true but inapplicable. Although organisms use energy, as I have pointed out (Yockey, 1974, 1977c), the entropy that is pertinent in the genetic system is the Shannon entropy or the Kolmogorov–Chaitin algorithmic entropy. Kitcher (1982), along with many others, is unaware that each probability space has its own peculiar entropy. Two probability spaces that are not isomorphic have no relation except in those cases where such a relation is established by a code. Patterson (1983) also makes the mistake of confusing order with complexity; he cites some references in which the Biblical creationists claim that evolution is impossible based on the false notion that evolution violates the second law. He believes that highly organized molecular assemblies have a low entropy. He thereby falls into the word-trap set by the creationists of using words like 'order' and 'complexity' without defining them mathematically. His explanation in other ways is rather lame.

Freske (1981) also gets caught in the same word-trap and includes the *negentropy* fallacy (section 2.4.4). Schidlowski (1976) regards the evolutionary trend towards higher (or differentiated) life to be a 'crime against entropy' (*sic*). 'Both maintenance of life as a non-equilibrium steady state with a marked negentropy and its evolutionary diversification are energy absorbing (endoergic) processes, driving "upstream" within the universal entropy flux'. He attributes this to: 'the utilization of the oxidative pathway of metabolism to play the role of the Maxwell demon which caused evolution to go in the direction of more order (or diversification) and less entropy'. Even a scientist as eminent as Eddington (1931a) believed there is a conflict between the second law of thermodynamics and evolution: 'We must of course drop the theory of evolution, or at least set alongside it a theory of anti-evolution as equally significant'. As usual Eddington could put his finger on the problem even if a solution was not available at the time. Later in that article he pointed out that it is *organization* that increases in evolution. But he had no mathematical definition of organization. This had to await the definition given by algorithmic entropy as discussed in section 2.4.1.

As has already been discussed in this book (section 2.2.2), the entropy to which the second law of thermodynamics applies is the Maxwell–Boltzmann–Gibbs entropy. This entropy addresses the energy efficiency of heat engines, the directions of chemical reactions and, in biology, the energy economy of the cell. Thus it is the 'orderliness' of energy that is addressed by the second law of thermodynamics. Organisms cannot defy the second law: there has never been any question in sober minds that organisms are not perpetual motion machines (although small children may seem to be). One should remember that classical thermodynamics deals only with energy and the Maxwell–Boltzmann–Gibbs entropy. It is even independent of the existence of atoms.

All probability spaces have an entropy peculiar to them (section 2.2.1). The entropy that is applicable to the case of the evolution of the genetic message is, as I believe the reader should now be convinced, the Shannon entropy of information theory or the Kolmogorov–Chaitin algorithmic entropy. In thermodynamics one has two invariants of the system, namely the mass of the system and the conservation of energy as expressed in the first law of thermodynamics. In communication systems one has only the conservation of the number of letters. There is no invariant corresponding to energy in the case of communication systems and consequently no analogue of temperature. Furthermore, the Shannon entropy and the Maxwell–Boltzmann–Gibbs entropy are based on probability spaces that are not isomorphic to each other. Consequently, for this reason also, as I have pointed out before (Yockey, 1977c), they have nothing to do with each other. Therefore it is easy to see that thermodynamics has nothing to do with Darwin's theory of evolution.

12.1.2 The increase in genetic algorithmic entropy during evolution

The Kolmogorov–Chaitin genetic algorithmic entropy is increased in evolution due to the duplications that occurred in DNA. For example, George *et al.* (1985) have reported evidence for a duplication from a study of ferredoxin alignments. The ferredoxins are regarded as very ancient proteins; therefore, this event must have occurred very early in the history of life. As the duplicated sections acquired new functions, the algorithm that describes the genome became longer. Thus the genetic algorithmic entropy increases with time just as the Maxwell–Boltzmann–Gibbs entropy does. Therefore creationists, who are fond of citing evolution as being in violation of the second law of thermodynamics (Wilder-Smith, 1981; Gish, 1989), are hoist by their own petard: evolution is not based on increasing *order*, it is based on increasing *complexity*. In fact, evolution requires an increase in the Kolmogorov–Chaitin algorithmic entropy of the genome in order to generate the complexity necessary for the higher organisms. Let us recall from section 2.4.3 that *highly organized* sequences, by the same token, have a large Shannon entropy and are embedded in the portion of the Shannon entropy scale also occupied by *random sequences*. Evolution is not in violation of the second law of thermodynamics. This is what any reasonable scientist believes; nevertheless, it is important to avoid word-traps and to reach the correct conclusion for the correct reasons.

12.2 The first meaning of the evolution of proteins: homologous sequences and the molecular clock

'Protein evolution' has a double meaning. Much confusion and useless debate has been engendered because of this. The first meaning is the evolution of *homologous* proteins, those that are defined as having a common function and evolution from a common ancestor (Reeck *et al.*, 1987; Lewin, 1987). The second meaning is the formation *de novo* of proteins with new functions and structures from older proteins. New morphology and new fitness arise from the *de novo* proteins. I shall take up the first case in this section and the second case in section 12.3.

In classical Darwinism, it is thought that any mutation must be either beneficial or deleterious. It is believed that neutral mutations are extremely rare and that substitutions must contribute some small increase in fitness to become fixed. Most mutations are believed to be deleterious, either lethal or *malefic*, to use a term coined by Fitch & Margoliash (1967) (Margoliash & Fitch, 1971). *Malefic* mutations are those that produce irreparable damage and must produce a lethal result either immediately or eventually.

The idea that most mutations are neutral was introduced independently by King & Jukes (1969) and Kimura (1968). Neutral mutations apply only to the evolution of homologous proteins. It is now generally accepted that mutations between amino acids in homologous proteins are almost always functionally neutral (Zuckerkandl, 1987). The metaphor for the Darwinian view of natural selection is that proteins or species ascend through gradual change to an adaptive peak in the evolutionary landscape (Wright, 1932). However, those who support the neutral theory believe that the activity landscape of homologous proteins is not like the Alps or even the rolling hills of Appalachia. Rather it is like the buttes of Monument Valley in Arizona (USA). The tops of these buttes have considerable area and are relatively flat, which is comparable to the ability of proteins to tolerate any change in which the amino acids remain functionally equivalent. However, like the buttes that tower precipitously above the plains, freedom within a certain area is bounded by a sharp limit: almost all changes outside this limit to amino acids that are not functionally equivalent reduce the activity by a factor of the order of 10^{-2} to 10^{-4}. Some changes do not reduce the activity so drastically, making them like the slopes of scree that lie at the foot of the buttes and blend the precipitous cliffs in with the plains. Thus homologous proteins tolerate a number of substitutions at certain variable sites with small fluctuations in activity, while certain other sites are invariant.

12.2.1 Attempts to fit the spectra of mutations to a probability distribution along the protein sequence

It seemed reasonable to a number of authors at the time the first protein sequences became available to attempt to fit the spectrum of mutation events along the protein sequence with a Poisson or other probability distribution. There has been considerable debate among various authors on the proper statistical methods to be used to represent the spectra of mutations along the protein sequence (Karon, 1979; Fitch, 1980, 1981; Holmquist & Jukes, 1981). Fitch & Margoliash (1967) concluded that the vast majority of the mutations are characterized by a Poisson distribution. Their analysis found 35 unmutated cytochrome c codons and predicted that, of these, 27–29 would remain unmutated, that is, invariant, as more cytochrome c sequences became available. Their method did not indicate which of the codons were invariant. Later Fitch & Markowitz (1970) proposed that there are 32 invariant codons and that the remainder could be represented by two groups of 65 and 16 codons in separate Poisson probability distributions.

Fitch (1976) found that, by designating a certain number of sites *a priori* to be invariant, he could fit the data if he used up to four different Poisson parameters representing four classes of variability some of which he called 'hypervariable'. The lowest estimate of the number of invariant sites in cytochrome c is that of Holmquist *et al.* (1983) who found only three.

Holmquist *et al.* (1983) found that they could get a better fit of the data by a negative binomial density as had been shown by Uzzell & Corbin (1971) much earlier. The negative binomial distribution has two adjustable parameters and includes the Poisson distribution which has only one such parameter as a special case. As is commonly known, the more adjustable parameters one has, the easier it is to fit experimental data.

By plotting the number of invariant codons, according to their statistical method, against the number of species and extrapolating to the ordinate, Fitch & Markowitz (1970) concluded that, at a given time in a given mammalian cytochrome c, less than 10% of the codons are capable of accepting a mutation. In addition, they interpreted the data to show that, as mutations are fixed, the sites that can accept mutations change so that the examination of a wide range of species reveals a wide range of mutated sites. Fitch & Markowitz (1970) gave the name *covarions* to the concomitantly variable sites in a protein chain. Fitch (1971) reported that the covarions change with time and with the organism from which they are taken; for example, the covarions of fungi are largely independent of those in insects and fish.

One sees from Tables 6.3 and 6.4 that, with many more sequences available today, there are 13 invariant sites. The rest are variable to a degree depending on the specific codon and the distance to the edge of the sphere that encloses the functionally equivalent amino acids. There is indeed little restriction to selection in some sites; that is, they have a low information content, or in Fitch's terminology are 'hypervariable'. As we see in Table 6.7, one finds two sites that are completely variable and consequently have zero information content. Thus it appears, from the discussion in Chapter 6, that there is a spectrum of variability among protein sites even wider than that found by Fitch. Furthermore, current knowledge of protein folding and structure indicates that each site has an individual function of greater or less specificity measured by its information content. The protein structure is remarkably flexible for acceptable amino acid replacements.

Functionally acceptable mutations that are actually found in evolution are apparently almost always independent and the effect on the activity of the protein is uncoordinated with mutations at other sites (Lesk & Chothia, 1980, 1982; Altschuh *et al.*, 1986; Alber *et al.*, 1987, 1988; Nagai, Luisi & Shih, 1988). Kantrowitz & Lipscomb (1988) report the effect of site-directed mutations on intrachain interaction, but no case where a mutation at one interacting site lost activity that was replaced by a second mutation at the other interacting site. Reidhaar-Olson & Sauer (1988) feel that most substitutions found in evolution are allowed as single substitutions without linkage. There appears to be no structural reason why linkage may not exist between the functionally equivalent amino acids at sites that are located near each other in the folded condition (Creighton & Chothia, 1989). A specific case has been reported by Brantly, Courtney & Crystal (1988) in a1-antitrypsin where a salt bridge is formed by a positive charge from Lys^{290} and a negative charge from Glu^{342}. These two amino acids are one Hamming distance apart so that a mutation $G \rightarrow A$ in GAG of Glu^{342} changes the residue to Lys and destroys the $+, -$ configuration and the activity of the protein. The salt bridge may be reformed by an $A \rightarrow G$ mutation either at site 342 or at site 290. Such sites have also been found in the hydrophobic core positions of the N-terminal domain of phage λ-repressor by Lim & Sauer (1989) using the technique of cassette mutagenesis. This linkage does not necessarily apply to other parts of the protein. Klig, Oxender & Yanofsky (1988) have found that mutants produced by hydroxylamine in the *trp* repressor of *E. coli* are 'corrected' at a second site by a subsequent treatment with hydroxylamine. These second site revertants were found to be more active than the parental mutant repressors. In the course of evolution in the natural state, to avoid passing through a non-viable intermediate, these two mutation events must be simultaneous. The probability of independent events is the product of the probability of each separate event. Since the probability of a mutational event is a number very much smaller than unity, simultaneous events are of the second order, i.e. extremely rare, and may be neglected (Lesk & Chothia, 1980, 1982; Altschuh *et al.*, 1986; Nagai, Luisi & Shih, 1988; Lim & Sauer, 1989). Therefore, it appears that current knowledge of protein structure, obtained by site-directed mutation since the covarion hypothesis was proposed, is contrary to the belief that *certain sites* are *incapable* of mutation only at *certain times* because such mutations would then be lethal or *malefic*. According to Lim & Sauer (1989), sites that are interdependent are always interdependent. The covarion hypothesis also appears to be inconsistent with evidence that supports the neutral theory of the evolution of homologous proteins and may be an artifact of the method of analysis of Fitch & Markowitz (1970).

In order to form phylogenetic evolutionary trees, various authors have tried to fit Poisson and other probability distributions to the sites forming the sequence. This cannot be done without specifying that there are some invariant sites that are not those predicted by the fitting procedure. Monod (1972) based his philosophy on the incorrect belief that the spatial selection along the chain of the sequences of amino acids was due to chance only. The fact that there is no perceived relation between succeeding sites in a protein simply reflects the fact that there is no intersymbol influence. It is clear from the discussion in Chapter 9 that the selection of amino acids is not due entirely to chance. The functionally acceptable amino acids at all sites are determined by the folding requirements and the biochemical requirements of the protein.

We now know that the application of the Poisson or other probability distribution to predict the number of mutations in the sequence is inappropriate since that would assume, incorrectly, that the amino acids that are the replacements at each site are equally probable and that the probability for each kind of mutation event is the same. It appears from

Tables 6.3 and 6.4 that the variation in mutability is not distributed statistically along the protein sequence.

The prescription for predicting functionally equivalent amino acids (section 6.3) has a bearing on the neutral theory of molecular evolution and on the efforts to establish a measure of the ability of different amino acids to replace one another as the result of mutation. It is frequently mentioned in the literature that if amino acid replacements are neutral mutations between similar amino acids would be favored over mutations between those less similar. An objective method of selecting only those amino acids with relatively similar chemical properties that may be expected to be suitable replacements at the given site is provided by the prescription given in Chapter 6 for selecting functionally equivalent amino acids at a given site. As we have seen in Chapter 6, the number and character of the functionally equivalent amino acids is peculiar to the site in the protein sequence. It was formerly thought that this is a manifestation of selectionism (Clarke, 1970; Uzzell & Corbin, 1971). In the light of the knowledge available today, one may regard mutations to functionally equivalent amino acids to be neutral mutations. The existence of well-known polymorphism in many organisms reflects the fixation of neutral mutations.

The pioneering work in this field was very useful but as more is learned about protein structure, folding and function, other ideas have been proposed. An experimental method is now available, namely, site-directed mutagenesis, which enables one to put the sites that now appear to be invariant to the test. An example of this work is that of Cupples & Miller (1988) who tested four sites in *E. coli* β-galactosidase and found that the substitution of 12 different amino acids at sites His[464] and Met[3] did not affect activity. On the other hand substitution of 13 different amino acids at sites Glu[461] and Tyr[503] reduces activity 2 to 4 orders of magnitude. Thus sites Glu[461] and Tyr[503] are found to be invariant. I suggest that the program of testing for invariance should proceed by testing the activity of replacements by site-directed mutagenesis, selecting first the amino acids one Hamming distance from the putative invariants and nearest in BH distance as shown in Table 6.2. One would then proceed in sequence, selecting the amino acids next in distance until conviction or fatigue sets in.

12.2.2 *The molecular clock and the construction of phylogenetic trees for homologous proteins*

The evolution of organisms is described by phylogenetic trees. Progress along the branches of these trees is determined by the rate that mutations occur and are fixed. The construction of phylogenetic trees is greatly simplified by the notion introduced by Zuckerkandl & Pauling (1965) that mutations occur at a reasonably constant rate in sidereal time and not in generation time. The analysis of certain homologous proteins by Jukes & Holmquist (1972) supported this proposal. This idea is revolutionary and has created some anguish among classical Darwinian selectionists because biologists were prone to think in terms of generation time. Papers are published from time to time to show that the molecular evolutionary clock runs fast or slow in one organism or in one protein or another. However, the fact that the rate of the molecular evolutionary clock is even remotely related to sidereal time is evidence that mutations are the result of purely physical and chemical molecular rearrangements and Gamow quantum mechanical tunnelling in the nucleotides and are not caused by the biological processes of cell division. The fact that the molecular evolutionary clock needs some calibration should not obscure its usefulness. After all, the ^{14}C clock must be calibrated because radiation from the Sun has varied in past time. The dating of rocks by radioactive methods must take into account the different half-lives of the isotopes being used as well as many other factors. Accurate dates can be obtained only by experts familiar with all the systematic errors. Zuckerkandl (1987) and Jukes (1987b) have reviewed this topic and brought it up to date.

Contemporary DNA, RNA and protein homologous sequences are messages derived from the time when the protein sequence originated and became fixed. In the case of the more ancient proteins that time is near the origin of life, at least 3.8×10^9 years ago, which provides evidence, independent of the fossil record and morphology, of evolutionary relationships between the organisms to which they belong. When methods to find these sequences were developed, attempts were begun to find the history of the evolution of species incorporated in these sequences. Thus organisms that are believed to be closely related on the grounds of morphology are expected to have proteins such as myoglobin, α and β hemoglobin and cytochrome c with many identical amino acids. The few differences in these amino acids in organisms that appear to be morphologically closely related are expected to be one Hamming distance apart. Organisms with remote common ancestors would be expected to differ by many mutations and one would expect some at two and even three Hamming distances. This is exactly what was found by Jukes & Holmquist (1972) from their REH (random evolutionary hit) theory.

The numerous major extinction events in the Earth's history must have restricted population size for many species causing bottlenecks in evolution. The rate of amino acid substitution is inversely correlated with population size (Ohta, 1987). Viewed backward in time, a bottleneck appears either like a source or like a common ancestor. Such bottlenecks may be as important in evolution as natural selection. The evolutionary lines of descent are represented by phylogenetic trees. The first evolutionary phylogenetic trees were made with the assumption of maximum parsimony (Camin & Sokal, 1965). That is, it was assumed that the most likely phylogenetic trees were those made with the minimum number of mutations. With the use of that algorithm and only those sequences of cytochrome c known at the time, Fitch & Margoliash (1967) built a comprehensive phylogenetic tree that largely coincided with trees developed from morphology and the fossil record. Baba *et al.* (1981) have reported a more extensive phylogenetic tree based on the maximum parsimony analysis of 87 cytochrome c sequences. The construction of phylogenetic trees is mathematically more complicated than it seemed at first because the total number of possible trees from which one must select the correct tree is extremely large. I shall consider the question of the intractable nature of such trees in the next section.

12.2.3 The total number of possible phylogenetic trees and the intractable and chaotic nature of the construction of phylogenetic trees

An intractable problem is one in which the length of the algorithm for its solution increases faster than e^n where n is the number of elements in the problem. A well-known example is that of the traveling salesman (Johnson & Papadimitriou, 1985; Lawler *et al.*, 1985; Packel & Traub, 1987; see also section 2.4.5). Such problems are soluble in principle but not in practice since the time and computer memory required escalate rapidly. For example, if there are n species of animals then the total number of *possible* rooted phylogenetic trees T_r is as follows (Fitch & Margoliash, 1968; Margoliash & Fitch, 1971):

$$T_r = \prod_{i=1}^{n-1} (2i-1) \tag{12.1}$$

The number of *possible* unrooted trees T_u is (Edwards & Cavalli-Sforza, 1964):

$$T_u = \frac{(2n-5)!}{(n-3)!\,2^{n-3}} \tag{12.2}$$

It is easy to show (see Exercise 12.2) that both of these functions increase faster than e^n. Therefore, as in the traveling salesman problem in section 2.4.5, the problem of examining each possible tree is *intractable*. This intractability, as well as the inaccurate knowledge one has of the mutation probabilities, results in the essentially *chaotic* nature of these calculations (Ornstein, 1989). This has been reflected in the essentially *ad hoc* nature of these methods that have been proposed to construct phylogenetic trees.

In our attempts to construct reasonable phylogenetic trees in spite of the intractability, undecidability and chaotic nature of the problem, we are in good company. Other disciplines also have difficulty due to the intractability or chaotic nature of their problem. For a dynamic system to be *chaotic* means that small errors in the initial conditions are amplified exponentially so that the final state is independent or nearly so of the initial conditions. A system that is *undecidable* is such that its long term behavior is unpredictable even if there is total knowledge of the initial conditions; (see sections 2.4.2, 2.4.3; see also section 2.4.6; Moore, 1990; Bennett, 1990). For example, astronomers have recently found that there are limitations in their ability to calculate the previous history of the solar system and by the same token the future motion of the planets. Although Newton's laws of motion are essentially deterministic, errors in the initial conditions grow exponentially with time, and even if they are tiny, large perturbations result in time. For example, Lasker (1989) has calculated the positions of the planets backward 200×10^6 years. He found that the planets' motion is chaotic and the orbits of the inner planets, including that of the Earth, are lost within a few tens of millions of years. Essentially what this means is that the traditional methods of calculating the orbits do not yield a unique solution: The astronomer simply loses track of his planet. Thus the boast of Laplace and Lagrange about the determinism of the solar system cannot be carried out except over a fraction of the age of the solar system.

The fossil record gives biologists some rather accurate dates of some events very early in evolution. This gives considerable help in forming phylogenetic trees. The chaotic nature of the problem is to be regarded with humility and is a fundamental limitation on our knowledge. Each of the standard methods of tree construction often gives an incorrect tree (Tateno & Tajima, 1986). For example, the phylogenetic position of rattle snake cytochrome c does not agree with the fossil record (Meyer, Cusanovich & Kamen, 1986). We can never know which of several reasonably probable pathways was in fact taken in evolution (Holmquist & Jukes, 1981; Meyer *et al.*, 1986). This is in contrast with the implied assumption of the deterministic approach in the early work, for example in the maximum parsimony papers, that the true evolutionary sequence of events could be found.

12.2.4 The formation of phylogenetic evolutionary trees by the method of maximum parsimony

Since examining all possible phylogenetic trees is intractable, it appeared reasonable to a number of authors that the path composed of branches with the minimum mutations would be the one most travelled by in evolution (Camin & Sokal, 1965; Fitch & Margoliash, 1967; Dayhoff, 1978). For times that are short between nodes in the tree that is a plausible assumption. Holmquist (1979) has given a detailed analysis of the maximum parsimony method. He found that in cases where there are few replacements, the maximum parsimony method often produces the correct phylogenetic tree. Several similar trees satisfy the condition of maximum parsimony, so the method does not always generate a unique tree. As the number of replacements increases, the back reactions and multiple replacements become appreciable, so that the maximum parsimony algorithm is correct only half the time. This means, of course, that the result is indeterminate and random. Sourdis & Nei (1988) have examined the reliability of the maximum parsimony method in comparison with five distance matrix methods, namely, Farris's distance-matrix (Wagner) method (Farris, 1972), Li's (1981) transformed distance method, Tateno's modified Farris method, Faith's method (1985) and Saitou & Nei's (1987) neighbor-joining method. They find that the phylogenetic trees developed from the maximum parsimony method are not always unique and that the maximum parsimony method is more affected by multiple, backward and parallel mutations than some other methods. As Holmquist & Jukes (1981) and Felsenstein (1978) have pointed out, the putative precision of the maximum parsimony method is misleading due to its *ad hoc* nature. Although a great deal has been accomplished by the maximum parsimony method and various related methods that have appeared in the literature, the stage is now set for a method of phylogenetic construction that is based on more fundamental biological and mathematical principles. If one accepts the neutral mutation theory of the evolution of homologous proteins, it is then clear that evolution is a random walk or a Markov chain of events. This is the subject of the following sections.

12.2.5 *The formation of phylogenetic evolutionary trees by non-ergodic Markov chains*

A Markov process is determined by the states of the system together with the transition probabilities between the states. The Markov states are not the sites along the protein chain because there is no intersymbol influence between sites. In the case of iso-1-cytochrome c, the probability sample space consists of 9.737×10^{93} sequences in the high probability group (section 9.2.2). Each of these sequences is transformed to another by a mutation event. Clearly this is far too many events to handle. A different approach to the problem is needed that maps what one now believes to be the actual sequence of events in the evolution of proteins.

King & Jukes (1969) pointed out that the gene frequency of a neutral allele executes a random walk that eventually reaches a fixation or a loss to an inactive sequence. The random walk, described by Markov chains of events, is consistent with the evolutionary process as it is now believed to occur. The codons that represent the functionally acceptable amino acids at each site are the states in a Markov process state diagram. Other codons are absorbing states and transitions or transversions to them are irreversible so that the process cannot return to those codons that represent the functionally equivalent amino acids. Di Giulio & Caldaro (1987) (see corrections by Curnow 1988) called the average number of transitions before the Markov chain encounters a termination codon the 'stay time'. Since the termination codons are absorbing states this process is not ergodic. Amino acids that have a larger number of codons have a larger number of paths that lead back to the same residue or to a functionally equivalent one before they mutate to an absorbing state. (See the comments in section 1.3.4). One notices in Table 6.4 that Trp, even when it is an allowed residue, is infrequently reported in the cytochrome sequences. Codons that are one Hamming distance from a termination codon are used less frequently in general (Shaw, Bloom & Bowman, 1977; Fitch, 1980; Golding & Strobeck, 1982). Trp, the single codon of which is one Hamming distance from a termination codon, has the shortest stay time, as calculated by Di Giulio & Caldaro (1987). The single codon of Met is two Hamming distances from a termination codon and this may contribute to the fact that Met has a somewhat longer stay time. On the other hand Leu, Arg and Ser, with their six codons, have the longest stay times.

In the case of homologous proteins where one considers the codons of the amino acids to be the Markov states, the process is not ergodic. Both the termination codons and the codons of amino acids that lie outside of the functionally equivalent sphere are absorbing states. It would be informative to carry out the study of Di Giulio & Caldaro (1987) at specific sites and regard the codons of the amino acids that are not functionally equivalent as absorbing states as well as the termination codons. When one of these codons is encountered, the result is an inactive protein. Thus mutations inside the sphere are the domain of the neutral theory and mutations that carry the amino acid outside the sphere are the domain of the selectionist theory (Jukes & Kimura, 1984). This is consistent with the five principles deduced by Kimura & Ohta (1974). The notion of neutral mutations applies only to homologous proteins. The neutral theory and the selectionist theory address different problems. There is no conflict between them and much wasted effort has been expended in debates about the relative merits of selectionism and neutralism.

12.2.6 *Specific application of the Markov process to the molecular evolution of homologous proteins*

The probability of occurrence of a mutation depends on the physical and chemical molecular rearrangements and Gamow quantum mechanical tunnelling in the nucleotides and not on the history of the sequence. The probability of fixation, however, depends on the site affected, the functional equivalence of the new amino acid, on conditions of the environment and on biological factors that exist at the time. This corresponds to the analogous property of a first-order Markov chain in which the transition probability is independent of the immediately preceding states.

The Markov chain or random walk method of constructing phylogenetic trees carries over much of the paradigm of the REH (random evolutionary hit) theory of Jukes & Holmquist (1972) (Holmquist, Cantor & Jukes, 1972; Holmquist & Jukes, 1981). Suppose we consider a node in a phylogenetic tree: only one sequence of amino acids, at the separation, was the common ancestor of each branch of the tree. The Markov chain proceeded from that starting point. It was not ergodic or stationary because the termination codons and the codons for amino acids outside the functionally equivalent sphere at each site are absorbing states. The transition probability matrix is not regular so that there is no equilibrium probability vector. However, one can find the transition probability matrix as time progresses by raising the first matrix to the appropriate power. The matrix elements give the probability of back mutations and multiple hits naturally, without introducing *ad hoc* assumptions or so-called corrections (section 1.3.4). One can put as much detail in the transition matrix as one's knowledge of the problem permits. One may be guided by a rich field of material on the subject of Markov chains and random walks (Rudnick & Gaspari, 1987). While the shapes of the average random walk contours are fairly *regular*, sample individual random walks produced by computer show that special cases may be extremely *irregular*. Accordingly, the real evolutionary tree may be much more irregular and complicated than has been found by methods used previously. One should not be surprised to find more than one reasonably probable path in the evolutionary tree. The maximum parsimony tree will be included among the others and its probability given. Furthermore, in constructing phylogenetic trees one must take into account the fact that major extinctions play a major role as well as classical Darwinian survival of the fittest. Each major extinction reduces the branches of the phylogenetic tree to a considerably reduced number of survivors (section 8.1.9).

An ergodic Markov process may be formed by making Markov states of the redundant or silent nucleotides. These four states make a closed set. There are no absorbing states so, accordingly, the Markov process is ergodic. The Perron–Frobenius theorem shows that such Markov processes will approach an equilibrium so that all knowledge of the starting conditions is lost. As two evolving homologous proteins approach this equilibrium the species to which they belong cannot be located on a phylogenetic tree. The redundant nucleotides are expected to be selectively neutral (Jukes, 1965, 1980; Jukes & King, 1979; Nichols & Yanofsky, 1979). Lipman & Wilbur (1983, 1985) have reported statistical evidence of textual constraints on synonymous codon selection in eukaryotes. They suggest no mechanism for these constraints, if they exist, and I shall consider all redundant nucleotides to be selectively neutral. This may be simply a matter of interpretation. There may not have been enough time for the Markov process to have become stationary.

A stationary Markov process was assumed by Lanave *et al.* (1984), Preparata & Saccone (1987) and Saccone *et al.* (1988). It was applied to the redundant third position of the codons in five mitochondrial genes coding for proteins in mouse, rat, cow and human. Their method was satisfactory for mouse, rat, and cow, but not for human. The rates for transitions exceeded those for transversions by an order of magnitude. The average substitution rate was found to be $(13.6 \pm 7.1) \times 10^{-9}$ per site per year. They obtained a phylogenetic tree from which they derived the divergence time for mouse–cow to be $(47.2 \pm 5.6) \times 10^6$ years. This is consistent with the rat–cow divergence that was found to be $(50.7 \pm 6.6) \times 10^6$ years. The rat–mouse divergence was 35×10^6 years. Lanave, Preparata & Saccone (1985) analyzed the first and second codon positions in the same mitochondrial genes and found them to satisfy the conditions of stationarity. They found consistent divergence times for rat, mouse and cow and a rodent–primate divergence time of 75 million years.

Preparata & Saccone (1987) showed that, with the use of the Markov method of analysis, phylogenetic trees of the indicated accuracy as well as molecular evolutionary clocks can be found at least for the evolution of mammals. This method addresses the question of the rate of the molecular clock without introducing *ad hoc* assumptions. Sequences with a high information content must have few functionally equivalent amino acids at each site. The histones, which have shown very little evolutionary change, are an example that comes to mind. I expect, therefore, that the rate of evolution of homologous proteins is inversely related to the information content.

This discussion has been applied to the evolution of homologous protein sequences and confined to the non-ergodic case where amino acids outside the sphere of functionally equivalent ones are absorbing states. In the following section, I shall consider the ergodic case where there are no absorbing states. This is applicable to the emergence of *de novo* functions in proteins where the new sequence is determined by Darwinian selectionism.

12.3 The second meaning of the evolution of proteins: the emergence of *de novo* functions from duplicated DNA by ergodic Markov chains

The second meaning of protein evolution is the formation of proteins with *de novo* functions. The ergodic Markov process is applicable here because there are no absorbing states as there are in the first meaning (section 12.2). Let us apply the ergodic Markov process to the theory that *de novo* proteins arise from duplicated DNA. It is well known that DNA may be duplicated from time to time and there is evidence that a number of proteins owe their genesis to evolution from duplicated genes (Ohno, 1970; Barker, Ketcham & Dayhoff, 1978; Jukes, 1980; Doolittle, 1981, 1987a). Let us imagine the amino acids at successive sites of a duplicated protein sequence. In one of the pairs of sequences, the amino acids remain in the spheres of those that are functionally equivalent at each site; that sequence retains its original activity. The evolution of this sequence is governed by the non-ergodic Markov process discussed in section 12.2. The other becomes inactive by genetic drift and is a potential *de novo* daughter sequence: none of its codons, including the termination codons, form absorbing states. This process is ergodic because each Markov state can be reached from another. The inactive DNA sequence is not under selection pressure so that the Markov processes of random walk allow a progression of codons at each site starting from the sequence of the first protein.

The progress of the initial probability vector representing the amino acid occupation of each site of the protein sequence as it was, immediately following duplication, is followed by multiplying that vector by the 64×64 transition probability matrix representing the mutation probabilities. According to the Perron–Frobenius theorem the probability of occurrence of all codons, including the non-sense codons, will reach an equilibrium and equality. Therefore, one does achieve in time a finite probability of finding codons designating all amino acids in the sphere of functionally equivalent amino acids at each site in the *de novo* protein.

There are two classes of outcome of this process. First, it may be that no new functional protein is found. The DNA sequence becomes 'junk' DNA, which is known to accumulate in the genomes of higher organisms. Second, a new protein with a new structure and function may be read off at some time in this process if the appropriate amino acids actually appear. (Remember we can deal only with probabilities. Actual predictions are credible only if the probabilities are near unity.) This new protein is now subject to Darwinian selectionism and must improve the fitness of the organism in which this happens. Of course, the requirements of population evolution must also be met so that the new gene can be fixed.

Experimental evidence for the saltation from one active protein to another is given by Waneck, Stein & Flavell (1988). They report that the conversion of Asp^{295} to Val by site-directed mutagenesis converts Qa-2, a cell-surface glycoprotein, to an integral membrane protein. Takahashi *et al.* (1989) report that the substitution of Tyr for Val was sufficient to interchange the specificities of two sequences that differed in six of 15 residues. Further experimental evidence is reported by Cantatore *et al.* (1987), who have found that the CUN gene for Leu in the mitochondrial DNA of the sea urchin came from a duplication event of the UUR tRNA gene followed by a selection for the new function.

Maynard Smith (1970, 1972) proposed that, if there are N proteins that can be formed by a step of one Hamming distance from the codon at each site and a fraction f of these is functional, then in order to form a network, $fN > 1$. His estimate of N for a protein (like cytochrome c) of 100 amino acids is 1900. In practice the genetic code limits this to about 1000. Following that argument he concluded that $f > 1/1000$. In consequence, the density of functional proteins in protein space must be quite high.

The basic idea is correct, but, in view of what has been discussed in this book, it is clear that the mathematics is much more complicated. As in the case of any message (Shannon, 1948), if there are n sites in a protein sequence, a given sequence may be regarded as a point in an n-dimensional probability space. The ensemble of homologous proteins forms a volume in that probability space. Other homologous protein sequence families also form volumes in that probability space. Those regions that are close together constitute to which the network that Maynard Smith (1970, 1972) referred.

In the case of the 110 site iso-1-cytochrome c sequence the total possible number of points is $20^{110} = 1.3 \times 10^{143}$. We learned in Exercise 9 of Chapter 9 that the effective number of amino acids is 17.621 so that the effective number of points is $(17.621)^{110} = 1.15 \times 10^{137}$. Of these, 2.316×10^{93} are occupied by iso-1-cytochrome c, in the high probability set.

Some of these points have Hamming distances closer to *de novo* proteins of the same number of sites than others. Assume for simplicity that the volumes enclosing a homologous protein family are spheres. The mathematical problem, then, is to estimate the number of spheres that can be packed in contact with the sphere enclosing a homologous protein family. That number is called the *kissing number* (Conway & Sloane, 1988).

It often happens in mathematics that two problems that appear at first sight to be quite different are in fact related. Because of its geometric character the problem of designing error correcting codes is related to the sphere-packing problem (Hamming, 1986; Hill, 1986). Noise produces a small region of uncertainty in hyperspace around each message. The object of designing error correcting codes is to find those codes such that these regions of hyperspace do not intersect. If they intersect the decoder cannot distinguish the correct message (Shannon, 1948). In the case of functionally equivalent amino acids the sphere in the *n*-dimensional hyperspace (not the BH space of Chapter 6) constitutes a region in which the message is decoded correctly as a member of the homologous protein family at that site. The mathematical problem is to find the kissing number of the spheres that are packed next to that sphere.

The mathematics of hyperspaces of very high dimensions is extremely complicated. For example, it is a curious fact that the volume of spheres in very high dimensional spaces is almost exclusively near the surface (Hamming, 1986; Conway & Sloane, 1988). No general means of calculating the kissing numbers for the general case have been found. The kissing numbers for certain lattices are known for dimensions up to 24. The kissing number for a lattice in 24 dimensions is 196560. Some kissing numbers are known for specific dimensions above 24, for example, the kissing number for a lattice in 128 dimensions is 1260230400. From this one may estimate for a protein of 128 sites that, in general, f need be only $> 7.9 \times 10^{-10}$.

Maynard Smith's condition for the continuous network required for evolution to be possible is much less strict than appeared to him at the time because the density of functional proteins in Maynard Smith's protein space may be very much more sparse than he thought. But clearly such a network exists and the sparseness of it suggests reasons why many of the same protein families are found in widely different organisms. There is evidence that proteins with *de novo* functions are more likely to be generated from genetically modified proteins rather than from sequences formed at random (Doolittle, 1981, 1987a). Notice, however, that in almost all cases more than one site must be changed to achieve a new activity. Therefore, the molecular clock must be applied with care to the generation of *de novo* protein sequences.

The Markov processes of evolution of *de novo* proteins can be tested by writing the algorithm that describes the scenario in detail and running the program on a computer with sufficient memory. The calculations can be applied to the rather long list of proteins which, because of the degree of similarity in the sequences, are believed to be related but are not homologous (Doolittle, 1981, 1987a). Such calculations will test the belief, for example, that the α and β hemoglobin chains diverged by gene duplication and evolved independently, perhaps beginning at the end of the Ordovician period (Kimura & Ohta, 1974). The duplication scenario, as described above, is not subject to the indeterminacy restrictions in the discussion of origin of life scenarios of Chapter 10 as long as only a handful of amino acids is changed along the chain in one event. Thus the information content involved in the duplication scenario is sufficiently small to allow it to proceed by selected mutations to bridge the gap between two proteins of different specificities.

The principle conclusion reached in this chapter is that the best model for evolution is a Markov process, whether the evolution is that in a homologous protein family or the evolution of *de novo* functional proteins.

Exercises

1. Calculate the mutual information for a number of sites in iso-1-cytochrome c as a function of time in evolution using the methods in section 12.2 and the knowledge of iso-1-cytochrome c given in Table 6.4. This exercise requires computing facilities at least equivalent to a personal computer with sufficient memory for a matrix computation program.
2. Show that expressions (12.1) and (12.2) increase faster than e^n. Hint: replace the factorials with Stirling's approximation.
3. In the Ala–Gln groups in Table 6.4, Val is the amino acid closest to the center of the sphere. Outline the transversions and transitions that are at one Hamming distance. List the functionally equivalent and non-functionally equivalent amino acids that result.

Epilogue

A brief summary will here be sufficient to recall to the reader's mind the more salient points in this work. Many of the views which have been advanced are highly speculative, and some no doubt will prove erroneous; but I have in every case given the reasons which have led me to one view rather than to another.
Charles Darwin (1874), *The Descent of Man* Chapter XXI, General Summary and Conclusions. Chicago and New York: Rand, McNally & Co.

We may now evaluate the progress that has been made in this monograph toward laying down salient points in the application of information theory and coding theory in molecular biology. I have shown that trying to construct a theory by squeezing meaning out of words leads at best nowhere, and usually to error. Altogether too often one finds arguments in the literature that are essentially a play on the different meanings of words. The meaning of words is in a sense arbitrary and usually there is no exact correspondence between languages. In certain cases the meaning of a given word is drastically different in a second language. Accordingly, in science, words must be considered only names for mathematical concepts and must get their meaning from the mathematics. Therefore, we must construct the mathematical foundations of molecular biology without resorting to a contrived and anecdotal series of *ad hoc* speculations. The foundations of theoretical molecular biology must be built on mathematics and developed by proofs from a basic set of axioms. The axioms of information theory and coding theory on which this book is based are themselves not speculations. Only the appropriateness of their application may be questioned. In this respect we have an advantage over Charles Darwin. These principles are applied to the scenario or problem under consideration to provide a test of its coherence and whether it is tenable or not in view of experimental knowledge. In this way one may begin to do in theoretical molecular biology what has been done in theoretical physics. It is only by producing useful explanations of experimental data and by helping to plan additional work for the experimentalist that theoretical molecular biology will take its place with theoretical physics (see Fisher, 1930).

In addition to providing highly accurate explanations for many

331

phenomena, the most important successes of theoretical physics have been to unify several fields that seem at first glance to be unrelated. For example, the axioms of Newton's three laws of motion and his expression for gravitational attraction of two bodies showed that the falling of the apple on the Newtonian pate was a result of the same phenomenon that holds the planets in their orbits. This was an astonishing notion at the time and Newton was widely criticized for it. Maxwell's equations unify electricity, magnetism and optics under one set of axioms. Thermodynamics is a third example of axioms that have extensive application outside the problem from which they originated, namely the operation of steam engines. The effort to unify several fields and, perhaps, to find a universal general theory is the dominant goal, today, in theoretical physics. This trend, of course, is simply the modern equivalent of the unification of geometry by the ancient Greek mathematicians as represented in the axioms of Euclid.

The first step in constructing a set of axioms is to define the basic ideas. The axioms then establish the relation between these ideas. The subsidiary concepts and quantities are derived from these basic primitive ideas. Energy and angular momentum are not mentioned in Newton's three laws of motion but are derived as theorems from these basic primitive axioms. And so the strange behavior of the gyroscope, for example, can be explained quantitatively and in detail, without contrived and *ad hoc* assumptions. In contrast with the axiomatic method, special contrived explanations often fail even in the domain for which they are invented.

The formalism of the theory of probability is generated from the primitive concepts of set theory. The theorems that are derived from the axiom of probability theory lead, by means of further theorems, to the concepts of information theory and coding theory. The dominant concepts in information theory and coding theory are Shannon entropy (section 2.1.3) and Kolmogorov and Chaitin algorithmic entropy (section 2.4.3). The reader should now have a good understanding of these concepts and the distinctions between them and the Maxwell–Boltzmann–Gibbs entropy of statistical mechanics. Each of these concepts of entropy has applications in molecular biology peculiar to the problem being discussed.

One may now reconsider the statement made in the Prologue regarding the issue of the sufficiency of physics and chemistry to deal with the complexities of biology. 'Molecular biology is consistent with chemical and physical processes, but *biological principles are not derivable from physics and chemistry alone*' (Monod, 1972; Mayr, 1982, 1988). It was foreseen by Bohr (1933) long ago that this insufficiency is of a fundamental nature. 'The recognition of the essential importance of fundamentally atomistic features in the function of living organisms is by no means sufficient, however, for a comprehensive explanation of biological phenomena, before we can reach an understanding of life on the basis of physical experience'.

Indeterminacy exists not only in physics but, astonishingly enough, in mathematics. Gödel (1931) has shown that the axioms of number theory are insufficient to prove certain theorems that are known to be true. The same mathematical problem was approached by Turing (1937). A sequence of binary digits, or bit string, can be defined as non-random if it can be calculated by a computer program and random if it cannot. That is, a sequence with some elements of pattern can be compressed to a shorter sequence. A random one cannot. Chaitin found a form of Gödel's theorem in algorithmic information theory. A message containing n bits of information can decide between 2^n equally probable events. By the same token, it cannot decide between more than 2^n equally probable events. As Chaitin (1982) put it so picturesquely: 'I would like to say that if one has ten pounds of axioms and a twenty pound theorem, then that theorem cannot be derived from those axioms'. Because of its mathematical isomorphism with computer programs, Turing machines, and mathematical logic, the genetic logic system as a whole cannot function unless it has sufficient information to conduct metabolism and reproduction.

The recognition that the biological sciences deal with phenomena unknown in inanimate objects is as old as Aristotle. Through the last quarter of the nineteenth century and the first half of this century the vitalist philosophers Henri Bergson and Hans Driesch believed that the complexity of living matter required an *élan vital* that was peculiar to living matter (Pauly, 1987). The complexity of enzymes and other proteins led many scientists to believe that eventually these molecules would be shown to have features unique to living systems (Watson *et al.*, 1987). On the other hand, materialists such as Jacques Loeb (1916) regarded calls for special biological laws as smacking of vitalism and dualism. Both schools are guilty of the fallacy of the *false alternative*. The sequence hypothesis has been widely accepted, *de facto*, as a new axiom in molecular biology. Mayr (1982), a philosopher of biology, states that it is unique to biology for there is no trace of a sequence determining the structure of a chemical or of a code between such sequences in the physical and chemical world. The ability of a sequence to determine a chemical compound is measured by the non-material Shannon entropy, Kolmogorov–Chaitin algorithmic entropy or information content. Like all new scientific axioms the sequence hypothesis meets the criterion that it is *consistent with but independent of* the laws of physics and chemistry. Or put it another way, the sequence hypothesis is in addition to but does not transcend the laws of physics and chemistry. This is quite different from the doctrines of vitalism and dualism whose proponents believed that there are processes unique to living organisms which are *contrary* to the laws of physics and chemistry (Mayr, 1982).

Information theory shows that the quantitative measure of molecular structure that is directly functional in the genetic information system is an abstract and non-material quantity, namely, the mutual entropy, information content or complexity of the informational biomolecules. I have considerably improved and brought up to date in Chapter 6 the methods that I published in 1977 (Yockey, 1977b) showing how to calculate this quantity for a homologous protein family. This measure may be used to test the faith most molecular biologists have in molecular structure as the ultimate explanation of life.

Regarding the protobiont as a von Neumann machine (von Neumann, 1966) it must have between 10^5 and 10^6 bits in its genome in order to metabolize and replicate (section 9.1.1). If one takes the information content calculated for iso-1-cytochrome c, for example, this means that the protobiont genome must be able to specify between 267 and 2670 proteins. I have shown, in Chapters 9 and 10, that none of the origin of life scenarios suggested at present comes even close to these figures.

A great deal of effort has been expended in finding theories (i.e. algorithms) for the origin of life without success. When a great deal of work has been applied to a problem without finding a solution the reason may not be that we are not smart enough or that we have not worked hard enough. The reason is sometimes that there is no structure or pattern that can be put into the terms of an algorithm of finite complexity. The reader who has solved Socrates' problem (discussed in Plato's dialogue *Meno*) of doubling the square will find his efforts to double the cube futile. This does not mean that cubes twice the size of any given cube do not exist. It means that the exact solution to the problem is undecidable; it is beyond human reasoning. In view of the isomorphisms discussed in section 2.4.6 and in sections 5.1.1 and 5.1.2 we may apply knowledge or theorems from one system to another isomorphic with it. Accordingly, the indeterminacy theorems of Gödel (1931), Turing (1937) and Chaitin (1987b) must apply. Gödel (1931) proved that there are theorems in number theory that are true but cannot be proved and propositions that are undecidable from the axioms of number theory. The isomorphism of the genetic logic system with the axioms of number theory shows that it is undecidable from the laws of physics and chemistry that life exists or that it does not exist. Contrary to the opinion expressed by Wald (1954), Simpson (1964) and Eigen (1971) it is not deterministic but undecidable, by the same token, whether life would almost inevitably arise spontaneously from non-living matter on young planets 'sufficiently similar' to Earth elsewhere in the universe. Of course the term 'sufficiently similar' is an escape clause since the conditions for the origin of life are not known. We must conclude that Bohr (1933) was right and that we must accept the existence of life as an axiom. The reason that there are principles of biology that cannot be deduced from the laws of physics and chemistry lies not in some esoteric philosophy but simply in the mathematical fact that the genetic information content of the genome for constructing even the simplest organisms is much larger than the information content of these laws. Chaitin (1985, 1987a) has examined the complexity of the laws of physics by actually programming them. He finds the complexity amazingly small.

It is universally the case that textbooks written for college under-graduates present the primeval soup paradigm as an established fact. In the discussion of chemical evolution in Chapter 8, I have emphasized that in science one must follow the results of experiments and mathematics and not one's faith, religion, philosophy or ideology. The primeval soup is unobservable since, by the paradigm, it was destroyed by the organisms from which it presumably emerged. It is most unsatisfactory in science to explain what is observable by what cannot be observed. Since creative skepticism and not faith is the cardinal virtue in science one would expect that proponents of the primeval soup paradigm would be actively searching for direct geological evidence of such a condition of the early ocean. The power of ideology to interpose a fact-proof screen (Hoffer, 1951) is so great that this has not been done (perhaps for fear that its failure may be exposed). For example, Ponnamperuma (1982), a proponent of the primeval soup paradigm, cites a paper that shows that the $\delta^{13}C$ in carbonaceous material found in meteorites is very different from that in very old terrestrial sediments. He fails to draw the conclusion that the extremely small amount of carbon in meteorites and from Moon rocks has nothing to do with the origin of life on Earth. McKay (1986) regards the credibility of the primeval soup paradigm as too well established to discuss, yet he cites a paper that contains data from the isotopic fractionation of carbon in the kerogen of early sediments (Schidlowski, Hayes & Kaplan, 1983). He fails to draw the obvious conclusion that the absence of abiotic carbon in the kerogen in these early sediments means that the primeval soup never existed. Furthermore, as I showed in Chapters 9 and 10, the paradigm does not perform the task its advocates claim. Although at the beginning the paradigm was worth consideration, now the entire effort in the primeval soup paradigm is self-deception based on the ideology of its champions.

The history of the 'life on Mars' effort which played a primary role in the Mariner and Viking space craft missions shows an intemperate *wish to believe* by leading scientists who saw in their data what they wished to see (Horowitz, 1986, 1990). The usual creative skepticism of scientists did not play a role until the enormous weight of evidence showed that belief that there is or was life on Mars is based on faith ('Now faith is the substance of things hoped for, the evidence of things unseen'; Hebrews 11:1). This has not yet manifested itself with regard to the belief in chemical evolution and the primeval soup paradigm.

The history of science shows that a paradigm, once it has achieved the status of acceptance (and is incorporated in textbooks) and regardless of its failures, is declared invalid only when a new paradigm is available to replace it. Nevertheless, in order to make progress in science, it is necessary to clear the decks, so to speak, of failed paradigms. This must be done even if this leaves the decks entirely clear and no paradigms survive. It is a characteristic of the true believer in religion, philosophy and ideology that he must have a set of beliefs, come what may (Hoffer, 1951). Belief in a primeval soup on the grounds that no other paradigm is available is an example of the logical *fallacy of the false alternative*. In science it is a virtue to acknowledge ignorance. This has been universally the case in the history of science as Kuhn (1970) has discussed in detail. There is no reason that this should be different in the research on the origin of life. The best advice that one could given to the alchemist would have been to study nuclear physics and astrophysics, although that would not have been helpful at the time. We do not see the origin of life clearly, but through a glass darkly. Perhaps the best advice to those who are interested in the origin of life would be to study biology. After dispensing with failed paradigms in Chapters 8, 9 and 10, I have shown that the problem to be solved in the origin of life and in evolution is the means by which complexity (as defined in this book) was generated. Half the solution to the problem is to define the question clearly.

A distinguished group of molecular biologists (Jukes & Bhushan, 1986; Reeck *et al.*, 1987; Lewin, 1987) has called attention to sloppy terminology in the misuse of 'homology' and 'similarity'. Nevertheless, editors still permit authors to qualify 'homology' and to confuse that word with similarity. *Mutual entropy* is the correct and robust concept and measure of similarity so that the sooner *per cent identity* disappears from usage the better. Mutual entropy is a mathematical idea that reflects the intuitive feeling that there is a quantity which we may call information content in homologous protein sequences. Clearly, the shortest message which describes at least one member of the family of sequences is what one would properly call the information content.

There is a great deal of discussion in the literature of conservative or non-conservative substitutions of amino acids in proteins. Often the judgements expressed are personal and non-quantitative. Accordingly, it is useful to replace these judgements with quantitative and objective procedures, for example, those based on the eigenvectors for each amino acid found by Borstnik & Hofacker (1985). Although the work of Borstnik & Hofacker (1985) is certainly not the last word, nevertheless, the prescription discussed in Chapter 6 is almost always consistent with experimental data. There are only two cases in which the prescription predicts a functionally equivalent amino acid that is not confirmed. I prefer not to tinker with the prescription since making it more complicated may reduce its usefulness. Comparison of the family of functionally equivalent amino acids at various sites in one protein as well as in different proteins may lead to determining a relationship or common function of these sites.

The probability of the occurrence of amino acids has been set proportional to the number of codons. This *Ansatz* may play a role similar to the chemist's and physicist's perfect gas. A perfect gas is satisfactory for many purposes but the engineer who is designing a steam turbine must have much better data especially near condensation of the working fluid. By the same token, the reader who has a special interest is given the equations by which he may calculate the information content using data which he regards as appropriate. Indeed, my choice of cytochrome c itself serves as a detailed example to make it clear how to apply the equations to calculate the information content of other homologous protein families.

In following the mathematics, we have found a number of theorems to which we would not be led by intuition and which are in fact contrary to intuition. It is a universal usage in the literature of molecular biology to base considerations on a calculation of the number of *possible* events without considering the fact that, except in certain games of chance, the probabilities of events are not equal. It is astonishing, at least from the present perspective, that there has been no thought of the effect of unequal probabilities especially by the authors who presume to apply information theory in molecular biology. Upon reading a few pages beyond the beginning of Shannon's paper, one finds the solution to this issue and that entropy solves the problem naturally, without being introduced in a contrived or *ad hoc* way (Theorem 2.2, the Shannon–McMillan theorem). One finds that often most of the *possible* events are too unlikely to merit attention. Indeed, as I mentioned, the appearance of some of the possible events, such as long sequence of heads in coin flipping, would be regarded by most people as evidence of fraud. We need concern ourselves only with those events in the high probability set.

It is frequently pointed out that the character of the genetic code was obtained by experiment, not by theory, in spite of considerable attention by George Gamow and other theoreticians of similar calibre. Although philosophers of science regard prediction as an important support of theory, nevertheless, they also regard the explanation of a phenomenon after it has been discovered to be significant, provided that the theory is not contrived or *ad hoc*. The reason that the character of the genetic code was not predicted is that inapplicable theories were used. As I discussed in Chapter 7, the assignments in the genetic code are primarily determined by coding theory. The evolution of the genetic code, as discussed in Chapter 7, led us to a description of the properties of the standard genetic code and the various non-standard genetic codes that has neither contrived or *ad hoc* assumptions nor anecdotal explanations. The first genetic codes could not have been 'error prone' as many authors have presumed. They must have been nearly as accurate as those of today, otherwise even short proteins could not have been transmitted in sufficient numbers. Thus it appears that the genetic code evolved to its present form guided by the principles of coding theory. The argument led to support of the endosymbiotic conjecture in a natural fashion and without being so contrived. The theory shows that the number of triplet codes that merit consideration is limited to the number of codes in the last two steps in the evolution of the doublet codes. The modern triplet genetic codes emerged from the bottleneck between the first extension doublet codes and the second extension triplet codes. It was by this means that the several modern triplet codes evolved without the necessity of trying the vast number of possible triplet codes.

Thus, as I have shown in Chapter 7, a theoretical explanation of the characteristics of the genetic code is provided.

Some authors regard the genetic codes in the mitochondria and other non-standard genetic codes as trivial deviations from the standard genetic code that they consider to be universal. I believe that Nature is giving us a clue to the evolution of the standard genetic code by the existence of the mitochondria genetic codes and other features of mitochondria. Furthermore, the eight cylinder codons remain as vestiges of their ancestry. Had the astronomers of the nineteenth century ignored the advance of the perihelion of Mercury by the tiny amount of 43 seconds of arc per century they would have ignored one of the clues that led to the general theory of relativity.

The fundamental axiom in molecular biology, which justifies the application of information and coding theory, is the *sequence hypothesis*. Although the sequence hypothesis is peculiar to living systems, I have given considerable attention to showing that the Central Dogma is a theorem from coding theory and is not peculiar to biology (Yockey, 1974, 1978). The fact that there must be a code between two sequences with different alphabets, if information is to be transmitted between them, follows from that axiom. There is no *élan vital* in biology so the genetic logic system must obey the same fundamental mathematical and physical laws as other logic operations. To believe otherwise is vitalism. The irreversibility of the genetic logic operation is unavoidable for two reasons. First, it is a consequence of thermodynamic irreversibility that occurs when the transition function does not have a single value. To allow reversible operation in the several-to-one mapping case the computer must record this information in its memory. The genetic logic system has no such memory. Second, irreversibility of information flow in the genetic logic system is the result of the several-to-one mapping of error detecting and error correcting codes. The genetic code has this property, which results in the source alphabet having a larger entropy than the receiving alphabet. No code exists such that messages can be sent from the receiver alphabet B to the source in alphabet A if the alphabet A has a larger entropy. Information cannot be transmitted from protein to DNA or mRNA for this reason. On the other hand, the information capacity of DNA may be used more fully by recording genetic messages in two or even three reading frames. Thus the phenomenon of the overlapping genes is related mathematically to the Central Dogma and the redundance of the genetic code.

One needs to be careful that one doesn't become beholden to terminology. There is evidence that shows that the stop codons may be assigned specificity by evolution of tRNA and its aminoacylation. They are not different *per se* from other codons. Thus the terminology 'suppression' seems not to be appropriate and may prejudice egregiously the study of this phenomenon. This is an example of the tendency to squeeze meaning from words themselves and not to stick to the mathematics. The term 'template' is often used to describe the relation between DNA, tRNA, mRNA and protein whereas the term 'genetic message' is the appropriate one. I have shown in Chapter 9 that there are 2.316×10^{93} different sequences for iso-1-cytochrome c. It is hard to believe that there could be that number of templates so the template terminology does not seem appropriate.

It has often been noted that the less a phenomenon is understood the more theories there are to explain it. Such is the case in senescence. Among the theories that presume to explain senescence are the error theories of aging. Unfortunately, the mathematical background of these theories is extremely unsatisfactory as I have pointed out previously (Yockey, 1974) and also here in Chapter 11. Regardless of the role that protein error may yet prove to play in aging it is important and useful at least to put the theory on a sound mathematical basis. I have made a beginning in previous papers and brought this up to date in Chapter 11 and in other parts of this book. In the case of the feedback theory of Orgel (1963, 1970) it is easy to speak lightly of feedback without calling into play the very large literature in this subject to ascertain whether feedback is playing a role in protein error correction. Feedback may be stable, unstable or oscillatory according to conditions of the system. I suggest that those who favor this theory study the subject with the mathematics of feedback theory in mind. An important puzzle in aging research is the so-called 'immortal' tumor cells. Error should be as important in their propagation as in any other cell. The solution of this puzzle should teach us much about aging in particular and cell reproduction in general. There is much more to be done to put this matter on a sound mathematical basis and this is beyond the scope of this book.

Much confusion and useless debate has occurred over the double meaning of protein evolution. Evolution proceeds in homologous protein families where the sequence of amino acids changes but the function does not. Since the late nineteenth century teleology has disappeared from science and appears only in religion; nevertheless, both Eigen & Schuster (1978a) and Dawkins (1986) regard proteins as having a 'master' or 'target' sequence that has resulted from fine tuning throughout the millions of years of evolution. On the other hand it is often remarked how flexible and tolerant to substitutions of amino acids proteins are. One concludes from the discussion in Chapters 6 and 9 that there is an enormous number of functionally equivalent master or target sequences, except for the few proteins, such as the histones, in which almost all sites are invariant. Mutations that exchange functionally equivalent amino acids among members of such homologous families are neutral. Mutations to non-functionally-equivalent amino acids are, of course, lethal or malefic. It is therefore useless to argue whether most mutations are neutral rather than lethal or malefic without making this distinction. There appear to be sites that have comparable functions in several unrelated proteins. These sites may be identified by the abstract sphere in Borstnik–Hofacker (BH) space that encloses the same amino acids.

It is well known that DNA may be duplicated from time to time. There is evidence that a number of proteins evolved their specificity from duplicated genes. Both functional and non-functional mutations may continue to occur. The duplicated sequences may continue to perform their function but in the case of a mutation to a non-functional amino acid the sequence loses its specificity. The emergence of *de novo* functions must occur in those sequences of DNA that have lost function. Except in the case of an immediate back mutation the sequence may become 'junk DNA' or it may acquire a new function. If the new functional protein is fixed the organism advances a step on the phylogenetic tree.

I hope that one accomplishment of this monograph has been to shoot down that *canard ancien*, now more than 100 years old, which states that evolution is incompatible with the second law of thermodynamics because evolution creates 'order' while the second law demands that 'disorder' increases with time in isolated systems. I have shown that this belief is merely a play on words. The correct explanation shows that an increase in the Kolmogorov–Chaitin algorithmic entropy is *required* for evolution to proceed.

The mathematical methods that have been used to calculate phylogenetic trees and thereby to acquire a picture of evolution have employed a number of *ad hoc* procedures. These procedures tend to bias the purported solution by putting into the problem what the investigator wishes to find. The maximum parsimony method may appear to be a reasonable way to construct a phylogenetic tree. Unfortunately, most authors who use it do not establish the limits of validity and it has often been applied, well outside its area of applicability, where back mutations are important. The method of using Markov chains to study evolution is based on first principles. Incomplete knowledge requires approximations to be made at the beginning of the calculation. Their effects are followed through. On the other hand, the effect of back and multiple mutations appears in the calculation in a natural fashion so that no 'corrections' are necessary. A few papers are beginning to appear in which the method of Markov chains has been used on homologous protein families to estimate the time of divergence of several related species such as rat and mouse. The effect of major extinctions may prove to have an important influence on evolution in addition to Darwinian selectionism.

The implied determinism in the construction of phylogenetic trees needs to be questioned and evaluated. The Perron–Frobenius theorem shows that Markov processes approach an equilibrium that is independent of the starting conditions. Thus the information in the protein sequences of modern organisms has some indeterminacy with regard to the evolutionary history. These sequences are not letters of gold carved in tablets of stone.

Information theory and coding theory have demonstrated the unity of molecular biology in several cases where problems appear to be unrelated. The discussions of particular problems, especially those of homologous protein families in Chapter 6, are intended to be regarded as examples of how information theory and coding theory can be applied to problems in molecular biology. In a number of cases, for example, aging and the theory of evolution, I have presented the basic approach with the intention of indicating the direction that the application of the mathematical methods presented in this book may take. This incompleteness must be understood from the fact that this book is a monograph, not a treatise. It is my hope that some readers will be inspired to use information theory and coding theory to pursue solutions of problems that I have only mentioned and, in addition, problems that have not been discussed.

References

Abelson, P. H. (1966). Chemical events on the primitive Earth. *Proceedings of the National Academy of Sciences USA* **55**, 1365–72.

Adelman, J. P., Bond, C. T., Douglas, J. & Herbert, E. (1987). Two mammalian genes transcribed from opposite strands of the same DNA. *Sciences* **235**, 1514–17.

Alber, T., Bell, J. A., Dao-pin, S., Nicholson, H., Wozniak, J. A., Cook, S. & Matthews, B. W. (1988). Replacements of Pro[86] in T4 lysozyme extend an α-helix but do not alter protein stability. *Science* **239**, 631–5.

Alber, T., Dao-pin, S., Wilson, K., Wozniak, J. A., Cook, S. & Matthews, B. W. (1987). Contributions of hydrogen bonds of Thr[157] to the thermodynamic stability of phage T4 lysozyme. *Nature* **330**, 41–6.

Alff-Steinberger, C. (1969). The genetic code and error transmission. *Proceedings of the National Academy of Sciences USA* **64**, 584–91.

Almasy, R. J. & Dickerson, R. E. (1978). *Pseudomonas* cytochrome c_{551} at 2.0 Å resolution: Enlargement of the cytochrome c family. *Proceedings of the National Academy of Sciences USA* **75**, 2674–8.

Alpher, R. & Herman, R. (1948). Evolution of the universe. *Nature* **162**, 774–5.

Alpher, R. & Herman, R. (1950). Theory of the origin and relative abundance distribution of the elements. *Reviews of Modern Physics* **22**, 153–212.

Alpher, R. & Herman, R. (1988). Reflections on early work on 'Big Bang' cosmology. *Physics Today* **41**, 24–34.

Alpher, R., Bethe, H. & Gamow, G. (1948). The origin of the chemical elements. *Physical Review* **73**, 803–4.

Altschuh, D., Lesk, D., Bloomer, A. C. & Klug, A. (1986). Correlation of co-ordinated amino acid substitutions with functions in viruses related to tobacco mosaic virus. *Journal of Molecular Biology* **136**, 225–70.

Alvarez, L. W. (1987). Mass extinctions caused by bolide impacts. *Physics Today* **40**, 24–33.

Alvarez, L. W., Alvarez, W., Asaro, F. & Michel, H. V. (1980). Extraterrestrial cause for the Cretaceous–Tertiary extinction. *Science* **208**, 1095–106.

Anders, R. (1989). Pre-biotic organic matter from comets and asteroids. *Nature* **342**, 255–7.

Anderson, P. W. (1983). Suggested model for prebiotic evolution: The use of chaos. *Proceedings of the National Academy of Sciences USA* **80**, 3386–90.

Anderson, S., de Bruijn, M. H. L., Coulson, A. R., Eperon, I. C., Sanger, F. & Young, I. G. (1982). Complete sequence of bovine mitochondrial DNA. *Journal of Molecular Biology* **156**, 683–717.

Anderson, S., Bankier, A. T., Barrell, B. G., de Bruijn, M. H. L., Coulson, A. R., Drouin, J., Eperon, I. C., Nierlich, D. P., Roe, B. A., Sanger, F., Schreier, P. H., Smith, A. J. H., Staden, R. & Young, I. G. (1981). Sequence and organization of the human mitochondrial genome. *Nature* **290**, 457–65.

Andini, S., Benedetti, E., Ferrara, L., Paolillo, L. & Temussi, P. A. (1975). NMR studies on prebiotic polypeptides. *Origins of Life* **6**, 147–53.

Anfinsen, C. B. (1973). Principles that govern the folding of protein chains. *Science* **181**, 223–30.

Apostolakis, G. (1990). The concept of probability in safety assessments of technological systems. *Science* **250**, 1359–64.

Argyle, E. (1977). Chance and the origin of life. *Origins of Life* **8**, 287–98.

Arp, H. C., Burbidge, G., Hoyle, F., Narlikar, J. V. & Wickramasinghe, N. C. (1990). The extragalactic universe: an alternative view. *Nature* **346**, 807–12.

Arthur, M. A., Dean, W. E. & Pratt, L. M. (1988). Geochemical and climatic effects of increased marine organic carbon burial at the Cenomanian/Turonian boundary. *Nature* **335**, 714–17.

Atkins, J. F. & Ryce, S. (1974). UGA and non-triplet suppressor reading of the genetic code. *Nature* **249**, 527–30.

Atlan, H. (1972). *L'organisation Biologique et la Theorie de l'Information.* Paris: Hermann.

Audouze, J. & Israël, G. (1985). *Cambridge Atlas of Astronomy.* Cambridge; New York: Cambridge University Press.

Baba, M. L., Darga, L. L., Goodman, M. & Czelusniak, J. (1981). Evolution of cytochrome c investigated by the maximum parsimony method. *Journal of Molecular Evolution* **17**, 197–213.

Bahcall, J. N. (1978). Solar neutrino experiments. *Reviews of Modern Physics* **50**, 881–903.

Bahcall, J. N. & Glashow, S. L. (1987). Upper limit on the mass of the electron neutrino. *Nature* **326**, 476–7.

Baltimore, D. (1970). Viral RNA-dependent DNA polymerase. *Nature* **226**, 1209–11.

Banin, A. & Navrot, J. (1975). Origin of life: Clues from relations between chemical compositions of living organisms and natural environments. *Science* **189**, 550–1.

Bar-Nun, A. & Shaviv, A. (1975). Dynamics of the chemical evolution of Earth's primitive atmosphere. *Icarus* **24**, 197–210.

Bar-Nun, A., Bar-Nun, N., Bauer, S. H. & Sagan, C. (1970). Shock synthesis of amino acids in simulated primitive environments. *Science* **168**, 470–3.

Bar-Nun, A., Bar-Nun, N., Bauer, S. H. & Sagan, C. (1971). Shock synthesis of amino acids in simulated primitive environments. In *Chemical Evolution and the Origin of Life*, eds R. Buvet & C. Ponnamperuma. Amsterdam; London: North-Holland.

Barker, W. C., Ketcham, L. K. & Dayhoff, M. (1978). Duplications in protein sequences. In *Atlas of Protein Sequence and Structure*, Vol. 5, Supplement 3, pp. 359–62. Washington DC: National Biomedical Research Foundation.

Barrell, B. G., Air, G. M. & Hutchinson, C. A. III (1976). Overlapping genes in bacteriophage ΦX174. *Nature* **264**, 34–41.

Barrell, B. G., Bankier, A. T. & Drouin, J. (1979). A different genetic code in human mitochondria. *Nature* **282**, 189–94.

Barrell, B. G., Anderson, S., Bankier, A. T., de Bruijn, M. H. L., Chen, E., Coulson, A. R., Drouin, J., Eperon, I. C., Nierlich, D. P., Roe, B. A., Sanger, F., Schreirer, P. H., Smith, A. J. H., Staden, R. & Young, I. G. (1980). Different pattern of codon recognition by mammalian mitochondrial tRNAs. *Proceedings of the National Academy of Sciences USA* **77**, 3164–6.

Barrow, J. D. & Silk, J. (1983). *The Left Hand of Creation*. New York: Basic Books.

Bartel, N., Rogers, A. E. E., Shapiro, I. I., Gorenstein, M. V., Gwinn, C. R., Marcaide, J. M. & Weiler, K. W. (1985). Hubble's constant determined using very-long baseline interferometry of a supernova. *Nature* **318**, 25–30.

Bartlett, M. S. (1945). Negative probability. *Proceedings of the Cambridge Philosophical Society* **41**, 71–73.

Batchinsky, A. G. & Ratner, V. A. (1976). Noise immunity of the genetic code. *Biomedical Zeitschift* **18**, 53–67.

Bayes, T. (1763). An essay towards solving a problem in the doctrine of chances. *Philosophical Transactions of the Royal Society* **53**, 370–418. Reprinted with introduction, G. A. Barnard (1958), *Biometrika* **45**, 295–315.

Beatty, J. K., O'Leary, B. & Chaikin, A., eds (1981). *The New Solar System*. Cambridge; London; New York: Cambridge University Press.

Bell, G. I. (1967). Cross sections for nucleosynthesis in stars and bombs. *Reviews of Modern Physics* **39**, 59–68.

Bellman, R. (1970). *Introduction to Matrix Analysis*, 2nd edn. New York, London: McGraw-Hill.

Benne, R. (1989). RNA editing in trypanosome mitochondria. *Biochemical Biophysica Acta* **1007**, 131–9.

Bennett, C. H. (1973). Logical reversibility of computation. *IBM Journal of Research and Development* **17**, 525–32.

Bennett, C. H. (1988). Notes on the history of reversible computation. *IBM Journal of Research and Development* **32**, 16–23.

Bennett, C. H. (1990). Undecidable dynamics. *Nature* **346**, 606–7.

Bennett, C. H. & Landauer, R. (1985). The fundamental physical limits of computation. *Scientific American* **253**, 48–56.

Berger, J. (1977). A comment on information theory and Central Dogma of molecular biology. *Journal of Theoretical Biology* **65**, 393–5.

Bergson, H. (1907). *Creative Evolution*. Translated from the French by Arthur Mitchell with a foreword by Irwin Edman (1944). New York: Random House Modern Library.

Bernal, J. D. (1951). *The Physical Basis of Life*. London: Routledge & Paul.

Bernal, J. D. (1960). *Nature* **186**, 694–5. (A reply to Hull, D. E. (1960), Thermodynamics and kinetics of spontaneous generation, *Nature* **186**, 693–4.)

Bernal, J. D. (1967). *The Origin of Life*. London: Weidenfeld & Nicolson.

Berry, M. J., Banu, L. & Larsen, P. R. (1991). Type I iodothyronine deiodinase is a selenocysteine-containing enzyme. *Nature* **349**, 438–40.

Bibb, M. J., van Etten, R. A., Wright, C. T., Walberg, M. W. & Clayton, D. A. (1981). Sequence and gene organization of mouse mitochondrian DNA. *Cell* **26**, 167–80.

Billingsley, P. (1965). *Ergodic Theory and Information* (see Theorem 15.1 in Chapter 5). New York; London; Sydney: John Wiley & Sons.

Blackburn, E. H. (1990). Copy editing – what's the U's. *Nature* **346**, 609–10.

Blahut, R. E. (1987). *Principles and Practices of Information Theory.* Reading, MA: Addison-Wesley.

Bludman, S. A. (1984). Thermodynamics and the end of a closed universe. *Nature* **308**, 319–22.

Blum, B., Bakalara, N. & Simpson, L. (1990). A model for RNA editing in kinetoplastid mitochondria: 'Guide' RNA molecules transcribed from maxicircle DNA provide the edited information. *Cell* **60**, 189–98.

Bohr, N. (1933). Light and life. *Nature* **131**, 421–3; 457–9.

Boltzmann, L. (1872). Weitere Studien über das Wärmegleichgewicht unter Gasmolekulen. *Königliche Academie der Wisschenschaft (Wien), Sitzungsberichte II, Abteilung* **66**, 275.

Boltzmann, L. (1877a). Über die Beziehung eines allgemeine Mechanischen Satzes zum zweiten Hauptsatzes der Wärmetheorie. *Königliche Academie der Wisschenschaft (Wien), Sitzungsberichte Mathematik-Naturwissenschaften* **75**, 67–73.

Boltzmann, L. (1877b). Über die Beziehung dem zweiten Hauptsatze der Mechanischen Wärmetheorie und der Wahrscheinlichkeitsrechnung, respective den Sätzen über das Wärmegleichgewicht. *Königliche Academie der Wisschenschaft (Wien), Sitzungsberichte II, Abteilung* **76**, 373–95.

Bondi, H. & Gold, T. (1948). The steady state theory of the expanding universe. *Monthly Notices of the Royal Astronomical Society* **108**, 252–70.

Bonitz, S., Berlani, R., Coruzzi, G., Li, M., Macino, G., Nobrega, F. G., Nobrega, M. P., Thalenfeld, B. E. & Tzagoloff, A. (1980). Codon recognition rules in yeast mitochodria. *Proceedings of the National Academy of Sciences USA* **77**, 3167–70.

Borstnik, B. & Hofacker, G. L. (1985). Functional aspects of the neutral patterns in protein evolution. In *Structure & Motion, Nucleic Acids & Proteins*, eds E. Clementi, G. Corongiu, M. H. Sarma & R. H. Sarma. New York; Guilderland: Adenine Press.

Borstnik, B., Pumpernik, D. & Hofacker, G. L. (1987). Point mutations as an optimal search process in biological evolution. *Journal of Theoretical Biology* **125**, 249–68.

Bové, J. M. (1984). Wall-less prokaryotes of plants. *Annual Reviews of Phytopathology* **22**, 361–96.

Bowers, R. L. & Deeming, T. (1984). *Astrophysics* I & *Astrophysics* II. Boston; Portola Valley, CA: Jones & Bartlett.

Bowie, J. U., Reidhaar-Olson, J. F., Lim, W. A. & Sauer, R. T. (1990). Deciphering the message in protein sequences: Tolerance to amino acid substitutions. *Science* **247**, 1306–10.

Brantly, M., Courtney, M. & Crystal, R. G. (1988). Repair of the secretion defect in the Z form of α 1-antitrypsin by addition of a second mutation. *Science* **242**, 1700–1.

Brans, C. & Dicke, R. H. (1961). Mach's principle and a relativistic theory of gravitation. *Physical Review* **124**, 925–35.

Brenner, S. A., Ellington, A. & Tauer, A. (1989). Modern metabolism as a palimpsest of the RNA world. *Proceedings of the National Academy of Sciences USA* **86**, 7054–8.

Bresch, C., Niesert, U. & Harnasch, D. (1980). Hypercycles, parasites and packages. *Journal of Theoretical Biology* **85**, 399–405.

Brillouin, L. (1953). The negentropy principle of information. *Journal of Applied Physics* **24**, 1153.

Brillouin, L. (1962). *Science and Information Theory*, 2nd edn. New York: Academic Press.

Brody, S. (1923). The kinetics of senescence. *Journal of General Physiology* **6**, 245–55.

Brooks, J. & Shaw, G. (1973). *Origin and Development of Living Systems*. London; New York: Academic Press.

Bruce, A. G., Atkins, J. F. & Gesteland, R. F. (1986). tRNA anticodon replacement experiments show that ribosomal frameshifting can be caused by doublet coding. *Proceedings of the National Academy of Sciences USA* **83**, 5062–6.

Bulmer, M. (1988). Codon usage and intragenic position. *Journal of Theoretical Biology* **133**, 67–71.

Burbidge, E. M., Burbidge, G. R., Fowler, W. A. & Hoyle, F. (1957). Synthesis of elements in the stars. *Reviews of Modern Physics* **29**, 547–650.

Burchfield, J. D. (1975). *Lord Kelvin and the Age of the Earth*. London: MacMillan.

Cairns-Smith, A. G. (1965). The origin of life and the nature of the primitive gene. *Journal of Theoretical Biology* **10**, 53–88.

Cairns-Smith, A. G. (1971). *The Life Puzzle*. Edinburgh: Oliver & Boyd.

Cairns-Smith, A. G. (1982). *Genetic Takeover and the Mineral Origin of Life*. Cambridge; London; New York: Cambridge University Press.

Calvin, M. (1961). Origin of life on Earth and elsewhere. In *The Logic of Personal Knowledge*. Glencoe, IL: The Free Press.

Calvin, M. (1969). *Chemical Evolution*. Oxford; New York: Oxford University Press.

Camin, J. H. & Sokal, R. R. (1965). A method for deducing branching sequences in phylogeny. *Evolution* **19**, 311–26.

Cantatore, P., Gadaleta, M. N., Roberti, M., Saccone, C. & Wilson, A. C. (1987). Duplication and remoulding of tRNA genes during the evolutionary rearrangement of mitochondrial genomes. *Nature* **329**, 853–5.

Canuto, V. M., Levine, J. S., Augustsson, T. R., Imhoff, C. L. & Giampapa, M. S. (1983). The young Sun and the atmosphere and photochemistry of the early Earth. *Nature* **305**, 281–6.

Capecchi, M. R., Capecchi, N. E., Hughes, S. H. & Wahl, G. M. (1974). Selective degradation of abnormal proteins in mammalian tissue. *Proceedings of the National Academy of Sciences USA* **71**, 4732–6.

Caron, F. & Meyer, E. (1985). Does *Paramecium primaurelia* use a different genetic code in its macronucleus? *Nature* **314**, 185–8.

Cech, T. R. (1986a). A model for the RNA-catalyzed replication of RNA. *Proceedings of the National Academy of Sciences USA* **83**, 4360–3.

Cech, T. R. (1986b). RNA as an enzyme. *Scientific American* **255**, 64–75.

Chaisson, E. (1981). *Cosmic Dawn*. Boston; Toronto: Little, Brown & Co.

Chaitin, G. J. (1966). On the length of programs for computing finite binary sequences. *Journal of the Association for Computing Machinery* **13**, 547–69.

Chaitin, G. J. (1969). On the length of programs for computing finite binary sequences: Statistical considerations. *Journal of the Association for Computing Machinery* **16**, 145–59.

Chaitin, G. J. (1970). Towards a mathematical definition of 'life'. *Association for Computing Machinery Special Interest Committee on Automata and Compatibility Theory News*, 12–18.

Chaitin, G. J. (1974). Information-theoretic computational complexity. *IEEE Transactions on Information Theory* **20**, 10–15.

Chaitin, G. J. (1975a). Randomness and mathematical proof. *Scientific American* **232**, 5, 47–52.

Chaitin, G. J. (1975b). A theory of program size formally identical to information theory. *Journal of the Association for Computing Machinery* **22**, 329–40.

Chaitin, G. J. (1977). Algorithmic information theory. *IBM Journal of Research and Development* **21**, 350–9.

Chaitin, G. J. (1979). Toward a mathematical definition of 'life'. In *The Maximum Entropy Formalism*, eds R. D. Levine & M. Tribus. Cambridge, MA: MIT Press.

Chaitin, G. J. (1982). Gödel's theorem and information. *International Journal of Theoretical Physics* **22**, 941–5.

Chaitin, G. J. (1985). An APL2 gallery of mathematical physics – a course outline. *Proceedings of the Japan '85 APL Symposium*, Publication N:GE18-9948-0 IBM Japan 1–56.

Chaitin, G. J. (1987a). Incompleteness theorems for random reals. *Advances in Applied Mathematics* **8**, 119–46.

Chaitin, G. J. (1987b). *Information, Randomness and Incompleteness*. Singapore: World Scientific.

Chaitin, G. J. (1988). *Algorithmic Information Theory* (revised second printing). Cambridge: Cambridge University Press.

Chambers, I., Frampton, J., Goldfarb, P., Affara, N., McBain, W. & Harrison, P. R. (1986). The structure of the mouse glutathione peroxidase gene: the selenocysteine in the active site is encoded by the 'termination' codon, TGA, *The EMBO Journal* **5**, 1221–7.

Chang, S. (1988). Planetary environments and the conditions of life. *Philosophical Transactions of the Royal Society of London A* **325**, 601–10.

Chang, S., DesMarais, D., Mack, R., Miller, S. L. & Streathearn, G. E. (1983). Prebiotic organic syntheses and the origin of life. In *Earth's earliest biosphere: Its origin and evolution*, ed. J. W. Schopf, pp. 53–92. Princeton, NJ: Princeton University Press.

Chernikov, A., Sagdeev, R. Z. & Zaslavsky, G. M. (1988). Chaos: How regular can it be? *Physics Today* **41**, 27–35.

Chyba, C. F. (1990). Impact delivery and erosion of planetary oceans in the early inner solar system. *Nature* **343**, 129–33.

Chyba, C. F. & Sagan, C. (1987). Cometary organics but no evidence for bacteria. *Nature* **329**, 208.

Chyba, C. F., Thomas, P. J., Brookshaw, L. & Sagan, C. (1990). Cometary delivery of organic molecules to the early Earth. *Science* **249**, 366–73.

Clare, J. & Farabaugh, P. (1985). Nucleotide sequence of a yeast Ty element: Evidence for an unusual mechanism of gene expression. *Proceedings of the National Academy of Sciences USA* **82**, 2829–33.

Clarke, B. (1970). Selective constraints on amino-acid substitutions during the evolution of proteins. *Nature* **228**, 159–60.

Clary, D. O. & Wolstenholme, D. R. (1985). The mitochondrial DNA molecule of *Drosophila yakuba*: Nucleotide sequence gene organization and genetic code. *Journal of Molecular Evolution* **22**, 252–71.

Clayton, D. D. & Woosley, S. E. (1974). Thermonuclear astrophysics. *Reviews of Modern Physics* **46**, 755–71.

Clayton, L. K., Goodman, M. F., Branscom, E. W. & Galas, D. J. (1979). Error induction and correction by mutant and wild type T4 DNA polymerases. *Journal of Biological Chemistry* **254**, 1902–12.

Compton, W. & Pidgeon, R. T. (1986). Jack Hills, Evidence of more very old detrital zircons in Western Australia. *Nature* **321**, 766–9.

Conway, J. H. & Sloane, N. J. A. (1988). *Sphere Packings; Lattices and Groups*. New York, Heidelberg, London: Springer-Verlag.

Covello, P. S. & Gray, M. W. (1989). RNA editing in plant mitochondria. *Nature* **341**, 662–6.

Coveney, P. V. (1988). The Second Law of Thermodynamics: entropy, irreversibility and dynamics. *Nature*, **333**, 409–15.

Cowan, J. J., Thielemann, F.-K. & Truran, J. W. (1987). Nuclear chronometers from the r-process and the age of the galaxy. *The Astrophysical Journal* **323**, 543–52.

Cox, D. R. (1955). A use of complex probabilities in the theory of stochastic processes. *Proceedings of the Cambridge Philosophical Society* **51**, 313–19.

Craigen, W. J., Fulton, S. M., Dobson, M. J., Wilson, W. & Kingsman, S. M. (1985). Bacterial peptide chain release factors: Conserved primary structure and possible frameshift regulation of release factor. *Proceedings of the National Academy of Sciences USA* **82**, 3616–20.

Creighton, T. E. & Chothia, C. (1989). Selecting buried residues. *Nature* **339**, 14–15.

Crick, F. (1968). The origin of the genetic code. *Journal of Molecular Biology* **38**, 367–79.

Crick, F. (1970). Central Dogma of molecular biology. *Nature* **227**, 561–3.

Crick, F. (1981). *Life Itself, Its Origin and Nature*. New York: Simon & Schuster.

Crick, F. H. C. & Orgel, Leslie, E. (1973). Directed panspermia. *Icarus* **19**, 341–6.

Crick, F., Griffith, J. S. & Orgel, L. E. (1957). Codes without commas. *Proceedings of the National Academy of Sciences USA* **43**, 416–21.

Crick, F., Barnett, L., Brenner, S. & Watts-Tobin, R. J. (1961). General nature of the genetic code for proteins. *Nature* **192**, 1227–32.

Cronin, J. R., Pizzarello, S. & Cruikshank, D. P. (1988). Organic matter in carbonaceous chondrites, planetary satellites, asteroids and comets. In *Meteorites and the Early Solar System*, eds J. F. Kerridge & M. S. Matthews, pp. 819–57. Tucson, AZ: University of Arizona Press.

Cullmann, G. (1981). A mathematical method for the enumeration of doublet codes. In *Origin of Life*, ed. Y. Wolman, pp. 405–13. Dordrecht; Boston; London: Reidel.

Cullmann, G. & Labouygues, J.-M. (1983). Noise immunity in the genetic code. *BioSystems* **16**, 9–29.

Cullmann, G. & Labouygues, J.-M. (1985). Le code génétique, instantané absolument optimal. *Compte Rendue* **301**, Série III, 157–60.

Cullmann, G. & Labouygues, J.-M. (1987). Evolution of proteins: An ergodic stationary chain. *Mathematical Modelling* **8**, 639–46.

Cupples, C. G. & Miller, J. H. (1988). Effects of amino acid substitutions at the active site in *Escherichia coli* β-galactosidase. *Genetics* **120**, 637–44.

Curnow, R. N. (1988). The use of Markov chain models in studying the evolution of the proteins. *Journal of Theoretical Biology* **134**, 51–7.

Darnell, J. E. & Doolittle, W. F. (1986). Speculations on the early course of evolution. *Proceedings of the National Academic of Sciences USA* **83**, 1271–5.

Darwin, C. (1982). *The Origin of Species*, pp. 443. New York; Scarborough, Ontario: Mentor Edition, reprint of sixth edition, The New American Library.

Das, G., Hickey, D. R., McLendon, D., McLendon, G. & Sherman, F. (1989). Dramatic thermostabilization of yeast iso-1-cytochrome c by an asparagine → isoleucine replacement at site 57. *Proceedings of the National Academy of Sciences USA* **86**, 496–9.

Davies, P. C. W. (1977). *Space and Time in the Modern Universe.* Cambridge; New York: Cambridge University Press.

Davis, M., Hut, P. & Muller, R. A. (1984). Extinction of species by periodic comet showers. *Nature* **308**, 715–17.

Dawkins, R. (1986). *The Blind Watchmaker.* New York, London: W. W. Norton.

Day, W. (1984). *Genesis on Planet Earth.* New Haven; London: Yale University Press.

Dayhoff, M. (1976). *Atlas of Protein Sequence and Structure*, Vol. 5, Supplement 2. Washington DC: National Biomedical Research Foundation.

Dayhoff, M. (1978). *Atlas of Protein Sequence and Structure*, Vol. 8, Supplement 3. Washington DC: National Biomedical Research Foundation.

Dayhoff, M. O., Eck, R. V. & Park, C. M. (1972). A model of evolutionary change in proteins. In *Atlas of Protein Sequence and Structure*, Vol. 5, ed. M. O. Dayhoff, pp. 89–100. Silver Spring, MD: National Biomedical Research Foundation.

de Duve, C. (1991). *Blueprint for a Cell: The Nature and Origin of Life.* Burlington, North Carolina, USA: Neil Patterson Publishers.

Delbrück, M. (1986). *Mind from Matter? An Essay on Evolutionary Epistemology*, eds G. S. Stent, E. P. Fischer, S. W. Golomb, D. Presti & H. Seiler. Palo Alto; Oxford; London: Blackwell.

Dickerson, R. E. (1978). Chemical evolution and the origin of life. *Scientific American* **239**, 70–86.

Dickerson, R. E., Takano, T., Eisenberg, D., Kallai, O., Samson, L., Cooper, A. & Margoliash, E. (1971). I. General features of the horse and bonito proteins at 2.8 Å resolution. *Journal of Biological Chemistry* **246**, 1511–71.

Di Guilio, M. & Caldaro, F. (1987). Absorbent Markov chains as a model for the study of the evolution of proteins. *Journal of Theoretical Biology* **124**, 485–90.

Doolittle, R. F. (1981). Similar amino acid sequences: Chance or common ancestry? *Science* **214**, 149–59.

Doolittle, R. F. (1983). Probability and the origin of life. In *Scientists*

Confront Creationism, ed. L. R. Godfrey. New York; London: W. W. Norton.

Doolittle, R. F. (1987a). The evolution of the vertebrate plasma proteins. *Biological Bulletin* **172**, 269–83.

Doolittle, R. F. (1987b). Of URFS and ORFS, A primer on how to analyze derived amino acid sequences. Mill Valley, CA: University Science Books.

Doolittle, R. F. (1988). More molecular opportunism. *Nature* **336**, 18.

Dose, K. (1975). Peptides and amino acids in the primordial hydrosphere. *BioSystems* **6**, 224–8.

Dose, K. (1976). Ordering processes and the evolution of the first enzymes. In *Protein Structure and Evolution*, eds J. L. Fox, Z. Deyl & A. Blazy, Chapter 9, pp. 149–84. New York: Dekker.

Dounce, A. L. (1952). Duplicating mechanism for peptide chain and nucleic acid synthesis. *Enzymologia* XV, 251–8.

Dufton, M. J. (1983). The significance of redundance in the genetic code. *Journal of Theoretical Biology* **102**, 521–6.

Dulbecco, R. (1987). In *The Design of Life*, pp. 440–6. New Haven; London: Yale University Press.

Durham, A. (1978). *New Scientist* **77**, 785–7.

Dyson, F. J. (1979a). *Disturbing the Universe*. New York: Harper & Row.

Dyson, F. J. (1979b). Time without end: Physics and biology in an open universe. *Reviews of Modern Physics* **51**, 447–60.

Dyson, F. J. (1982). A model for the origin of life. *Journal of Molecular Evolution* **18**, 344–50.

Dyson, F. J. (1985). *Origins of Life*. Cambridge; London; New York: Cambridge University Press.

Ebling, W. & Jiminez-Montano, M. A. (1980). On grammars, complexity, and information measures of biological macromolecules. *Mathematical Bioscience* **52**, 53–71.

Ebling, W. & Mahnke, R. (1978). On the complexity of cytochrome c and the influence of the genetic code. *Studia Biophysica* **71**, 173–80.

Ebling, W., Feistel, R. & Herzel, H. (1987). Dynamics and complexity of biomolecules. *Physica Scripta* **35**, 761–8.

Ebringer, A. (1975). Information theory and the Central Dogma (a comment on code degeneracy). *Journal of Theoretical Biology* **53**, 243.

Eddington, A. S. (1926). *The Internal Constitution of the Stars*, Dover edition (1959), New York: Dover Publications.

Eddington, A. S. (1928). *The Nature of the Physical World*, pp. 84–5. Cambridge: Cambridge University Press.

Eddington, A. S. (1930). On the instability of Einstein's spherical world. *Monthly Notices of the Royal Astronomical Society* **90**, 668–78.

Eddington, A. S. (1931a). The end of the world: From the standpoint of mathematical physics. *Nature* **127**, 447–53.

Eddington, A. S. (1931b). The evolution of the universe. *Nature* **128**, 709.

Eddington, A. S. (1933). *The Expanding Universe*. Cambridge: Cambridge University Press.

Edwards, A. W. F. & Cavalli-Sforza, L. L. (1964). *Phenetic and phylogenetic classification*, eds V. H. Heywood & J. McNeill, pp. 67–76. London: Systematics Association, Publication No. 6.

Ehrenberg, M., Kurland, C. G. & Blomberg, C. (1986). Kinetic costs of accuracy in translation. In *Accuracy in Molecular Processes*, eds T. B. L. Kirkwood & R. F. Rosenberg, pp. 329–60. London; New York: Chapman & Hall.

Eigen, M. (1971). Self-organization of matter and the evolution of biological macromolecules. *Naturwissenschaften* **58**, 465–523.

Eigen, M. & Schuster, P. (1977). The hypercycle: A principle of natural self-organization. Part A: Emergence of the hypercycle. *Naturwissenschaften* **64**, 541–65.

Eigen, M. & Schuster, P. (1978a). The hypercycle: A principle of natural self-organization. Part B: The abstract hypercycle *Naturwissenschaften* **65**, 7–41.

Eigen, M. & Schuster, P. (1978b). The hypercycle: A principle of natural self-organization. Part C: The realistic hypercycle *Naturwissenschaften* **65**, 341–69.

Eigen, M. & Schuster, P. (1982). Stages of emerging life – Five principles of early organization. *Journal of Molecular Evolution* **19**, 47–61.

Eigen, M. & Winkler-Oswatitsch, R. (1981). Transfer-RNA, an early gene? *Naturwissenschaften* **68**, 282–92.

Eigen, M., Gardiner, W. C. Jr. & Schuster, P. (1980). Hypercycles and compartments. *Journal of Theoretical Biology* **85**, 407–11.

Eigen, M., Winkler-Oswatitsch, R. & Dress, A. (1988). Statistical geometry in sequence space: A method of quantitative comparative sequence analysis. *Proceedings of the National Academy of Sciences USA* **85**, 5913–17.

Einstein, A. (1905). Über die von molecular-kinetischen Theorie der Wärme geforderte Bewegung in ruhenden flüssigkeiten suspendierten Teilchen. *Annalen der Physik* **17**, 549–60.

Einstein, A. (1906). Zur Theorie der Brownischen Bewegung. *Annalen der Physik* **19**, 371–81.

Einsten, A. (1915a). Zur allgemeinen Relativitätstheorie. *Sitzungsberichte der Königliche Preussische Akademie der Wisschenschaften* **XLIV**, 778–86.

Einstein, A. (1915b). Zur allgemeinen Relativitätstheorie (Nachtrag). *Sitzungsberichte der Königliche Preussische Akademie der Wisschenschaften* **XLVI**, 799–801.

Einstein, A. (1915c). Erklärung der Perihelbebewegung des Merkur aus der allgemeinen Relativatätstheorie. *Sitzungsberichte der Königliche Preussische Akademie der Wisschenschaften* **XLVII**, 831–9.

Einstein, A. (1915d). Die Feldgleichungen der Gravitation. *Sitzungsberichte der Königliche Preussische Akademie der Wisschenschaften* **XLVIII**, 884–47.

Einstein, A. (1916). Die Grundlage der Allgemeine Relativitätstheorie. *Annalen der Physik* **49**, 769–822.

Einstein, A. (1917). Kosmological Betrachtungen zur allgemeine Relativitätstheorie. *Sitzungsberichte der Königliche Preussische Akademie der Wisschenschaften Physikalische–Mathematische Klasse* **8**, February, 142–52.

Einstein, A. (1931). Zum kosmologischen Problem der allgemeine Relativitätstheorie. *Sitzungsberichte der Königliche Preussische Akademie der Wisschenschaften Physikalische–Mathematische Klasse* **16**, April, 235–7.

Einstein, A. & de Sitter, W. (1932). On the relation between the expansion and the mean density of the universe. *Proceedings of the National Academy of Sciences USA* **18**, 213–14.

Epstein, C. J. (1966). Role of the amino-acid 'code' and of selection for conformation in the evolution of proteins. *Nature* **210**, 25–8.

Epstein, C. J. (1967). Non-randomness of amino-acid changes in the evolution of homologous proteins. *Nature* **215**, 355.

Erikson, G. J. & Smith, C. R., eds. (1988). *Maximum-Entropy and Bayesian Methods in Science and Engineering, Volume 1: Foundations*. Dordrecht: Kluwer Academic Publishers.

Essene, E. J. & Fisher, D. C. (1986). Lightning strike fusion: extreme reduction and metal-silicate liquid immiscibility. *Science* **234**, 189–93.

Euripides (485–407 BC). For the character of Aphrodite see in Euripides' play *Hippolytus* a speech by the goddess herself. For the characters of Hera, Athene and Aphrodite see a speech by Helen of Troy on the Judgement of Paris in Euripides' play *The Women of Troy*. Each goddess offered a bribe to Paris to choose her as the fairest. Aphrodite offered to Paris, Helen, the wife of Menelaus, King of Sparta. That started the Trojan War.

Eves, H. (1966). *Elementary Matrix Theory*. New York: Dover Publications.

Fadiev, D. A. (1956). On the notion of the entropy of a finite probability space (in Russian). *Uspekhi Matematicheskikh Nauk* **11**, 227–31.

Failla, G. (1958). The aging process and cancerogenesis. *Annals of the New York Academy of Sciences* **721**, 1124–40.

Failla, G. (1960). The aging process and somatic mutations. In *The Biology of Aging*, ed. B. L. Strehler. Washington DC: American Institute of Biological Sciences, Publication No. 6.

Faith, D. P. (1985). Distance methods and approximation of most parsimonious trees. *Systematic Zoology* **34**, 312–25.

Farina, M., Esquivel, D. M. S. & de Barros, H. G. P. L. (1990). Magnetic iron–sulfur crystals from a magnetotactic microorganism. *Nature* **343**, 256–8.

Farris, J. S. (1972). Estimating phylogenetic trees from distance matrices. *American Naturalist* **106**, 645–68.

Feely, D. E., Erlandson, S. L. & Chase, D. G. (1984). In *Giardia and Giardiasis*, eds S. L. Erlandson & E. A. Meyer, pp. 3–31. New York: Plenum.

Feferman, S., Dawson, J. W. Jr., Kleene, S. C., Moore, G. H., Soloway, R. M. & van Heijennoort I. (1986). *Kurt Gödel: Collected Works I: Publications 1929–1936*. New York: Oxford University Press.

Feinstein, A. (1958). *Foundations of Information Theory*. New York: McGraw-Hill.

Feld, B. & Szilard, G. W. (1972). *The Collected Works of Leo Szilard*. Cambridge, MA: MIT Press.

Feller, W. (1968). *An Introduction to Probability Theory and Its Applications*, 3rd edn, Vol. 1. New York: John Wiley & Sons.

Feller, W. (1971). *An Introduction to Probability Theory and its Applications*, 2nd edn, Vol. 2. New York: John Wiley & Sons.

Felsenstein, J. (1978). Cases in which parsimony or compatibility methods will be positively misleading. *Systematic Zoology* **27**, 401–9.

Feng, D. F., Johnson, M. S. & Doolittle, R. F. (1985). Aligning amino acid sequences: Comparison of commonly used methods. *Journal of Molecular Evolution* **21**, 112–25.

Ferscht, A. R. & Dingwall, C. (1979). Cysteinyl-tRNA synthetase from *Escherichia coli* does not need an editing mechanism for the rejection of Ser and Ala: High binding energy of small groups in specific molecular interactions. *Biochemistry* **18**, 1245–50.

Ferscht, A. R., Schindler, J. S., Tsui, W.-C. (1980). Probing the limits of protein-amino side chain recognition with the aminoacyl-tRNA synthetases. Discrimination against Phe by tyrosyl-tRNA synthetases. *Biochemistry* **19**, 5520–4.

Fiddes, J. C. (1977). The nucleotide sequence of a viral DNA. *Scientific American* **237**, 55–67.

Fiers, W., Contreras, R., Haegman, G., Rogiers, R., van der Voorde, A., van Heuverswyn, H., van Herreweghe, J., Volckaert, G. & Ysebaert, M. (1978). Complete sequence of SA 40 DNA. *Nature* **273**, 113–20.

Figureau, A. & Labouygues, J.-M. (1981). The origin and evolution of the genetic code. In *Origin of Life*, ed. Y. Wolman. Dordrecht; Boston; London: Reidel.

Finch, C. E. & Schneider, E. L., eds (1985). *Handbook of the Biology of Aging*, 2nd edn. New York: Van Nostrand Reinhold.

Fischer, A. G. & Arthur, M. A. (1977). Secular variations in the pelagic realm. The Society of Economic Paleontologists and Mineralogists, Special Publication No. 25, 19–50. **TITLE** Soc. Econ. Paleontology Min. Spe. Publ 25, 19–50.

Fisher, D. E. (1976). Rare gas clues to the origin of the terrestrial atmosphere. In *The Early History of the Earth*, ed. B. F. Windley. London; New York: John Wiley & Sons.

Fisher, R. A. (1930). *The Genetical Theory of Natural Selection*, 2nd revised edn, pp. 1–12. Oxford: Oxford University Press. (Republished 1958, New York: Dover Publications.)

Fitch, W. M. (1971). Rate of change of concomitantly variable codons. *Journal of Molecular Evolution* **1**, 84–96.

Fitch, W. M. (1976). Molecular evolution of cytochrome c in eukaryotes. *Journal of Molecular Evolution* **8**, 13–40.

Fitch, W. M. (1980). Estimating the total number of nucleotide substitutions since the common ancestor of a pair of homologous genes: Comparison of several methods and three β-hemoglobin messenger RNAs. *Journal of Molecular Evolution* **16**, 153–209.

Fitch, W. M. (1981). The old REH theory remains unsatisfactory and the new REH theory is problematical – A reply to Holmquist and Jukes. *Journal of Molecular Evolution* **18**, 60–7.

Fitch, W. M. & Margoliash, E. (1967). A method for estimating the number of invariant amino acid coding positions in a gene using cytochrome c as a model case. *Biochemical Genetics* **1**, 65–71.

Fitch, W. M. & Margoliash, E. (1968). The construction of phylogenetic trees II. How well do they reflect past history? *Brookhaven Symposium on Biology* **21**, 217–42.

Fitch, W. M. & Margoliash, E. (1970). The usefulness of amino acid and nucleotide sequences in evolutionary studies. In *Evolutionary Biology*, eds T. Dobzhansky, M. K. Hecht & W. C. Steere, Vol. 4, pp. 67–109.

Fitch, W. M. & Markowitz, E. (1970). An improved method for determining codon variability in a gene and its application to the rate of fixation of mutations in evolution. *Biochemical Genetics* **4**, 579–93.

Folsome, C. E. (1976). Synthetic organic microstructures and the origin of cellular life. *Die Naturwissenschaften* **7**, 303–6.

Fougère, P. F. (1988). Maximum entropy calculations on a discrete probability space. In *Maximum-entropy and Bayesian Methods in Science and Engineering*, Vol. 1, eds G. J. Erickson & C. R. Smith, pp. 205–34. Dordrecht: Kluwer Academic Publishers.

Fox, S. W. (1973). Molecular evolution to the first cells. *Pure & Applied Chemistry* **34**, 641–69.

Fox, S. W. (1975). The matrix for the protobiological quantum: Cosmic casino or shapes of molecules? *International Journal of Quantum Chemistry: Quantum Biology Symposium No. 2*, 307–20.

Fox, S. W. (1976a). The evolutionary significance of ordering phenomena in thermal proteinoids and proteins. In *Protein structure and evolution*, eds J. L. Fox, Z. Deyl & A. Blazej. New York; Basel: Dekker.

Fox, S. W. (1976b). Response to comments on thermal polypeptides by Temussi *et al.*, *Journal of Molecular Evolution* **8**, 301–4.

Fox, S. W. (1984a). Proteinoid experiments and evolution theory. In *Beyond neo-Darwinism*, eds M. W. Ho & P. T. Saunders, pp. 15–60. London; New York: Academic Press.

Fox, S. W. (1984b). Creationism and evolutionary protobiogenesis. In *Science and Creationism*, ed. A. Montagu, pp. 194–239. Oxford; New York: Oxford University Press.

Fox, S. W. (1986). Selforganization in evolution. In *Selforganization*, ed. S. W. Fox, pp. 35–56. Guilderland; New York: Adenine Press.

Fox, S. W. & Dose, K. (1972). *Molecular Evolution and the Origin of Life*. San Francisco: W. H. Freeman & Co.; revised edition, 1977.

Fox, S. W. & Matsuno, K. (1983). Self-organization of the protocell was a forward process. *Journal of Theoretical Biology* **101**, 321–3.

Fox, S. W. & Nakashima, T. (1980). The assembly and properties of protobiological structures: The beginnings of cellular peptide synthesis. *BioSystems* **12**, 155–66.

Fox, T. D. (1987). Natural variations in the genetic code. In *Annual Reviews of Genetics*, eds A. Campbell, I. Herskowitz & L. M. Sander, pp. 67–91.

Fox, T. D. & Leaver, C. J. (1981). The Zea mays mitochondrial gene coding cytochrome oxidase subunit II has an intervening sequence and does not contain TGA codons. *Cell* **26**, 315–23.

Frautschi, S. (1982). Entropy in an expanding universe. *Science* **217**, 593–9.

Freeman, K. H., Hayes, J. M., Trendel, J.-M. & Albrecht, P. (1990). Evidence from carbon isotope measurements for diverse origins of sedimentary hydrocarbons. *Nature* **343**, 254–6.

Freifelder, D. (1987). *Molecular Biology*. Boston; Portola Valley, CA: Jones & Bartlett.

Freske, S. (1981). Creationist misunderstanding, misrepresentation, and misuse of the Second Law of Thermodynamics. *Creation/Evolution* **IV**, 8–16.

Freter, R. R. & Savageau, M. A. (1980). Proofreading system of multiple stages for improved accuracy of biological discrimination. *Journal of Theoretical Biology* **85**, 99–123.

Fricke, H. (1988). Coelacanths, the fish that time forgot. *National Geographic* **173**, 824–38.

Fricke, H. & Plante, R. (1988). Habitat requirements of the living coelacanth *Latimeria chalumnae* at Grande Comore, Indian Ocean. *Naturwissenschaft* **75**, 149–51.

Friday, A. & Ingram, D. S. eds. (1985). *The Cambridge Encyclopedia of Life Sciences.* Cambridge; London; New York: Cambridge University Press.

Friebele, E., Shimoyama, A. & Ponnamperuma, C. (1980). Adsorption of protein and non-protein amino acids on clay mineral: A possible role of selection in chemical evolution. *Journal of Molecular Evolution* 16, 269–78.

Friedmann, A. (1922). Über die Krümmung des Raumes. *Zeitschrift für Physik* 10, 377–86.

Friedmann, A. (1924). Über die Möglichkeit einer Welt mit konstanter negativer Krümmung des Raumes. *Zeitschrift für Physik* 21, 326–32.

Friedmann, K. & Shimony, A. (1971). Jaynes's maximum entropy prescription and probability. *Journal of Statistical Physics* 3, 381–4.

Frobenius, G. (1908). Über Matrizen aus positiven Elementen. *Sitzungsberichte der Königliche Preussische Akademie der Wisschenschaft* (Berlin), 471–6.

Frobenius, G. (1909). Über Matrizen aus positiven Elementen. *Sitzungsberichte der Königliche Preussische Akademie der Wisschenschaft* (Berlin), 514–18.

Frobenius, G. (1912). Über Matrizen aus nicht negativen Elementen. *Sitzungsberichte der Königliche Preussische Akademie der Wisschenschaft* (Berlin), 456–77.

Gabel, N. W. & Ponnamperuma, C. (1972). Primordial organic chemistry. In *Exobiology*, ed. C. Ponnamperuma. Amsterdam; London: North-Holland.

Galas, D. J. & Branscom, E. W. (1978). Enzymatic determinants of DNA polymerase accuracy: Theory of coliphage T4 polymerase mechanisms. *Journal of Molecular Biology* 124, 653–87.

Gamow, G. (1946). Expanding universe and the origin of the elements. *Physical Review* 70, 572–3.

Gamow, G. (1954). Possible relation between deoxyribonucleic acid and protein structure. *Nature* 173, 318.

Gamow, G. (1970). *My World Line – An Informal Autobiography.* New York: Viking Press.

Gamow, G. & Ycas, M. (1958). The crytographic approach to the problem of protein synthesis. In *Symposium on Information Theory in Biology*, eds H. P. Yockey, R. L. Platzman & H. Quastler, pp. 63–9. New York; London: Pergamon Press.

Gardell, S. J., Craik, C. S., Hilvert, D., Urdea, M. S. & Rutter, W. J. (1985). Site-directed mutagenesis shows that tyrosine 248 of carboxypeptide A does not play a crucial role in catalysis. *Nature* 317, 551–5.

Garrison, W. M., Morrison, D. C., Hamilton, J. G., Benson, A. A. & Calvin, M. (1951). Reduction of carbon dioxide in aqueous solutions by ionizing radiation. *Science* 114, 416–18.

Garwin, L. (1988). Of impacts and volcanoes. *Science* 336, 714–16.

Gatlin, L. L. (1966). The information content of DNA, I. *Journal of Theoretical Biology* 10, 281–300.

Gatlin, L. L. (1968). The information content of DNA, II. *Journal of Theoretical Biology* 18, 181–94.

Gatlin, L. L. (1972a). *Information Theory and the Living System.* New York; London: Columbia University Press.

Geller, A. I. & Rich, A. (1980). A UGA termination suppression tRNA[Trp] active in rabbit reticulocytes. *Nature* 283, 41–6.

George, D. G., Hunt, L. T., Yeh, L.-S., & Barker, W. C. (1985). New perspectives on bacterial ferrodoxin evolution. *Journal of Molecular Evolution* **22**, 20–31.

Gericke, H. & Wäsche, H. (1974). Linear algebra. *In Fundamentals of Mathematics*, Vol. 1, eds H. Benke, F. Bachmann, K. Fladt & W. Süss, pp. 233–90. Cambridge, MA; London: MIT Press.

Gibbons, G. W., Hawking, S. & Siklos, S. T. C. (1983). *The Very Early Universe*. Cambridge; New York: Cambridge University Press.

Gilbert, W. (1986). The RNA world. *Nature* **319**, 618.

Gish, D. T. (1989). In a discussion of the origin of life on radio station KKLA, Los Angeles, CA, on 29 June, 1989 with Dr. H. P. Yockey, Dr. Gish repeatedly insisted that evolution was in contradiction of the Second Law of Thermodynamics in spite of my explanation to the contrary.

Gödel, K. (1931). Über formal unentscheidbare Sätze der Principia Mathematica und verwandte System I (On formal undecidable propositions of *Principia Mathematica* and related systems I). *Monatshefte für Mathematik und Physik* **38**, 173–98. (Reprinted in S. Feferman *et al.* (1986), New York: Oxford University Press.)

Godson, G. N., Barrell, B. G., Staden, R. & Fiddes, J. C. (1978). Nucleotide sequence of bacteriophage G4 DNA. *Nature* **276**, 236–47.

Goldberg, A. L. & Dice, J. F. (1974). Intracellular protein degradation in mammalian and bacterial cells, Part I. *Annual Reviews of Biochemistry* **43**, 835–69.

Goldberg, A. L. & St John, A. C. (1976). Intracellular protein degradation in mammalian and bacterial cells, Part II. *Annual Reviews of Biochemistry* **45**, 747–803.

Goldberg, A. L. & Wittes, R. E. (1966). Genetic code: Aspects of organization. *Science* **153**, 420–4.

Goldenberg, D. P., Frieden, R. W., Haack, J. A. & Morrison, T. B. (1989). Mutational analysis of a protein-folding pathway. *Nature* **338**, 127–32.

Goldhaber, M. (1983). Physics and the world in 1982. *Physics Today* **36**, 35–8.

Golding, G. B. & Strobeck, C. (1982). Expected frequencies of codon use as a function of mutation rates and codon fitnesses. *Journal of Molecular Biology* **18**, 379–86.

Gompertz, B. (1825). On the nature of the function expressive of the law of human mortality and on a new mode determining life contingencies. *Philosophical Transactions of the Royal Society of London*, II, 513–85.

Goodman, M. F. & Branscom, E. W. (1986). DNA replication fidelity and base mispairing mutagenesis. In *Accuracy in Molecular Processes*, eds T. B. L. Kirkwood, R. F. Rosenberger & C. J. Galas. London; New York: Chapman & Hall.

Goodman, M. F., Hopkins, R. L., Watanabe, S. M., Clayton, L. K. & Guidotti, S. (1980). On the molecular basis of mutagenesis: Enzymological and genetic studies with the bacteriophage T4 system. In *Mechanistic Studies of DNA Replication and Genetic Recombination*, ICN–UCLA Symposia on Molecular and Cellular Biology, Vol. XIX, eds B. Alberts & C. R. Fox, pp. 685–705. New York: Academic Press.

Gough, D. O. (1981). Solar interior structure and luminosity variations. *Solar Physics* **74**, 21–34.

Gould, S. J. (1984). Toward the vindication of functional change. In

Catastrophes and Earth History, eds W. A. Bergren & J. A. van Couvering. Princeton, NJ: Princeton University Press.

Grantham, R. (1974). Amino acid difference formula to help explain protein evolution. *Science* **185**, 862–85.

Greeley, R. (1987). Release of juvenile water on Mars: Estimated amounts and timing associated with volcanism. *Science* **236**, 1653–7.

Greenberg, J. M. & Zhao, N. (1988). Cometary organics. *Nature* **331**, 124.

Gribbin, J. (1986). *In Search of the Big Bang*. New York; London: Bantam Books.

Griffith, J. S. (1967). Self-replication and scrapie. *Nature* **215**, 1043–4.

Grosjean, H. J., de Henau, S. & Crothers, D. M. (1978). On the physical basis for ambiguity in genetic coding interactions. *Proceedings of the National Academy of Sciences USA* **75**, 610–14.

Gualberto, J. M., Lamattina, L., Bonnard, G., Weil, J.-H. & Grienenberger, J.-M. (1989). RNA editing in wheat mitochondria results in the conservation of protein sequences. *Nature* **341**, 660–2.

Gull, S. F. & Daniell, G. J. (1978). Image reconstruction from incomplete and noisy data. *Nature* **272**, 686–90.

Gull, S. F. & Skilling, J. (1984). The maximum entropy method. In *Indirect Imaging*, ed. J. A. Roberts. Cambridge: Cambridge University Press.

Haldane, J. B. S. (1929). The origin of life. In *Rationalist Annual*. London: C. A. Watts & Co. Also in *The Origin of Life*, ed. J. D. Bernal (1967). London: Weidenfeld & Nicolson and in *On Being the Right Size*, ed. J. Maynard Smith (1985). Oxford, New York: Oxford University Press.

Haldane, J. B. S. (1931). *The Philosophical Basis of Biology*, p. 36. London; Aylesbury, UK: Hodder & Stoughton.

Hallam, A. (1987). End-Cretaceous mass extinction events: Argument for terrestrial causation. *Science* **238**, 1237–42.

Hamming, R. W. (1950). Error detecting and error correcting codes. *Bell System Technical Journal* **29**, 147–60.

Hamming, R. W. (1986). *Coding and Information Theory*, 2nd edn. Englewood Cliffs, NJ: Prentice-Hall.

Hampsey, D. M., Das, G. & Sherman, F. (1986). Amino acid replacements in yeast iso-1-cytochrome c. *The Journal of Biological Chemistry* **261**, 3259–71.

Hampsey, D. M., Das, G. & Sherman, F. (1988). Yeast iso-1-cytochrome c: genetic analysis of structural requirements. *FEBS Letters* **231**, 275–83.

Hanyu, N., Kuchino, Y. & Nishimura, S. (1986). Dramatic events in ciliate evolution: alteration of UAA and UAG termination codons to glutamine codons due to anticodon mutations in two *Tetrahymena* tRNAs[Gln]. *The EMBO Journal* **5**, 1307–11.

Hart, M. H. (1978). The evolution of the atmosphere of the Earth. *Icarus* **33**, 23–9.

Hart, M. H. (1979). Habitable zones about main sequence stars. *Icarus* **37**, 351–7.

Hart, M. H. & Zuckerman, B., eds. (1982). *Extraterrestrials, Where are they?* New York; Oxford: Pergamon Press.

Hartle, J. B. & Hawking, S. W. (1983). Wave function of the universe. *Physical Review D* **28**, 2960–75.

Hartley, R. V. L. (1928). Transmission of information. *Bell System Technical Journal*, July, 535.

Hartmann, J. M., Brand, M. C. & Dose, K. (1981). Formation of specific amino acid sequences during thermal polymerization of amino acids. *BioScience* **13**, 141–7.

Hawking, Stephen W. (1983). The cosmological constant. *Philosophical Transactions of the Royal Society of London A* **310**, 303–10.

Hawking, S. W. (1984). The quantum state of the universe. *Nuclear Physics B* **239**, 257–76.

Hawking, S. W. (1988). *A Brief History of Time*. New York: Bantam.

Hawking, S. W. & Ellis, G. F. R. (1968). The cosmic black body radiation and the existence of singularities in our universe. *Astrophysical Journal*, **152**, 25–36.

Hawking, S. W. & Penrose, R. (1970). The singularities of gravitational collapse and cosmology. *Proceedings of the Royal Society of London, Series A* **314**, 529–48.

Hawks, W. C. & Tappel, A. L. (1983). In vitro synthesis of glutathione peroxidase from selenite translational incorporation of selenocysteine. *Biochemica et Biophysica Acta* **793**, 225–34.

Heckman, J. E., Sarnoff, J., Alzner-DeWeerd, B., Yin, S. & RajBhandrary, U. L. (1980). Novel features in the genetic code and codon reading patterns in *Neurospora crassa* mitochondria based on sequences of six mitochondrial tRNAs. *Proceedings of the National Academy of Sciences USA* **77**, 3159–63.

Henderson-Sellers, A. & Henderson-Sellers, B. (1988). Equable climate in the early Archaen. *Nature* **336**, 117.

Henikoff, S., Keene, M. A., Fechtel, K. & Fristrom, J. (1986). Gene within a gene: Nested Drosophila genes encode unrelated proteins on opposite DNA strands. *Cell* **44**, 33–42.

Hermes, H. & Markwald, W. (1974). Foundations of mathematics. In *Fundamentals of Mathematics*, Vol. 1, eds H. Benke, F. Bachmann, K. Fladt & W. Süss, pp. 1–80. Cambridge, MA: MIT Press.

Hernquist, L. (1989). Tidal triggering of starbursts and nuclear activity in galaxies. *Nature* **340**, 687–91.

Hiesel, R. & Brennicke, A. (1983). Cytochrome oxidase subunit II gene in mitochondria of *Oenothera* has no intron. *The EMBO Journal* **2**, 2173–8.

Hiesel, R., Wissinger, B., Schuster, W. & Brennicke, A. (1989). RNA editing in plant mitochondria. *Science* **246**, 1632–4.

Hilbert, D. (1902). Mathematical Problems. *Bulletin of the American Mathematical Society* **2**, 337–479.

Hill, R. (1986). *A First Course in Coding Theory*. Oxford, New York, Toronto: Oxford University Press.

Himeno, H., Masaki, H., Kawai, T., Ohta, T., Kumagai, I., Miura, K. & Watanabe, K. (1987). Unusual genetic codes and a novel gene structure for $tRNA_{AGY}^{Ser}$ in starfish mitochondrial DNA. *Gene* **56**, 219–30.

Hirata, K. S., Kajita, T., Kifune, K., Nakahata, M., Nakamura, K., Ohara, S., Oyama, Y., Sato, N., Totsuka, Y., Yaginuma, Y., Mori, M., Suzuki, A., Takahashi, K., Tanimori, T., Yamada, M. & Koshiba, M., Miyano, K., Miyata, H., Takei, H., Kaneyki, K., Nagashima, Y., Suzki, Y., Beier, E. W., Feldscher, L. R., Frank, E. D., Frati, W., Kim, S. B., Mann, A. K., Newcomer, F. M., Van Berg, R. & Zhang, W. (1989). Observation of ^{8}B solar neutrinos in the Kamiokande-II detector. *Physical Review Letters* **63**, 16–19.

Hobson, A. (1969). A new theorem of information theory. *Journal of Statistical Physics* **1**, 383–91.

Hochstim, A. R. (1971). The role of shock waves in the formation of organic compounds in the primeval atmosphere. In *Chemical Evolution and the Origin of Life*, eds R. Buvet & C. Ponnamperuma. Amsterdam; London: North-Holland.

Hoffer, E. (1951). *The True Believer.* New York: Harper & Row. (This book gives a profile of the true believer primarily in politics and economics but the profile is germane in science as well.)

Hogan, C. J. (1990). Experimental triumph. *Nature* **344**, 107.

Holland, H. D., Lazar, B. & McCaffrey, M. (1986). Evolution of the atmosphere and oceans. *Nature* **320**, 27–33.

Holliday, R., ed. (1986). *Genes, Proteins and Cellular Ageing.* New York: Van Nostrand Reinhold.

Holliday, R. & Kirkwood, T. B. L. (1981). Predictions of the somatic mutation and mortalization theories of cellular ageing are contrary to experimental observations. *Journal of Theoretical Biology* **93**, 627–42.

Holliday, R. & Kirkwood, T. B. L. (1983). Theories of cell ageing: A case of mistaken identity. *Journal of Theoretical Biology* **103**, 329–30.

Holmquist, R. (1979). The method of maximum parsimony: An experimental test and theoretical analysis of the adequacy of molecular restoration studies. *Journal of Molecular Biology* **135**, 939–58.

Holmquist, R. & Jukes, T. H. (1981). The current status of REH theory. *Journal of Molecular Evolution* **18**, 47–59.

Holmquist, R. & Cimino, J. B. (1980). A general method for biological inference: Illustrated by the estimation of gene nucleotide transition probabilities. *BioSystems* **12**, 1–22.

Holmquist, R. & Pearl, D. (1980). Theoretical foundations for quantitative paleogenetics, Part III. *Journal of Molecular Evolution* **16**, 211–67.

Holmquist, R., Cantor, C. & Jukes, T. H. (1972). Improved procedures for comparing homologous sequences in molecules of proteins and nucleic acids. *Journal of Molecular Biology* **64**, 145–61.

Holmquist, R., Goodman, M., Conroy, T. & Czelusniak, J. (1983). The spatial distribution of fixed mutations within genes coding for proteins. *Journal of Molecular Evolution* **19**, 437–48.

Homer, (*c.* 850 BC). *The Iliad*, a, Book III, lines 314–325; b, Book VII, lines 170–199; c, Book XV, lines 158–217; d, Book III, lines 373–446 (a conversation between Aphrodite and Helen of Troy). *The Odyssey*, Book X, XIV & XVI.

Hopfield, J. J. (1974). Kinetic proofreading: A new mechanism for reducing errors in biosynthetic processes requiring high specificity. *Proceedings of the National Academy of Sciences USA* **71**, 4135–9.

Hopfield, J. J. (1980). The energy relay: A proofreading scheme based on dynamic cooperativity and lacking all characteristic symptoms of kinetic proofreading in DNA replication and protein synthesis. *Proceedings of the National Academy of Sciences USA* **77**, 5248–52.

Hopfield, J. J., Yamane, T., Yue, V. & Coutts, S. M. (1976). Direct experimental evidence for kinetic proofreading in amino acylation of tRNA^{Ile}. *Proceedings of the National Academy of Sciences USA* **73**, 1164–8.

Horowitz, N. H. (1986). *To Utopia and Back: The Search for Life in the Solar System.* New York: W. H. Freeman & Co.

Horowitz, N. H. (1990). Mission impractical. *The Sciences*, March–April, 44–9.

Horowitz, N. H. & Hubbard, J. S. (1974). The origin of life. In *Annual Reviews of Genetics* 3, 393–410.

Horowitz, S. & Gorovsky, M. A. (1985). An unusual genetic code in nuclear genes of *Tetrahymena*. *Proceedings of the National Academy of Sciences USA* **82**, 2452–5.

Hoyle, F. (1948). A new model for the expanding universe. *Monthly Notices of the Royal Astronomical Society* **108**, 372–82.

Hoyle, F. (1949). Stellar evolution and the expanding universe. *Nature* **163**, 196–7.

Hoyle, F. (1980). *Steady-State Cosmology Re-visited.* Cardiff, UK: University of Cardiff Press.

Hoyle, F. (1982). The universe: Past and present reflections. *Annual Reviews of Astronomy and Astrophysics* **20**, 1–35.

Hoyle, F. & Taylor, R. J. (1964). The mystery of the cosmic abundance of helium. *Nature* **203**, 1108–10.

Hoyle, F. & Wickramasinghe, N. C. (1977a). Prebiotic molecules and interstellar grain clumps. *Nature* **266**, 241–3.

Hoyle, F. & Wickramasinghe, N. C. (1977b). Polysaccharides and infrared spectra of galactic sources. *Nature* **268**, 610–12.

Hoyle, F. & Wickramasinghe, N. C. (1977c). Origin and nature of carbonaceous material in the galaxy. *Nature* **270**, 701–3.

Hoyle, F. & Wickramasinghe, N. C. (1978). *Lifecloud: The Origin of Life in the Universe.* London; Toronto; Melbourne: Dent. (1979) New York: Harper & Row.

Hoyle, F. & Wickramasinghe, N. C. (1987). Organic dust in Comet Halley. *Nature* **328**, 117.

Hoyle, F. & Wickramasinghe, N. C. (1988a). Cometary organics. *Nature* **331**, 123–4.

Hoyle, F. & Wickramasinghe, N. C. (1988b). Cometary organics. *Nature* **331**, 666.

Hoyle, F., Wickramasinghe, N. C., Brooks, J. & Shaw, G. (1977). Prebiotic polymers and infrared spectra of galactic sources. *Nature* **269**, 674–6.

Hubble, E. (1929). A relation between distance and radial velocity among extra-galactic nebulae. *Proceedings of the National Academy of Sciences USA* **15**, 168–73.

Hubble, E. & Humason, M. (1931). Velocity–distance relationships among extra-galactic nebulae. *Astrophysical Journal* **74**, 43–80.

Hulett, H. R. (1969). Limitations on prebiological synthesis. *Journal of Theoretical Biology* **24**, 56–72.

Hull, H. E. (1960). Thermodynamics and kinetics of spontaneous generation. *Nature* **186**, 693–4.

Hull, D. L. (1973). *Darwin and his Critics.* Cambridge, MA: Harvard University Press.

Huszagh, V. A. & Infante, J. P. (1989). The hypothetical way of progress. *Nature* **338**, 109.

Imbrie, J. & Imbrie, J. Z. (1980). Modelling the climatic response to orbital variations. *Science* **207**, 943–53.

Inarine, V. (1981). Mutation simulations in present and primitive genetic codes. In *Proceedings of the First International Conference on Applied Modelling and Simulation*, Lyons, France, September 7–11, Vol. 5.

Ingels, F. M. (1971). *Information and Coding Theory*. San Francisco; Toronto; London: Intext Educational Publishers.

Jankowski, J. M., Krawetz, S. A., Walcyzk, E. & Dixon, G. (1986). In vitro expression of two proteins from overlapping reading frames in a eukaryotic DNA sequence. *Journal of Molecular Evolution* **24**, 61–71.

Jaynes, E. T. (1957a). Information theory and statistical mechanics. *Physical Review* **106**, 620–30.

Jaynes, E. T. (1957b). Information theory and statistical mechanics II. *Physical Review* **108**, 171–90.

Jaynes, E. T. (1963). Information theory and statistical mechanics. In *Statistical Physics*, Vol. 3, eds G. E. Uhlenbeck, N. Rosenzwieg, A. J. F. Siegert, E. T. Jaynes & S. Fujita. New York; Amsterdam: W. A. Benjamin.

Jaynes, E. T. (1968). Prior probabilities. *IEEE Transactions on Systems Science and Cybernetics* **SSC-4**, 227–41.

Jaynes, E. T. (1979). Where do we stand on maximum entropy? In *The Maximum Formalism*, eds R. D. Levine & M. Tribus, pp. 15–118. Cambridge, MA: MIT Press.

Jaynes, E. T. (1988a). How does the brain do plausible reasoning? In *Maximum-entropy and Bayesian Methods in Science and Engineering*, Vol. 1, eds G. J. Erickson & C. R. Smith, pp. 1–24. Dordrecht, Boston, London: Kluwer Academic Publishers.

Jaynes, E. T. (1988b). The relation of Bayesian and maximum entropy methods. In *Maximum-entropy and Bayesian Methods in Science and Engineering*, Vol. 1, eds. G. J. Erickson & C. R. Smith, pp. 25–29. Dordrecht, Boston, London: Kluwer Academic Publishers.

Jenkin, H. C. F. (1867). The origin of species. *The North British Review* **46**, 277–318. (Reprinted in D. L. Hull (1973), *Darwin and his Critics*, Cambridge: Harvard University Press.)

Johnson, D. S. & Papadimitriou, C. H. (1985). Computational complexity. In *The Traveling Salesman Problem*, eds E. L. Lawler, J. K. Lenstra, A. H. G. Rinnooy Kan & D. B. Shmoys. New York; Brisbane; Toronto; Singapore: John Wiley & Sons.

Johnson, H. A. (1970). Information theory in biology after 18 years. *Science* **168**, 1545–50.

Jona, F. & Shirane, G. (1962). *Ferroelectric Crystals*. New York: Pergamon Press.

Joyce, G. F. (1989). RNA evolution and the origins of life. *Nature* **338**, 217–24.

Joyce, G. F., Schwartz, A. W., Miller, S. L. & Orgel, L. E. (1987). The case for an ancestral genetic system involving simple analogues of the nucleotides. *Proceedings of the National Academy of Sciences USA* **84**, 4398–402.

Joyce, G. F., Visser, G. M., van Boeckel, C. A. A., van Boom, J. H., Orgel, L. E. & van Westrenen, J. (1984). Chiral selection in poly(C)-directed synthesis of oligo(G). *Nature* **310**, 602–4.

Jukes, T. H. (1965). Coding triplets and their possible evolutionary implications. *Biochemical and Biophysical Research Communications* **19**, 391–6.

Jukes, T. H. (1966). *Molecules and Evolution*. New York: Columbia University Press.

Jukes, T. H. (1973). Possibilities for the evolution of the genetic code from a preceding form. *Nature* **246**, 22–6.

Jukes, T. H. (1974). On the possible origin and evolution of the genetic code. *Origins of Life* **5**, 331–50.

Jukes, T. H. (1980). Silent nucleotide substitutions and the molecular evolutionary clock. *Science* **210**, 973–8.

Jukes, T. H. (1981). Amino acid codes in mitochondria as possible clues to primitive codes. *Journal of Molecular Evolution* **18**, 15–17.

Jukes, T. H. (1983a). Evolution of the amino acid code: Inferences from mitochondrial codes. *Journal of Molecular Evolution* **19**, 219–25.

Jukes, T. H. (1983b). Mitochondrial codes and evolution. *Nature* **301**, 19–20.

Jukes, T. H. (1985). A change in the genetic code in *Mycoplasma caprolium*. *Journal of Molecular Evolution* **22**, 361–2.

Jukes, (Doubting) Thomas H. (1987a). Letter to the editor, *Science* **238**, 732: 'The idea that Martian bugs are dangerous is based on faith rather than on science. ...The absence of detectable life on Mars is regarded by the devout as a test of faith in their belief, rather than as scientific disproof of the existence of danger.
> *How blest are they who have not seen,*
> *But yet whose faith has constant been,*
> *For they eternal life shall win,*
> *Alleluia!*

Jukes, T. H. (1987b). Transitions, transversions, and the molecular evolutionary clock. *Journal of Molecular Evolution* **26**, 87–98.

Jukes, T. H. & Bhushan, V. (1986). Silent nucleotide substitutions and G + C content of some mitochrondrial and bacterial genes. *Journal of Molecular Evolution* **24**, 39–44.

Jukes, T. H. & Gatlin, L. L. (1970). Recent studies concerning the coding mechanism. *Progress in Nucleic Acid Research* **11**, 324–31.

Jukes, T. H. & Holmquist, R. (1972). Estimation of evolutionary changes in certain homologous polypeptide chains. *Journal of Molecular Biology* **64**, 163–79.

Jukes, T. H. & Kimura, M. (1984). Evolutionary constraints and the neutral theory. *Journal of Molecular Evolution* **21**, 90–2.

Jukes, T. H. & King, J. L. (1979). Evolutionary nucleotide replacements in DNA. *Nature* **281**, 605–6.

Jukes, T. H. & Margulis, L. (1980). Letter to the editor, Getting started. *The Sciences* **20**, 2–3.

Jukes, T. H., Holmquist, R. & Moise, H. (1975). Amino acid composition of proteins: Selection against the genetic code. *Science* **189**, 50–1.

Jukes, T. H., Osawa, S., Muto, A. & Lehman, N. (1987). Evolution of anticodons: Variations in the genetic code. *Cold Spring Harbor Symposia on Quantitative Biology* **52**, 769–76.

Jungck, J. R. (1978). The genetic code as a periodic table. *Journal of Molecular Evolution* **11**, 211–24.

Jurka, J. W. (1977). On replication of nucleic acids in relation to the evolution of the genetic code and of proteins. *Journal of Theoretical Biology* **68**, 515–20.

Jurka, J., Kolasza, Z. & Roterman, I. (1982). Globular proteins, GU wobbling, and the evolution of the genetic code. *Journal of Molecular Evolution* **19**, 20–7.

Kampis, G. & Csányi, V. (1987). Notes on order and complexity. *Journal of Theoretical Biology* **124**, 111–21.

Kantrowitz, E. R. & Lipscomb, W. N. (1988). *Escherichia coli* aspartate transcarbamylase: The relation between structure and function. *Science* **241**, 699–74.

Kaplan, R. W. (1971). The problem of chance in formation of protobionts by random aggregation of macromolecules. In *Chemical Evolution and the Origin of Life*, eds. R. Buvet & C. Ponnamperuma. Amsterdam: North-Holland.

Kaplan, R. W. (1974). Theoretical considerations on probabilities of biopolymers and life's origin. *Radiation and Environmental Biophysics* **40**, 31–40.

Karon, J. M. (1979). The covarion model for the evolution of proteins: Parameter estimates and comparison with Holmquist, Cantor and Jukes' stochastic model. *Journal of Molecular Evolution* **12**, 197–218.

Kasting, J. F. (1988). Runaway and moist greenhouse atmospheres and the evolution of Earth and Venus. *Icarus* **74**, 472–94.

Katchalsky, A. (1973). Prebiotic synthesis of biopolymers on inorganic templates. *Naturwissenschaften*, **60**, 215–20.

Katchalsky, E. (1951). Poly-α-amino acids. In *Advances in Protein Chemistry* **6**, 123–85.

Kaula, W. M. (1990). Venus: A contrast in evolution to Earth. *Science* **247**, 1191–6.

Kawaguchi, Y., Honda, H., Taniguchi-Morimura, J. & Iwasaki, S. (1989). The codon CUG is read as serine in an asporogenic yeast *Candida cylindracea*. *Nature* **341**, 164–6.

Kemeny, J., Snell, J. L. & Knapp, A. W. (1976). *Denumerable Markov Chains*. New York; Heidelberg; Berlin: Springer-Verlag.

Kerr, R. A. (1988a). Huge impact is favored K-T boundary killer. *Science* **242**, 865–7.

Kerr, R. A. (1988b). Snowbird II: Clues to Earth's impact history. *Science* **242**, 1380–2.

Khinchin, A. I. (1957). *Mathematical Foundations of Information Theory*, translated by R. A. Silverman & M. D. Friedman. New York: Dover Publications.

Kimberlin, R. H. (1982). Scrapie agent: prions or virinos? *Nature* **297**, 107–8.

Kimura, M. (1968). Evolutionary rate at the molecular level. *Nature* **217**, 624–6.

Kimura, M. (1983). *The Neutral Theory of Molecular Evolution*. Cambridge; London; New York: Cambridge University Press.

Kimura, M. & Ohta, T. (1974). On some principles governing molecular evolution. *Proceedings of the National Academy of Sciences USA* **71**, 2848–52.

King, J. L. & Jukes, T. H. (1969). Non-Darwinian evolution. *Science* **164**, 788–9.

Kirkwood, T. B. L., Rosenberger, R. F. & Galas, D. J., eds (1986). *Accuracy in Molecular Processes*. London; New York: Chapman & Hall.

Kiseleva, E. V. (1989). Secretory protein synthesis in *Chironomus* salivary gland cells is not coupled with protein translocation across endoplasmic reticulum membrane. *FEBS Letters* **257**, 251–3.

Kitcher, P. (1982). *Abusing Science. The Case Against Creationism*. Cambridge, MA; London: MIT Press.

Klig, L. S., Oxender, D. L. & Yanofsky, C. (1988). Second-site revertants of *Escherichia coli trp* repressor mutants. *Genetics* **120**, 651–5.

Klug, W. S. & Cummings, M. R. (1986). *Concepts of Genetics*, 2nd edn. Columbus, Toronto, London: Merrill Co.

Kochen, M. (1959). Extension of Moore–Shannon model for relay circuits. *IBM Journal*, April, 169–86.

Kolata, G. B. (1977). Overlapping genes: More than anomalies? *Science* **196**, 1187–8.

Kolmogorov, A. N. (1933). Grundbegriffe der Wahrscheinlichkeitsrechnung. *Ergebnisse der Mathematik* **2**, No. 3. Berlin: Springer-Verlag. English edition (1956), New York: Chelsea.

Kolmogorov, A. N. (1958). A new metric of invariants of transitive dynamical systems and automorphisms in Legesgue spaces. *Doklady Akademica Nauk SSSR* **119**, 861–4.

Kolmogorov, A. N. (1959). Entropy per unit time as a metric invariant of automorphisms. *Doklady Akademica Nauk SSSR* **124**, 754–5.

Kolmogorov, A. N. (1965). Three approaches to the concept of the amount of information. *IEEE Problems on Information Transmission* **1**, No. 1, 1–7.

Kolmogorov, A. N. (1968). Logical basis for information theory and probability theory. *IEEE Transactions on Information Theory* **14**, 662–4.

Kraft, L. G. (1949). *A Device for Quantizing, Grouping, and Coding Amplitude Modulated Pulses*. M. S. Thesis, Electrical Engineering Department, Massachusetts Institute of Technology.

Kruger, K., Grabowski, P. J., Zaug, A. J., Sands, J., Gottschling, D. E. & Cech, T. R. (1982). Self-splicing RNA: Autoexcision and autocyclization of ribosomal RNA intervening sequences of *Tetrahymna*. *Cell* **31**, 147–57.

Kuchino, Y., Beier, H., Akita, N. & Nishimura, S. (1987). Natural UAG suppressor glutamine tRNA is elevated in mouse cells infected with Moloney murine leukemia virus. *Proceedings of the National Academy of Sciences USA* **84**, 2668–72.

Kuchino, Y., Hanyu, N., Tashiro, F. & Nishimura, S. (1985). *Tetrahymena thermophilia* glutamine tRNA and its gene that corresponds to UAA termination codon. *Proceedings of the National Academy of Sciences USA* **82**, 4758–62.

Kuhn, T. S. (1957). *The Copernican Revolution*. Cambridge, MA; London: Harvard University Press.

Kuhn, T. S. (1970). *The Structure of Scientific Revolutions*. Chicago; London: University of Chicago Press.

Kullback, S. (1968). *Information Theory and Statistics*. New York: Dover Publications.

Kunisawa, T., Horimoto, K. & Otsuka, J. (1987). Accumulation pattern of amino acid substitutions in protein evolution. *Journal of Molecular Evolution* **24**, 357–65.

Küppers, B.-O. (1983). *Molecular Theory of Evolution*, translated by P. Woolley. Berlin; Heidelberg; New York: Springer-Verlag.

Küppers, B.-O. (1986). *Der Ursprung biologischer Information: Zur Naturphilosophie der Lebensentstehung*. Munich; R. Piper GmbH & Co.

Küppers, B.-O. (1990). *Information and the Origin of Life*, p. 166. Cambridge, MA; London; MIT Press. (An English translation of Küppers (1986) by P. Woolley.)

Kurland, C. G. (1986). The error catastrophe: A molecular Fata Morgana? *BioEssays* **6**, 33–5.

Kyte, F. T., Zhou, L. & Wasson, J. T. (1988). New evidence on the size and possible effects of a late pliocene oceanic asteroid impact. *Science* **241**, 63–5.

Labouygues, J.-M. & Figureau, A. (1982). L'Origine et l'évolution du code génétique. *Reviews of Canadian Biology Experiments* **41**, 209–16.

Labouygues, J.-M. & Figureau, A. (1984). The logic of the genetic code: Synonyms and optimality against effects of mutations. *Origins of Life* **14**, 405–13.

Labouygues, J.-M. & Gourgand, J.-M. (1981). Optimation contre la survenue de terminateurs pour des codes théoriques doublets. *Compte Rendue Académie Scientifique Paris* **292**, Série III, 1105–8.

Lacey, J. C. & Mullin, D. W., Jr. (1983). Experimental studies related to the origin of the genetic code and the process of protein synthesis: A review. *Origins of Life* **13**, 3–42.

Lada, C. J. & Shu, F. H. (1990). The formation of Sun-like stars. *Science* **248**, 564–72.

Lahav, N., White, D. & Chang, S. (1978). Peptide formation in the prebiotic era: Thermal condensation of glycine in fluctuating clay environments. *Science* **201**, 67–9.

Lanave, C., Preparata, G. & Saccone, C. (1985). Mammalian genes as molecular clocks? *Journal of Molecular Evolution* **21**, 346–50.

Lanave, C., Preparata, G., Saccone, C. & Serio, G. (1984). A new method for calculating evolutionary substitution rates. *Journal of Molecular Evolution* **20**, 86–93.

Lancaster, P. & Tismenetsky, M. (1985). *The Theory of Matrices*, 2nd edn. New York; London: Academic Press.

Landauer, R. (1961). Irreversibility and heat generation in the computing process. *IBM Journal of Research and Development* **5**, 183–91.

Landauer, R. (1988). Dissipation and noise immunity in computation and communication. *Nature* **335**, 779–84.

Langton, C. G. (1984). Self-reproduction in cellular automata. *Physica* **10D**, 135–44.

Laplace, P. S. de (1796). *Essai Philosophique sur les probabilities: A Treatise on Probability*, translated by F. W. Truscott & F. L. Emory (1951). New York: Dover.

Lasaga, A. C., Holland, H. D. & Dwyer, M. J. (1971). Primordial oil slick. *Science* **174**, 53–5.

Lasker, J. (1989). A numeric experiment on the chaotic behavior of the solar system. *Nature* **338**, 237–8.

Lawler, E. L., Lenstra, J. K., Ronooy Kan, A. H. G. & Shmyos, D. B. (1985). *The Traveling Salesman Problem, A Guided Tour of Combinational Optimization*. Somerset, NJ: John Wiley & Sons.

Lehman, N. & Jukes, T. H. (1988). Genetic code development by stop codon takeover. *Journal of Theoretical Biology* **135**, 203–14.

Leinfelder, W., Zehelein, E., Mandrand-Berthelot, M.-A. & Bock, A. (1988). Gene for a novel tRNA species that accepts L-serine and cotranslationally inserts selenocysteine. *Nature* **331**, 723–5.

Lemaître, G. (1931a). A homogeneous universe of constant mass and increasing radius accounting for the radial velocity of extra-galactic nebulae. *Monthly Notices of the Royal Astronomical Society* **91**, 483–90. (This is an English translation of his paper which appeared in *Annales de la Société Scientifique de Bruxelles* **XLVII**, série A, première partie, April 1927.)

Lemaître, G. (1931b). The expanding universe. *Monthly Notices of the Royal Astronomical Society* **91**, 490–501.

Lemaître, G. (1931c). The evolution of the universe. *Nature* **128**, 704–6.

Lempel, A. & Ziv, J. (1976). On the complexity of finite sequences. *IEEE Transactions on Information Theory* **IT-22**, 75–81.

Lesk, A. M. (1988). Introduction: Protein engineering. *BioEssays* **8**, 51–2.

Lesk, A. M. & Chothia, C. (1980). How different amino acid sequences determine similar protein structures: The structure and evolutionary dynamics of the globins. *Journal of Molecular Biology* **136**, 225–70.

Lesk, A. M. & Chothia, C. (1982). Evolution of proteins formed by β-sheets. II, The core of the immunoglobin domains. *Journal of Molecular Biology* **160**, 325–42.

Levin, L. A. (1984). Randomness conservation inequalities: Information and independence in mathematical theories. *Information and Control* **61**, 15–37.

Lewin, R. (1982). *Thread of Life*. Washington DC: Smithsonian Books, distributed by W. W. Norton & Co., New York.

Lewin, R. (1987). When does homology mean something else? *Science* **237**, 1570.

Li, W.-H. (1981). Simple method for constructing phylogenetic trees from distance matrices. *Proceedings of the National Academy of Sciences USA* **78**, 1085–9.

Li, M. & Tzagoloff, A. (1979). Assembly of the mitochondrial membrane system: Sequences of yeast mitochondrial valine and an unusual threonine tRNA gene. *Cell* **18**, 47–53.

Liang, N., Mauk, A. G., Pielak, G. J., Johnson, J. A., Smith, M. & Hoffman, B. M. (1988). Regulation of interprotein electron transfer by residue 82 of yeast cytochrome c. *Science* **240**, 311–13.

Liang, N., Peilak, G. J., Mauk, A. G., Smith, M. & Hoffman, B. M. (1987). Yeast cytochrome c with phenylalanine or tyrosine at position 87 transfers electrons to (zinc cytochrome c peroxidase)+ at a rate ten thousand times that of the serine-87 or glycine-87 variants. *Proceedings of the National Academy of Sciences USA* **84**, 1249–52.

Lim, W. A. & Sauer, R. T. (1989). Alternative packing arrangements in the hydrophobic core of λ-repressor. *Nature* **339**, 31–6.

Lindley, D. V. (1970). *Introduction to Probability and Statistics from a Bayesian Viewpoint*, Part 1. Cambridge, England: Cambridge University Press.

Lindley, D. (1990). An excess of perfection. *Nature* **343**, 207.

Lindley, D. V. (1965). *Introduction to Probability and Statistics*, Part 1. Cambridge: Cambridge University Press.

Lipman, D. J. & Wilbur, W. J. (1983). Contextual constraints on synonymous codon choice. *Journal of Molecular Biology* **163**, 363–76.

Lipman, D. J. & Wilbur, W. J. (1985). Interaction of silent and replacement changes in eukaryotic coding sequences. *Journal of Molecular Evolution* **21**, 161–7.

Loeb, J. (1912). The mechanistic conception of life. In *The Mechanistic Conception of Life*. Chicago, IL: University of Chicago Press.

Loeb, J. (1916). *The Organism as a Whole*, p. 39. New York; London: G. P. Putnam's Sons.

Loeb, W. (1913). Über das Verhalten des Formamids unter der Wirkung stillen Entlandung. *Berichte der chemische Gesellschaft* **46**, 684–97.

Loftfield, R. B. (1963). The frequency of errors in protein synthesis. *Biochemical Journal* **89**, 82–7.

Loftfield, R. B. & Vanderjagt, D. (1972). The frequency of errors in protein biosynthesis. *Biochemical Journal* **128**, 1353–6.

Lowe, D. R., Byerly, G. R., Asaro, F. & Kyte, F. J. (1989). Geological and geochemical record of 3400-million-year-old terrestrial meteorite impacts. *Science* **245**, 959–62.

Lyle, M. (1988). Climatically forced organic carbon burial in equatorial Atlantic and Pacific Oceans. *Nature* **335**, 529–32.

MacDougall, J. D. (1988). Seawater strontium isotopes, acid rain, and the Cretaceous-Tertiary boundary. *Science* **239**, 485–7.

Macino, G., Coruzzi, G., Nobrega, F. G., Li, M. & Tzagoloff, A. (1979). Use of UGA terminator as a tryptophan codon in yeast. *Proceedings of the National Academy of Sciences USA* **76**, 3784–5.

MacKay, A. L. (1967). Optimization of the genetic code. *Nature* **216**, 159–160.

MacLane, S. (1986). *Mathematics Form and Function*. New York; Heidelberg: Springer-Verlag.

Maddox, J. (1983). Is biology now a part of physics? *Nature* **306**, 311.

Maddox, J. (1986). Can chance be less than zero? *Nature* **320**, 481.

Maddox, J. (1988). Finding wood among the trees. *Nature* **333**, 11.

Maddox, J. (1989a). Where next with peer-review? *Nature* **339**, 11.

Maddox, J. (1989b). Down with the Big Bang. *Nature* **340**, 425.

Maher, K. A. & Stevenson, D. J. (1988). Impact frustration of the origin of life. *Nature* **331**, 612–14.

Mandelbrot, B. B. (1982). *The Fractal Geometry of Nature*. New York: Freeman.

Mann, S., Sparks, N. H. C., Frankel, R. B., Bazylinski, D. A. & Holger, W. J. (1990). Biomineralization of ferrimagnetic greigite (Fe_3S_4) and iron pyrite (FeS_2) in a magnetostatic bacterium. *Nature* **343**, 258–61.

Maréchal, L., Guillemaut, P., Grienenberger, J.-M., Jeannin, G. & Weil, J.-H. (1985). Sequence and codon recognition of bean mitochondria and chloroplast tRNAs[Trp]: Evidence for a high degree of homology. *Nucleic Acids Research* **13**, 4411–16.

Margoliash, E. (1972). Functional evolution of cytochrome c. In *The Harvey Lectures*, Series 66. New York: Academic Press.

Margoliash, E. & Fitch, W. M. (1971). The evolutionary information content of protein amino acid sequences, *Comptes Rendue de la Second Conférence International de la Physique Théorique et Biologie*, ed. M. Marois, pp. 208–24. Paris: Centre National de la Recherche Scientifique.

Margoliash, E., Barlow, G. H. & Byers, V. (1970). Differential binding properties of cytochrome c: Possible relevance for mitochondrial ion transport. *Nature* **228**, 723–6.

Margoliash, E., Fitch, W. M., Markowitz, E. & Dickerson, R. E. (1972). In *Oxidation Reduction Enymes*, eds A. Akeson & A. Ehrenberg. Oxford; New York: Pergamon Press.

Margoliash, E., Ferguson-Miller, S., Tullos, J., Kang, C. H., Feinberg, B. A.,

Brautigan, D. L. & Morrison, M. (1973). Separate intramolecular pathways for reduction and oxidation of cytochrome c in electron transport chain reaction. *Proceedings of the National Academy of Sciences, USA* **70**, 3245–9.

Margulis, L. (1970). *Origin of Eucaryote Cells.* New Haven: Yale University Press.

Marks, C. B., Naderi, H., Kosen, P. A., Kuntz, I. D. & Anderson, S. (1987). Mutants of bovine pancreatic trypsin inhibtor lacking cysteines 14 and 38 can fold properly. *Science* **235**, 1370–3.

Martin, N. F. G. & England, J. W. (1981). *Mathematical Theory of Entropy.* London; Reading, MA: Addison-Wesley.

Martin-Löf, P. (1966). The definition of randomness. *Information and Control* **9**, 602–19.

Matthews, G. J. & Cowan, J. J. (1990). New insights into the astrophysical r-process. *Nature* **345**, 491–4.

Matsui, T. & Abe, Y. (1986). Impact-induced atmospheres and oceans on Earth and Venus. *Nature* **322**, 526–8.

Matsumara, M., Signor, G. & Matthews, B. W. (1989). Substantial increase of protein stability by multiple disulphide bonds. *Nature* **342**, 291–3.

Maxwell, C. (1871). *The Theory of Heat*, 1st edn. London: Longmans Green.

Maynard Smith, J. (1970). Natural selection and the concept of a protein space. *Nature* **225**, 563–4.

Maynard Smith, J. (1972). *On Evolution.* Edinburgh: Edinburgh University Press.

Maynard Smith, J. (1979). Hypercycles and the origin of life. *Nature* **280**, 445–6.

Mayr, E. (1982). The place of biology in the sciences and its conceptional structure. In *The growth of biological thought*, pp. 21–82. Cambridge, MA; London: Harvard University Press.

Mayr, E. (1988). Introduction, pp. 1–7; Essay 1, Is biology an autonomous science?, pp. 8–23. In *Toward a New Philosophy of Biology*, Part I. Cambridge, MA; London: Harvard University Press.

McCarthy, E. D. & Calvin, M. (1967). Organic geochemical studies. I. Molecular criteria for hydrocarbon genesis. *Nature* **216**, 642–7.

McEliece, R. J. (1984). *The Theory of Information and Coding.* Cambridge: Cambridge University Press.

McKay, C. (1986). Exobiology and future Mars missions: The search for Mars' earliest biosphere. *Advances in Space Research* **6**, 269–85.

McLachlan, A. D. (1971). Tests for comparing related amino-acid sequences, cytochrome c and cytochrome c551. *Journal of Molecular Biology* **61**, 409–24.

McMillan, B. (1953). The basic theorems of information theory. *Annals of Mathematical Statistics* **24**, 196–219.

McPherson, A. & Schlichta, P. (1988). Heterogeneous and epitaxial nucleation of protein crystals on mineral surfaces. *Science* **239**, 385–7.

Medvedev, Z. A. (1966). Protein biosynthesis and problems of heredity, development and ageing (translated by Ann Synge). New York: Plenum Press.

Melosh, H. J. & Vickery, A. M. (1989). Impact erosion of the primordial atmosphere of Mars. *Nature* **338**, 487–9.

Meyer, F. Schmidt, H. J., Plümper, E. H., Andrej, M. G., Meyer, H. E., Engstrom, A. & Heckman, K. (1991). UGA translated as cysteine in pheromone 3 of *Euplotes octocarinatus*. *Proceedings of the National Academy of Sciences USA* **88**, 3758–61.

Meyer, T. E., Cusanovich, M. A. & Kamen, M. D. (1986). Evidence against use of bacterial amino acid sequence data for construction of all-inclusive phylogenetic trees. *Proceedings of the National Academy of Sciences USA* **83**, 217–20.

Mikelsaar, R. (1987). A view of early cellular evolution. *Journal of Molecular Evolution* **25**, 168–183.

Mildvan, A. S. & Strehler, B. (1960). A critique of theories of mortality. In *The Biology of Aging, A Symposium*, ed B. Strehler. Washington DC: American Institute of Biological Sciences, Publication No. 6.

Miller, S. L. (1953). A production of amino acids under possible primitive Earth conditions. *Science* **117**, 528–9.

Miller, S. L. (1955). Production of some organic compounds under possible primitive Earth conditions. *Journal of the American Chemical Society* **77**, 2351–61.

Miller, S. L. & Bada, J. L. (1988). Submarine hot springs and the origin of life. *Nature* **334**, 609–11.

Miller, S. L. & Bada, J. L. (1989). Origin of life. *Nature* **337**, 23. (A reply to Nisbet (1989), *Nature* **337**, 23.)

Miller, S. L. & Orgel, L. E. (1974). *The Origins of Life on Earth*. Englewood Cliffs, NJ; Prentice-Hall.

Miller, S. L. & Urey, H. C. (1959). Organic compound synthesis on the primitive Earth. *Science* **130**, 245–51.

Miyata, T. & Yasunaga, T. (1978). The evolution of overlapping genes. *Nature* **272**, 532–5.

Miyata, T., Hayashida, H., Kikuno, R., Hasegawa, M., Kobayashi, M. & Koike, K. (1982). Molecular clock of silent substitution: At least six-fold preponderance of silent changes in mitochondrial genes over those in nuclear genes. *Journal of Molecular Evolution* **19**, 28–35.

Mizutani, T. & Hitaka, T. (1988). The conversion of phosphoserine residues to selenocysteine residues on an opal suppressor tRNA and casein. *Federation of European Biochemical Societies Letters* **232**, 243–8.

Monod, J. (1972). *Chance and Necessity*. New York: Knopf.

Montoya, J., Ojala, D. & Attardi, G. (1981). Distinctive features of the 5'-terminal sequences of the human mitochondrial mRNAs. *Nature* **290**, 465–70.

Moore, C. (1990). Unpredictability and undecidability in dynamic systems. *Physical Review Letters* **64**, 2354–7.

Moore, E. F. & Shannon, C. E. (1956). Reliable circuits using less reliable relays. Part I, *Journal of the Franklin Institute* **262**, 191–208; Part II, *Journal of the Franklin Institute* **262**, 281–97.

Moran, P. A. P. (1986). *An Introduction to Probability Theory*. London; Oxford: Oxford University Press.

Morrison, P. (1981). Reflections. In *Life in the Universe*, ed. J. Billingham. Cambridge, MA; London: MIT Press.

Mueckenheim, W. (1986). A review of extended probabilities. *Physics Reports* **166**, 337–410.

Mullenbach, G., Tabrizi, A., Irvine, B. D., Bell, G. I. & Hallewell, R. A. (1987). Sequence of a cDNA coding for human glutathione peroxidase confirms TGA encodes active site selenocysteine. *Nucleic Acids Research* **15**, 5484.

Muller, R. (1989). *Nemesis: The Death Star*. New York: Weidenfeld & Nicolson.

Muto, A., Yamao, F., Hori, H. & Osawa, S. (1985). Gene organization of *Mycoplasma caprolium*. *Advances in Biophysics* **21**, 49–56.

Nagai, K., Luisi, B. & Shih, D. (1988). Evolution of haemoglobin studied by protein engineering. *BioEssays* **8**, 79–82.

Nakashima, T., Jungck, J. R., Fox, S. W., Lederer, E. & Das, B. C. (1977). A test for randomness in peptides isolated from a thermal polyamino acid. *International Journal of Quantum Chemistry: Quantum Biology Symposium* **4**, 65–72.

Nature, unsigned (1970). Central Dogma reversed. *Nature* **226**, 1198–9.

Nauta, D. (1972). *The Meaning of Information*. The Hague; Paris: Mouton.

Needleman, S. B. & Wunsch, C. D. (1970). A general method applicable to the search for similarities in the amino acid sequence of two proteins. *Journal of Molecular Biology* **48**, 443–53.

Newman, M. J. & Rood, R. T. (1977). Implications of solar evolution for the Earth's early atmosphere. *Science* **198**, 1035–937.

Newsom, H. E. & Taylor, S. R. (1989). Geochemical implications of the formation of the Moon by a single giant impact. *Nature* **338**, 29–34.

Ney, E. P. (1977). Star dust. *Science* **195**, 541–6.

Nichols, B. P. & Yanofsky, C. (1979). Nucleotide sequences of *trpA* of *Salmonella typhimurium* and *Escherichia coli*: An evolutionary comparison. *Proceedings of the National Academy of Sciences USA* **76**, 5244–8.

Niesert, U. (1987). How many genes to start with? A computer simulation about the origin of life. *Origins of Life* **17**, 155–69.

Niesert, U., Harnasch, D. & Bresch C. (1981). Origin of life between Scylla and Charybdis. *Journal of Molecular Evolution* **17**, 348–53.

Ninio, J. (1975). Kinetic amplification of enyme discrimination. *Biochemie* **57**, 587–95.

Nisbet, E. G. (1985). The geological setting of the earliest life forms. *Journal of Molecular Evolution* **21**, 289–98.

Nisbet, E. G. (1986a). RNA and hot-water springs. *Nature* **322**, 206.

Nisbet, E. G. (1986b). RNA, Hydrothermal systems, zeolites and the origin of life. *Episodes* **9**, 83–90.

Nisbet, E. G. (1987). *The Young Earth*. Boston: Allen & Unwin.

Nisbet, E. G. (1989). Origin of life. *Nature* **337**, 23. (A reply to Miller & Bada (1989), *Nature* **334**, 609–11.)

Noren, C. J., Anthony-Cahill, S. J., Griffith, M. C. & Schult, P. G. (1989). A general method for site-specific incorporation of unnatural amino acids into proteins. *Science* **244**, 182–8.

Oberbeck, V. R. & Fogelman, G. (1989). Impacts and the origin of life. *Nature* **339**, 434.

Officer, C. B., Hallam, A., Drake, C. L. & Devine, J. D. (1987). Late Cretaceous and paroxysmal Cretaceous/Tertiary extinctions. *Nature* **326**, 143–9.

Ohno, S. (1970). *Evolution by Gene Duplication*. Heidelberg; Berlin; New York: Springer-Verlag.

Ohta, T. (1987). Very slightly deleterious mutations and the molecular clock. *Journal of Molecular Evolution* **26**, 1–6.

Ohtsuki, K., Nakagawa, Y. & Nakazawa, K. (1988). Growth of the Earth in nebular gas. *Icarus* **75**, 552–65.

O'Keefe, J. D. & Ahrens, T. J. (1989). Impact production of CO_2 by the Cretaceous/Tertiary extinction bolide and the resultant heating of the Earth. *Nature* **338**, 247–9.

Oort, J. H. (1950). The structure of the cloud of comets surrounding the solar system, and a hypothesis concerning its origin. *Bulletin of the Astronomical Institute of the Netherlands* **11**, 91–110.

Oparin, A. I. (1924). Proiskhozdenic Zhizny, Moscow: Izd Moskovski Rabochii. Published in English translation in J. D. Bernal (1967), *The Origin of Life*. London: Weidenfeld & Nicolson.

Oparin, A. I. (1938). *The Origin of Life*. New York: MacMillan.

Oparin, A. I. (1957). *The Origin of Life on Earth*, 3rd revised edn (translated by Ann Synge). Edinburgh; London: Oliver & Boyd.

Oparin, A. I. (1972). The appearance of life in the Universe. In *Exobiology*, ed. C. Ponnamperuma. Amsterdam; London: North-Holland.

Orgel, L. E. (1963). The maintenance of the accuracy of protein synthesis and its relevance to ageing. *Proceedings of the National Academy of Sciences USA* **49**, 517–21.

Orgel, L. E. (1968). Evolution of the genetic apparatus. *Journal of Molecular Biology* **38**, 381–93.

Orgel, L. E. (1970). The maintenance of the accuracy of protein synthesis and its relevance to ageing: A correction. *Proceedings of the National Academy of Sciences USA* **67**, 1476.

Orgel, L. E. (1973a). Ageing of clones of mammalian cells. *Nature* **243**, 441–5.

Orgel, L. E. (1973b). *The Origin of Life: Molecules and Natural Selection*. New York: John Wiley & Sons.

Orgel, L. E. (1986). RNA catalysis and the origin of life. *Journal of Theoretical Biology* **123**, 127–49.

Orgel, L. E. (1990). Adding to the genetic alphabet. *Nature* **343**, 18–20.

Ornstein, D. S. (1970). Bernoulli shifts with the same entropy are isomorphic. *Advances in Mathematics* **4**, 337–52.

Ornstein, D. S. (1974). *Ergodic Theory, Randomness, and Dynamical Systems*, Yale Mathematical Monographs 5. New Haven, CT: Yale University Press.

Ornstein, D. S. (1989). Ergodic theory, randomness and 'chaos'. *Science* **243**, 182–7.

Osawa, S., Jukes, T. H., Muto, A., Yamao, F., Ohama, T. & Andachi, Y. (1987). Role of directional mutation pressure in the evolution of the eubacterial genetic code. *Cold Spring Harbor Symposia on Quantitative Biology* **52**, 777–89.

Othmer, M. (1989). Missing formula. *Nature* **341**, 380.

Overbeck, V. R. & Fogelman, G. (1989). Impacts and the origin of life. *Nature* **339**, 434.

Owen, T., Cess, R. D. & Ramanathan, V. (1979). Enhanced CO_2 greenhouse to compensate for reduced solar luminosity. *Science* **277**, 640–2.

Packel, E. W. & Traub, J. F. (1987). Information-based complexity. *Nature* **328**, 29–33.

Paecht-Horowitz, M. (1971). Polymerization of amino acid phosphate anhydrates in the presence of clay minerals. In *Chemical Evolution and the Origin of Life*, eds R. Buvet & C. Ponnamperuma. Amsterdam; London: North-Holland.

Paecht-Horowitz, M., Berger, J. & Katchalsky, A. (1970). Prebiotic synthesis of polypeptides by heterogeneous polycondensation of amino-acid adenylates. *Nature* **228**, 636–9.

Pais, A. (1982) '*Subtle is the Lord.' The Science and the Life of Albert Einstein.* Oxford: Clarendon Press.

Parker, J. (1989). Errors and alternatives in reading the universal genetic code. *Microbiological Reviews* **53**, 273–98.

Parthia, R. K. (1962). A statistical study of randomness among the first 10000 digits of π. *Mathematics of Computation* **16**, 188–97.

Patterson, J. W. (1983). Thermodynamics and evolution. In *Scientists Confront Creationism*, ed. L. R. Godfrey, pp. 99–116. New York, London: W. W. Norton & Co.

Pauling, L. (1957). The probability of errors in the process of synthesis of protein molecules. In *Festschrift for Prof. Dr. Arthur Stoll zum Siebsigsten Geburtstag*, pp. 597–602. Basel: Birkhäser Verlag.

Pauly, P. J. (1987). *Controlling Life: Jacques Loeb and the Engineering Ideal in Biology*. Oxford; New York: Oxford University Press.

Peebles, P. J. E. & Silk, J. (1990). A cosmic book of phenomena. *Nature* **346**, 233–9.

Pelc, S. R. & Welton, M. G. E. (1966). Steriochemical relationship between coding triplets and amino-acids. *Nature* **209**, 868–70.

Penrose, R. (1989). *The Emperor's New Mind*. Oxford; New York; Melbourne: Oxford University Press.

Penzias, A. A. & Wilson, R. W. (1965). A measurement of excess antenna temperature at 4080 Mc/s. *Astrophysical Journal* **142**, 419–21.

Perlwitz, M. D., Burks, C. & Waterman, M. S. (1988). Pattern analysis of the genetic code. *Advances in Applied Mathematics* **9**, 7–21.

Perron, O. (1907). Zur Theorie der Matrizen. *Mathematische Annalen* **64**, 248–63.

Perutz, M. F. (1987). Physics and the riddle of life. *Nature* **326**, 555–8.

Petersen, K. (1983). *Ergodic Theory*. Cambridge: Cambridge University Press.

Peterson, W. W. & Weldon, E. J., Jr. (1972). *Error-correcting Codes*, 2nd edn. Cambridge, MA; London: MIT Press.

Pflug, H. D. (1984). Early Earth geological record and the origin of life. *Naturwisschenschaften* **71**, 63–8.

Piccirilli, J. A., Krauch, T., Moroney, S. E. & Benner, S. A. (1990). Enzymatic incorporation of a new base pair into DNA and RNA extends the genetic alphabet. *Nature* **343**, 33–7.

Pickert, G. & Görke, L. (1974). Construction of the system of real numbers. In *Fundamentals of Mathematics*, Vol. 1, eds H. Behnke, F. Bachmann, K. Fladt & W. Süss, pp. 93–153. Cambridge, MA: MIT Press.

Planck, M. (1906). *Vorlesungen über die Theorie der Wärmestrahlung*. Leipzig: Barth.

Pocklington, R. (1971). Free amino-acids dissolved in North Atlantic Ocean waters. *Nature* **230**, 374–5.

Ponnamperuma, C. (1972). Organic compounds in the Murchison meteorite. *Annals of the New York Academy of Science* **194**, 56–70.

Ponnamperuma, C. (1982). Primordial organic chemistry. In *Extraterrestrials: Where are they?*, eds M. H. Hart & B. Zuckerman. New York; Oxford: Pergamon Press.

Ponnamperuma, C. (1983). Cosmochemistry and the origin of life. In

Cosmochemistry and the Origin of Life, ed. C. Ponnamperuma. Dordrecht; Boston; London: Reidel.

Ponnamperuma, C., Shimoyama, A. & Friebele, E. (1982). Clay and the origin of life. *Origins of Life* **12**, 9–40.

Popper, K. R. (1959). *The Logic of Scientific Discovery*. London: Hutchinson & Co.

Popper, K. R. (1963). *Conjectures and Refutation: The Growth of Scientific Knowledge*. London: Routledge & Kegan Paul.

Popper, K. R. (1972). *Objective Knowledge: An Evolutionary Approach*. Oxford: Clarendon.

Popper, K. R. (1990). Pyrite and the origin of life. *Nature* **344**, 387.

Pounds, K. A., Nandra, K., Stewart, G. C., George, I. M. & Fabian, A. C. (1990). X-ray reflection from cold matter in the nuclei of active galaxies. *Nature* **344**, 132–3.

Powell, L. M. W., Pease, R. J., Edwards, Y. H., Knott, T. J. & Scott, J. (1987). A novel form of tissue-specific RNA processing produces apolipoprotein-B48 in intestine. *Cell* **50**, 831–40.

Preer, J. R., Preer, L. B., Rudman, B. M. & Barnett, A. J. (1985). Deviation from the universal code shown by the gene for surface protein 51A in *Paramecium*. *Nature* **314**, 188–90.

Preparata, G. & Saccone, C. (1987). A simple quantitative model of the molecular clock. *Journal of Molecular Evolution* **26**, 7–15.

Prigogine, I. & Nicolis, G. (1971). Biological order, structure and instabilities. *Quarterly Review of Biophysics* **4**, 2 and 3, 107–48.

Prigogine, I., Nicolis, G. & Babloyantz, A. (1972a). Thermodynamics of evolution: Part I. *Physics Today* **25**, 23–8.

Prigogine, I. Nicolis, G. & Babloyantz, A. (1972b). Thermodynamics of evolution: Part II. *Physics Today* **25**, 38–44.

Project Cyclops (1973). Moffett Field, California, NASA CR114445.

Prusiner, S. B. (1982). Novel proteinaceous infectious particles cause scrapie. *Science* **216**, 136–44.

Pütz, J. P., Puglis, J. D., Florentz, C. & Giegé, R. (1991). Identity elements for specific aminoacylation of yeast tRNA[Asp] by cognate aspartyl-tRNA synthetase. *Science* **252**, 1696–9.

Quastler, H. (1953). *Information Theory in Biology*. Urbana, IL: University of Illinois Press.

Quastler, H. (1964). *The Emergence of Biological Organization*. New Haven; London: Yale University Press.

Raff, R. & Mahler, H. R. (1972). The non symbiotic origin of mitochondria. *Science* **177**, 575–82.

Raff, R. & Mahler, H. R. (1973). Origin of mitochondria. *Science* **180**, 517.

Rampino, M. R. & Stothers, R. B. (1984). Terrestrial mass extinctions and the Sun's motion perpendicular to the galactic plane. *Nature* **308**, 709–12.

Rampino, M. R. & Stothers, R. B. (1988). Flood basalt volcanism during the past 250 million years. *Science* **241**, 663–8.

Raup, D. M. (1986). *The Nemesis Affair: A Story of the Death of Dinosaurs and the Ways of Science*. New York: W. W. Norton & Co.

Raup, D. M. & Seproski, J. J., Jr. (1984). Periodicity of extinctions in the geologic past. *Proceedings of the National Academy of Sciences USA* **81**, 801–5.

Raup, D. M. & Seproski, J. J., Jr. (1986). Periodic extinction of families and genera. *Science* **231**, 833–6.

Raup, D. M. & Seproski, J. J., Jr. (1988). Testing for periodicity of extinctions. *Science* **241**, 94–6.

Raup, D. M. & Valentine, J. W. (1983). Multiple origins of life. *Proceedings of the National Academy of Sciences USA* **80**, 2981–4.

Reddy, V. B., Themmappaya, B., Dhar, R., Subramanian, K. N., Zain, B. S., Pan, J., Ghosh, P. K., Celma, M. L. & Weissman, S. M. (1978). The genome of simian virus 40. *Science* **200**, 494–502.

Reeck, G. R., de Haën, C., Teller, D. C., Doolittle, R. F., Fitch, W. M., Dickerson, R. E., Chambon, P., McLachan, A. D., Margoliash, E., Jukes, T. H. & Zuckerkandl, E. (1987). Homology in proteins and nucleic acids: A terminological muddle and a way out of it. *Cell* **50**, 667.

Rees, M. J. (1988). Tidal disruption of stars by black holes of 10^6–10^8 solar masses in nearby galaxies. *Nature* **333**, 523–8.

Reidhaar-Olson, J. F. & Sauer, R. T. (1988). Combinatorial cassette mutagenesis as a probe of the informational content of protein sequences. *Science* **241**, 55–7.

Rein, R., Nir, S. & Stamadiadou, M. N. (1971). Photochemical survival principle in molecular evolution. *Journal of Theoretical Biology* **33**, 309–18.

Richards, M. A., Duncan, R. A. & Courtillot, V. E. (1989). Flood basalts and hot-spot tracks: Plume heads and tails. *Science* **246**, 103–7.

Richardson, J. S. & Richardson, D. C. (1988). Amino acid preferences for specific locations at the ends of α helices. *Science* **240**, 1648–52.

Riddle, D. & Carbon, J. (1973). Frameshift suppression: a nucleotide addition in the anticodon of a glycine transfer RNA. *Nature New Biology* **242**, 230–4.

Riddle, D. & Roth, J. R. (1972). Frameshift suppression II. Genetic mapping and dominance studies. *Journal of Molecular Biology* **66**, 483–93.

Reichert, T. A. (1972). The amount of information stored in proteins and other short biological code sequences. In *Proceedings of the Sixth Berkeley Symposium on Mathematical Statistics and Probability*. Berkeley; Los Angeles: University of California Press.

Rissanen, J. (1986). Complexity of strings in the class of Markov sources. *IEEE Transactions on Information Theory* **IT-32**, 526–32.

Root-Bernstein, R. S. (1983). Protein replication by amino acid pairing. *Journal of Theoretical Biology* **100**, 99–106.

Rosenberger, R. F. & Kirkwood, T. B. L. (1986). Errors and the integrity of genetic information transfer. In *Accuracy in Molecular Processes*, eds T. B. L. Kirkwood, R. F. Rosenberger & D. J. Galas. London; New York: Chapman & Hall.

Roth, J. R. (1981). Frameshift suppression. *Cell* **24**, 601–2.

Rowlinson, J. S. (1970). Probability, information and entropy. *Nature* **225**, 1196.

Rudnick, J. & Gaspari, G. (1987). The shape of random walks. *Science* **237**, 384–338.

Russell, M. J., Hall, A. J. & Gize, A. P. (1990). Pyrite and the origin of life. *Nature* **344**, 387.

Russell, M. J., Hall, A. J., Cairns-Smith, A. G. & Braterman, P. S. (1988). Submarine hot springs and the origin of life. *Nature* **336**, 117.

Ryan, J. P. (1972). Information, entropy and various systems. *Journal of Theoretical Biology* **36**, 139–46.

Ryan, J. P. (1980). Information–entropy interfaces and different levels of biological organization. *Journal of Theoretical Biology* **84**, 31–48.

Saccone, C., Lanave, C., Pesole, G. & Preparata, G. (1990). Influence of base composition on quantitative estimates of gene evolution. In *Methods in Enzymology*, ed. R. F. Doolittle, pp. 570–83. San Diego, CA: Academic Press.

Saccone, C., Attimonelli, M., Lanave, C., Gallerani, R. & Pesole, G. (1988). The evolution of mitochondrially coded cytochrome genes: A quantitative estimate. In *Cytochrome Systems, Molecular Biology and Bioenergetics*, eds S. Papa, B. Chance & L. Ernster. New York; London: Plenum Press.

Sagan, C. (1965). Organic matter and the Moon. In *The Origins of Preobiological Systems*, ed S. W. Fox. New York: Academic Press.

Sagan, C. (1973). Ultraviolet selection pressure on the earliest organisms. *Journal of Theoretical Biology* **39**, 195–200.

Sagan, C. (1977). Reducing greenhouses and the temperature history of Earth and Mars. *Nature* **269**, 224–6.

Sagan, C. & Khare, B. N. (1971). Longwavelength ultraviolet photoproduction of amino acids on the primitive Earth. *Science* **173**, 417.

Sagan, C. & Mullen, G. (1972). Earth and Mars: Evolution of atmospheres and surface temperatures. *Science* **177**, 52–6.

Sagan, C. & Newman, W. T. (1983). The solipsist approach to extraterrestrial intelligence. *Quaterly Journal of the Royal Astronomical Society* **24**, 113–21. (Reprinted in *Extraterrestrials – Science and Alien Intelligence*, ed. E. Regis, Jr. (1987), Cambridge; New York: Cambridge University Press.)

Saitou, N. & Nei, M. (1987). The neighbor-joining method: A new method for reconstructing phylogenetic trees. *Molecular Biological Evolution* **4**, 406–25.

Sandage, A. & Tamman, G. A. (1984). The Hubble constant as derived from 21 cm linewidths. *Nature* **307**, 326–9.

Sander, C. & Schulz, G. (1979). Degeneracy of the information contained in amino acid sequences: Evidence from overlaid genes. *Journal of Molecular Evolution* **13**, 245–52.

Sanger, F., Air, G. M., Barrell, B. G., Brown, N. L., Coulson, A. R., Fiddes, J. C., Hutchison, C. A., III, Slocombe, P. M. & Smith, M. (1977). Nucleotide sequence of bacteriophage ΦX174. *Nature* **268**, 687–95.

Saunders, P. T. & Ho, M. W. (1976). On the increase of complexity in evolution. *Journal of Theoretical Biology* **63**, 375–84.

Saunders, P. T. & Ho, M. W. (1981). On the increase in complexity in evolution II. The relativity of complexity and the principle of minimum increase. *Journal of Theoretical Biology* **90**, 515–30.

Schidlowski, M. (1976). Archean atmosphere and evolution of the terrestrial oxygen budget. In *The Early History of the Earth*, ed. B. F. Windley. London; New York: John Wiley & Sons.

Schidlowski, M. (1983). Biologically mediated isotope fractionations: Biochemistry, geochemical significance and preservation in the Earth's oldest sediments. In *Cosmochemistry and the origin of life*, ed. C. Ponnamperuma, pp. 277–322. Dordrecht; Boston; London: Reidel.

Schidlowski, M. (1988). A 3,800-million-year isotopic record of life from carbon in sedimentary rocks. *Nature* **333**, 313–18.

Schidlowski, M., Hayes, J. M. & Kaplan, I. R. (1983). Isotopic inferences of ancient biochemistries: Carbon, sulphur, hydrogen, and nitrogen. In *Earth's Earliest Biosphere: Its Origin and Evolution*, ed. J. W. Schopf, pp. 149–85. Princeton, NJ: Princeton University Press.

Schlesinger, G. & Miller, S. L. (1983a). Prebiotic synthesis in atmospheres containing CH_4, CO and CO_2. I. Amino acids. *Journal of Molecular Evolution* **19**, 376–82.

Schlesinger, G. & Miller, S. L. (1983b). Prebiotic synthesis in atmospheres containing CH_4, CO and CO_2. II. Hydrogen cyanide, formaldehyde and ammonia. *Journal of Molecular Evolution* **19**, 383–90.

Schneider, S. U., Leible, M. B. & Yang, X.-P. (1989). Strong homology between the small subunit of ribulose-1,5-bisphosphate carboxylase/oxygenase of two species of Acetabularia and the occurrence of unusual condon usage. *Molecular General Genetics* **218**, 445–52.

Schön, A., Kannangara, C. G., Gough, S. & Söll, D. (1988). Protein biosynthesis in organelles require misaminoacylation of tRNA. *Nature* **331**, 187–90.

Schopf, J. W. (1972). Precambrian paleobiology. In *Exobiology*, ed. C. Ponnamperuma. Amsterdam; London: North-Holland.

Schopf, J. W. & Packer, B. M. (1987). Early Archean (3.3-billion to 3.5-billion-year-old) microfossils from Warrawoona Group, Australia. *Science* **237**, 70–3.

Schrödinger, E. (1944). *What is Life? The Physical Aspects of the Living Cell*. Cambridge: Cambridge University Press. (See also comments on Schrödinger's book and the origin of the ideas in it in M. F. Perutz (1987), Physics and the riddle of life, *Nature* **326**, 555–8.)

Schwartz, A. & Bakker, C. G. (1989). Was adenine the first purine? *Science* **245**, 1102–4.

Schwartz, R. D. & James, P. B. (1984). Periodic mass extinctions and the Sun's oscillation about the galactic plane. *Nature* **308**, 712–5.

Schwartzman, D. W. & Volk, T. (1989). Biotic enhancement of weathering and the habitability of Earth. *Nature* **340**, 457–60.

Seneta, E. (1973). *Non-negative Matrices, An Introduction to Theory and Application*. London: George Allen & Unwin.

Shannon, C. E. (1948). A mathematical theory of communication. *Bell System Technical Journal* **27**, 379–423, 623–56.

Shannon, C. E. (1951). Prediction and entropy of printed English. *Bell System Technical Journal* **30**, 50–64.

Shannon, C. E. (1956). The zero capacity of a noisy channel *IRE Transactions on Information Theory* **IT-2**, 8–19.

Shannon, C. E. & Weaver, W. (1949). *The Mathematical Theory of Communication*. Urbana, IL: University of Illinois Press.

Shapiro, R. (1986). Doubt and uncertainty; Bubbles, ripples and mud. In *Origins, A Skeptic's Guide to the Creation of Life on Earth*, Chapter 8, pp. 190–224. New York: Summit Books.

Shapiro, S. L. & Teukolsky, S. A. (1988). Building black holes: Supercomputer cinema. *Science* **241**, 421–5.

Sharma, B. D., Mitter, J. & Mohan, M. (1987). On measures of 'useful' information. *Information and Control* **39**, 323–36.

Shaw, D. C., Bloom, R. W. & Bowman (1977). Hemoglobin and the genetic code. Evolution of protection against somatic mutations. *Journal of Molecular Evolution* **9**, 225–30.

Shaw, D. C., Walker, J. E., Northrop, F. D., Barrell, B. G., Godson, G. N. & Fiddes, J. C. (1978). Gene K, a new overlapping gene in bacteriophage G4. *Nature* **272**, 510–15.

Shaw, J. M., Feagin, J., Stuart, K. & Simpson, L. (1988). Editing of kinetoplastid mitochondrial mRNAs by uridine addition and deletion generates conserved amino acid sequences and AUG initiation codons. *Cell* **53**, 401–11.

Sherman, F., Stewart, J. W., Parker, J. H., Putterman, G. J., Agrawal, B. B. L. & Margoliash, E. (1970). The relationship of gene structure and protein structure of iso-1-cytochrome c from yeast. In *Symposia of the Society for Experimental Biology*, XXIV, *Control of Organelle Development*, pp. 85–107. Cambridge: Cambridge University Press.

Shields, P. C. (1973). *The Theory of Bernoulli Shifts*. Chicago, IL: University of Chicago Press.

Shklovskii, I. S. & Sagan, C. (1966). *Intelligent Life in the Universe*. New York: Dell Publications.

Shore, J. E. & Johnson, R. W. (1980). Axiomatic derivation of the principle of maximum entropy and the principle of minimum cross-entropy. *IEEE Transactions of Information Theory* **IT-26**, 26–37.

Shore, J. E. & Johnson, R. W. (1983). Comments on and correction to 'Axiomatic derivation of the principle of maximum entropy and the principle of minimum cross-entropy.' *IEEE Transactions of Information Theory* **IT-29**, 942–3.

Sibler, A.-P., Dirheimer, G. & Martin, R. (1981). Nucleotide sequence of a yeast mitochondrial threonine-tRNA able to decode the C-U-N leucine codons. *Federation of European Biochemical Societies Letters* **132**, 344–8.

Sik, K., Leung-Yan-Cheung & Cover, T. M. (1978). Some equivalences between Shannon entropy and Kolmogorov complexity. *IEEE Transactions on Information Theory* **IT-24**, 331–8.

Sillén, L. G. (1965). Oxidation state of Earth's ocean and atmosphere, *Arkiv för Kemi* **24**, 431–56.

Simpson, G. G. (1964). The nonprevalence of humanoids. *Science* **143**, 769–75.

Simpson, L. (1990). RNA editing – A novel genetic phenomena? *Science* **250**, 512–13.

Simpson, L. & Shaw, J. (1989). RNA editing and the mitochondrial cryptogenes of kinetoplasmid protozoa. *Cell* **57**, 355–66.

Skilling, J. (1984). The maximum entropy method. *Nature* **309**, 748–9.

Sleep, N. H., Zhanle, K. J., Kasting, J. F. & Morowitz, H. J. (1989). Annihilation of ecosystems by large asteroid impacts on the early Earth. *Nature* **342**, 139–42.

Slepian, D. (1973). *Key Papers in the Development of Information Theory*. New York: IEEE Press.

Smith, L., Davies, H., Reichlin, M. & Margoliash, E. (1973). Separate oxidase and reductase reaction sites on cytochrome c demonstrated with purified site-specific antibodies. *Journal of Biological Chemistry* **248**, 237–43.

Smith, M., Brown, G. M., Air, G. M., Barrell, B. G., Coulson, A. R., Hutchinson, C. A. III & Sanger, F. (1977). DNA sequence at the C termini of the overlapping genes A and B in bacteriophage ΦX174. *Nature* **265**, 702–5.

Smith, S. (1974). A derivation of entropy and the maximum entropy criterion in the context of decision problems. *IEEE Transactions on Systems, Man and Cybernetics* **SMC-4**, 157–63.

Smith, T. F. & Waterman, M. S. (1980). Protein constraints induced by multiframe encoding. *Mathematical BioSciences* **49**, 17–26.

Sneath, P. H. A. (1966). Relations between chemical structure and biological activity in proteins. *Journal of Theoretical Biology* **12**, 157–95.

Sogin, M. L., Gunderson, J. H., Elwood, H. J., Alonso, R. A. & Peattie, D. A. (1989). Phylogenetic meaning of the kingdom concept: An unusual ribosomal RNA from *Giardia lamblia*. *Science* **243**, 75–7.

Solomonoff, R. J. (1964). A formal theory of inductive inference. *Information and Control* **7**, 1–22; 224–54.

Sonneborn, T. M. (1965). Degeneracy of the genetic code: Extent, nature, and genetic implications. In *Evolving Genes and Proteins*, eds V. Bryson & H. J. Vogel, pp. 377–97. New York: Academic Press.

Sourdis, J. & Nei, M. (1988). Relative efficiencies of the maximum parsimony and distance-matrix methods in obtaining the correct phylogenetic tree. *Molecular Biological Evolution* **5**, No. 3, 298–311.

Spencer, C. A., Gietz, R. D. & Hodgetts, R. B. (1986). Overlapping transcription units in the dopa decarboxylase region of *Drosophila*. *Nature* **322**, 279–81.

Squyres, S. W. (1989). Urey Prize Lecture: Water on Mars. *Icarus* **79**, 229–88.

Stent, G. S. (1988). Light and life: Niels Bohr's legacy to contemporary biology. In *Niels Bohr: Physics and the World*, eds H. Feshbach, T. Matsui & A. Oleson, pp. 231–44. London; Paris; New York: Harwood Academic Publishers.

Stevenson, D. J. (1988). Greenhouses and magma oceans. *Nature* **335**, 587–8.

Stewart, I. (1988). The ultimate in undecidability. *Nature* **332**, 115–16.

Stigler, S. M. & Wagner, M. J. (1987). A substantial bias in nonparametric tests for periodicity in geophysical data. *Science* **238**, 940–4.

Stigler, S. M. & Wagner, M. J. (1988). Response to D. M. Raup & J. W. Seproski (1988), *Science* **241**, 94–6. *Science* **241**, 96–9.

Stoneham, R. G. (1965). A study of 60000 digits of the transcendental '*e*'. *American Mathematical Monthly* **72**, 483–500.

Stribling, R. & Miller, S. L. (1987). Energy yields for hydrogen cyanide and formaldehyde syntheses: The HCN and amino acid concentrations in the primitive ocean. *Origins of Life* **17**, 261–73.

Strigunkova, T. F., Lavrentiev, G. A. & Otroshchenko, V. A. (1986). Abiogenic synthesis of oligonucleotides on kaolinite under the action of ultraviolet radiation. *Journal of Molecular Evolution* **23**, 290–3.

Sunde, R. A. & Evenson, J. K. (1987). Serine incorporation into the selenocysteine moiety of glutathione peroxidase. *The Journal of Biological Chemistry* **262**, 933–7.

Suppes, P. (1972). *Axiomatic Set Theory*. New York: Dover Publications.

Szekely, M. (1977). Phi X 174 sequenced. *Nature* **265**, 685.

Szekely, M. (1978). Triple overlapping genes. *Nature* **272**, 492.

Szilard, L. (1959). On the nature of the aging process. *Proceedings of the National Academy of Sciences USA* **45**, 30–45.

Takahashi, H., Merli, S., Putney, S. D., Houghten, R., Moss, B., Germain, R.

N. & Berzofsky, J. A. (1989). A single amino acid interchange yields reciprocal CTL specificities for HIV-1 group 160. *Science* **246**, 118–212.

Tateno, Y. & Tajima, X. (1986). Statistical properties of molecular tree construction methods under the neutral mutation model. *Journal of Molecular Evolution* **23**, 354–61.

Taylor, J. H., Fowler, L. A. & McCulloch, P. M. (1979). Measurements of general relativistic effects in the binary pulsar PSR1913 + 16. *Nature* **277**, 437–40.

Taylor, W. R. (1986). The classification of amino acid conservation. *Journal of Theoretical Biology* **119**, 205–18.

Temin, H. M. & Mizutani, S. (1970). RNA-dependent DNA polymerase in virions of rous sarcoma virus. *Nature* **226**, 1211–13.

Temussi, P. A. (1976). Reply (to S. W. Fox (1976b), *Journal of Molecular Evolution* **8**, 301–4). *Journal of Molecular Evolution* **8**, 305.

Temussi, P. A., Paolillo, L., Ferrara, L., Benedetti, E. & Andini, S. (1976). Structural characterization of thermal prebiotic polypeptides. *Journal of Molecular Evolution* **7**, 105–10.

Tennyson, G. E., Sabatos, C. A., Higuchi, K., Meglin, N. & Brewer, H. B., Jr. (1989a). Expression of apoliprotein B mRNAs encoding higher- and lower-molecular weight isoproteins in rat liver and intestine. *Proceedings of the National Academy of Sciences USA* **86**, 500–4.

Tennyson, G. E., Sabatos, C. A., Eggerman, T. L. & Brewer, H. B., Jr. (1989b). Characterization of single base substitutions in edited apoliprotein B transcripts. *Nucleic Acids Research* **17**, 691–8.

Thaddeus, P. & Chanan, G. (1985). Cometary impacts, molecular clouds, and the motion of the Sun perpendicular to the galactic plane. *Nature* **314**, 73–5.

Thielemann, F.-K., Metzinger, J. & Klapdor, H. V. (1983). Beta-delayed fission and neutron emission: Consequences for the astrophysical r-process and the age of the galaxy. *Zeitschrift für Physik A – Atoms and Nuclei* **309**, 301–17.

Thompson, C. J. & McBride, J. L. (1974). On Eigen's theory of the self-organization of matter and the evolution of biological macromolecules. *Mathematical Biosciences* **21**, 127–42.

Thwaites, W. & Awbrey, F. (1981). Biological evolution and the second law. *Creation/Evolution* **IV**, 5–7.

Tikochinsky, T., Tishby, N. Z. & Levine, R. D. (1984). Consistent inference of probabilities for reproducible experiments. *Physical Review Letters* **52**, 1357–60.

Tipler, F. J. (1985a). Extraterrestrial intelligent beings do not exist. In *Extraterrestrials – Science and Alien Intelligence*, ed. E. Regis, Jr. Cambridge; New York: Cambridge University Press.

Tipler, F. J. (1985b). Extraterrestrial intelligent beings do not exist. In *Frontiers of Modern Physics*, ed. A. Rothman, pp. 155–89. New York: Dover Publications.

Tipler, F. J. (1988). Review of *The Search for Extraterrestrial Life: Recent Developments* (ed. M. Papagiannis, Boston: Riedel). *Physics Today*, December 1987. See also the replies in the letters column, *Physics Today* (1988), **41**, 14–15; 142–6.

Tomarev, S. I. & Zinovieva, R. D. (1988). Squid major lens polypeptides are homologous to glutathione S-transferases subunits. *Nature* **336**, 86–8.

Traub, J. F., Wasilkowski, G. W. & Wazniakowski, H. (1983). *Information, Uncertainty, Complexity.* Reading, MA: Addison-Wesley.

Trefil, J. S. (1983). *The Moment of Creation.* New York: Charles Scribner's Sons.

Tribus, (1979). *The Maximum Entropy Formalism.* Cambridge, MA; London: MIT Press.

Tribus, M. & Motroni, H. (1972). Comments on the paper 'Jaynes's maximum entropy prescription and probability theory.' *Journal of Statistical Physics*, **4**, 227–8.

Tully, R. B. (1988). Origin of the Hubble constant controversy. *Nature* **334**, 209–12.

Turing, A. M. (1937). On computable numbers, with an application to the Entscheidungsproblem. *Proceedings of the London Mathematical Society* **42**, 230–65. (A correction, **43**, 544–6.)

Uy, R. & Wold, F. (1977). Postranslational covalent modification of protein. *Science* **198**, 890–6.

Uzzell, T. & Corbin, K. W. (1971). Fitting discrete probability distributions to evolutionary events. *Science* **172**, 1089–96.

Uzzell, T. & Spolsky, C. (1973). Origin of mitochondria. *Science* **180**, 516–17.

Uzzell, T. & Spolsky, C. (1974). Mitochondria and plastids as endosymbionts: a revival of Special Creation? *American Scientist* **62**, 334–43.

Varga, R. S. (1962). *Matrix Iterative Analysis.* Englewood Cliffs, NJ: Prentice-Hall.

Venn, J. (1888). *The Logic of Chance.* London: MacMillan. (Reprinted 1963, New York: Chelsea.)

Vessot, R. R. C., Levine, M. W., Mattison, E. M., Blomberg, E. L., Hoffman, T. E., Nystrom, G. U. & Farrel, B. F. and Decher, R., Eby, P. B., Baugher, C. R., Watts, J. W., Teuber, D. E. & Wills, F. D. (1980). Test of gravitation with a space-borne hydrogen maser. *Physical Review Letters* **45**, 2081–4.

von Helmholtz, H. (1854). *On the Interaction of the Natural Forces.* Reprinted in *Popular Scientific Lectures*, ed. M. Klein (1961). New York: Dover Publications.

von Neumann, J. (1966). *Theory of Self-Reproducing Automata.* Urbana; London: University of Illinois Press.

Wächtershäuser, G. (1988a). An all-purine precursor of nucleic acids. *Proceedings of the National Academy of Sciences USA* **85**, 1134–35.

Wächtershäuser, G. (1988b). Pyrite formation, the first energy source for life: A hypothesis. *Systems and Applied Microbiology* **10**, 207–10.

Wächtershäuser, G. (1988c). Before enzymes and templates: Theory of surface metabolism. *Microbiological Reviews* **52**, 452–84.

Wächtershäuser, G. (1990). Evolution of the first metabolic cycles. *Proceedings of the National Academy of Sciences USA* **87**, 200–4.

Wald, G. (1954). The Origin of Life. *Scientific American* **191**, 44–53.

Waldrop, J. L. (1988). Supernova 1987A: Facts and fancies. *Science*, **239**, 460–2.

Waldrop, M. M. (1988). After the fall. *Science* **239**, 977.

Waldrop, M. M. (1990). Spontaneous order, evolution and life. *Science* **247**, 1543–5.

Walker, J. C. G. (1976). Implications for atmospheric evolution of the inhomogeneous accretion model of the origin of the Earth. In *The Early*

History of the Earth, ed. B. F. Windley. London; New York: John Wiley & Sons.

Walker, G. C., Marsh, L. & Dodson, L. A. (1985). Genetic analysis of DNA repair: Inference and extrapolation. *Annual Review of Genetics* **19**, 103–26.

Walker, J. C. G., Klein, C., Schidlowski, M., Schopf, J. W., Stevenson, D. J. & Walter, M. R. (1983). Environmental evolution of the Archean-Early proterozoic Earth. In *Earth's Earliest Biosphere: Its Origin and Evolution*, ed. J. W. Schopf, pp. 260–89. Princeton, NJ: Princeton University Press.

Waneck, G. L., Stein, M. E. & Flavell, R. A. (1988). Conversion of a PI-anchored protein to an integral membrane protein by a single amino acid mutation. *Science* **241**, 697–9.

Watson, J. D. & Crick, F. H. C. (1953a). Molecular structure of nucleic acids. *Nature* **171**, 737–8.

Watson, J. D. & Crick, F. H. C. (1953b). Genetical implications of the structure of deoxyribose nucleic acid. *Nature* **171**, 964–7.

Watson, J. D. & Crick, F. H. C. (1953c). The structure of DNA. *Cold Spring Harbor Symposium on Quantitative Biology* **18**, 123–31.

Watson, J. D., Hopkins, N. H., Roberts, J. W., Steitz, J. A. & Weiner, A. M., eds. (1987). *Molecular Biology of the Gene*, 4th edn. Menlo Park, CA: Benjamin/Cummings.

Weber, A. L. & Miller, S. L. (1981). Reasons for the occurrence of the twenty coded protein amino acids. *Journal of Molecular Evolution* **17**, 273–84.

Wehrl, A. (1978). General properties of entropy. *Reviews of Modern Physics* **50**, 221–58.

Weinberg, S. (1977). *The First Three Minutes*. New York: Basic Books.

Weiner, A. M. (1987). The origins of life. In *Molecular Biology of the Gene*, J. D. Watson, N. H. Hopkins, J. W. Roberts, J. A. Steitz & A. M. Weiner, Chapter 28, pp. 1098–163. Menlo Park, CA; Reading, MA Benjamin/Cummings.

Weiss, R. B. (1984). Molecular model of ribosome frameshifting. *Proceedings of the National Academy of Sciences USA* **81**, 5797–801.

Weissman, P. R. (1990). The Oort cloud. *Science* **344**, 825–30.

Weissmann, C., Cattaneo, R. & Billeter, M. A. (1990). Sometimes an editor makes sense. *Nature* **343**, 697–9.

Welton, M. G. E. & Pelc, S. R. (1966). Specificity of the stereochemical relationship between ribonucleic acid-triplets and amino-acids. *Nature* **209**, 870–2.

Wendoloski, J. J., Matthew, J. B., Weber, P. C. & Salemme, F. R. (1987). Molecular dynamics of a cytochrome c-cytochrome b_5 electron transfer complex. *Science* **238**, 794–6.

Westheimer, F. H. (1986). Polyribonucleic acids as enzymes. *Nature* **319**, 534–6.

Whitmire, D. P. & Jackson, A. A. IV (1984). Are periodic mass extinctions driven by a distant solar companion? *Nature* **308**, 713–15.

Wicken, J. S. (1984). On the increase of complexity in evolution. In *Beyond Neo-Darwinism*, eds M. W. Ho & P. T. Saunders, pp. 89–112. London; New York: Academic Press.

Wickramasinghe, D. T. & Allen, D. A. (1986). Discovery of organic grains in comet Halley. *Nature* **323**, 44–6.

Wiener, N. (1948). *Cybernetics or Control and Communication in the Animal and the Machine.* New York: John Wiley & Sons.

Wiener, N. (1949). *Extrapolation, Interpolation, and Smoothing of Stationary Time Series.* New York: John Wiley & Sons.

Wilder-Smith (1981). *The Natural Sciences Know Nothing of Evolution.* San Diego: Creation Life Publishers.

Wilkening, L. L. (1978). Carbonaceous chondritic material in the solar system. *Naturwissenschaften* **65**, 73–9.

Wilkins, M. H. F., Stokes, A. R. & Wilson, H. R. (1953). Molecular structure of deoxypentose nucleic acids. *Nature* **171**, 738–40.

Will, C. M. (1990). General relativity at 75: How right was Einstein? *Science* **250**, 770–6.

Williams, R. J. P. (1990). Iron and the origin of life. *Nature* **343**, 213–14.

Williams, T. & Fried, M. (1986). A mouse locus at which transcription from both DNA strands produces mRNAs complementary at their 3' ends. *Nature* **322**, 275–8.

Wills, P. R. (1986). Scrapie, ribosomal proteins and biological information. *Journal of Theoretical Biology* **122**, 157–78.

Wills, P. R. (1989). Genetic information and the determination of functional organization in biological systems. *Systems Research* **6**, 219–26.

Woese, C. R. (1965a). Order in the genetic code. *Proceedings of the National Academy of Sciences USA* **54**, 71–5.

Woese, C. R. (1965b). On the evolution of the genetic code. *Proceedings of the National Academy of Sciences USA* **54**, 1546–52.

Woese, C. R. (1967). *The Genetic Code.* New York: Harper & Row.

Woese, C. R. (1970). The problem of evolving a genetic code. *BioScience* **20**, 471–80.

Woese, C. R. (1979). A proposal concerning the origin of life on the planet Earth. *Journal of Molecular Biology* **13**, 95–101.

Woese, C. R. (1990). Evolutionary questions: The 'Progenote'. *Science* **247**, 789.

Wolfenden, R. V., Cullis, P. M. & Southgate, C. C. F. (1979). Water, protein folding and the genetic code. *Nature* **206**, 575–7.

Wolfenstein, L. & Beier, E. W. (1989). Neutrino oscillations and solar neutrinos. *Physics Today,* **42**, 28–36.

Wong, J. T.-F. (1975). A co-evolution theory of the genetic code. *Proceedings of the National Academy of Science USA* **72**, 1909–12.

Wong, J. T.-F. (1976). The evolution of the genetic code. *Proceedings of the National Academy of Science USA* **73**, 2336–40.

Wong, J. T.-F. (1981). Co-evolution of genetic code and amino acid biosynthesis. *Trends in Biochemical Science* **6**, February, 33–5.

Wong, J. T.-F. (1983). Membership mutation of the genetic code: Loss of fitness by tryptophan. *Proceedings of the National Academy of Sciences USA* **80**, 6303–6.

Wong, J. T.-F. (1988). Evolution of the genetic code. *Microbiological Sciences* **5**, 174–81.

Wong, J. T.-F. & Cedergren, R. (1986). Natural selection versus primitive gene structure as determinant of codon usage. *European Journal of Biochemistry* **159**, 175–80.

Woosley, S. E. & Phillips, M. M. (1988). Supernova 1987! *Science* **240**, 750–9.

Wright, S. (1932). The roles of mutation, inbreeding, crossbreeding and selection in evolution. In *Proceedings of the Sixth International Congress on Genetics*, Vol. 1, pp. 356–6.

Wu, C.-I., Li, W.-H., Shen, J. J., Scarpulla, R. C., Limbach, K. J. & Wu, R. (1986). Evolution of cytochrome c genes and pseudogenes. *Journal of Molecular Evolution* **23**, 61–75.

Yamane, T. & Hopfield, J. J. (1977). Experimental evidence for kinetic proofreading in the aminoacylation of tRNA by synthetase. *Proceedings of the National Academy of Science USA* **74**, 2246–50.

Yamao, F., Muto, A., Kawauchi, Y., Iwami, M., Iwagami, S., Azumi, Y. & Osawa, S. (1985). UGA is read as trytophan in *Mycoplasma capricolum*. *Proceedings of the National Academy of Science USA* **82**, 2306–9.

Yang, J., Turner, M. S., Steigman, G., Schramm, D. N. & Olive, K. A. (1984). Primordial nucleosynthesis: A critical comparison of theory and observation. *Astrophysical Journal* **281**, 493–511.

Ycas, M. (1970). On earlier states of the biochemical system. *Journal of Theoretical Biology* **44**, 145–60.

Yockey, H. P. (1956). An application of information theory to the physics of tissue damage. *Radiation Research* **5**, 146–55.

Yockey, H. P. (1958). A study of aging, thermal killing and radiation damage by information theory. In *Symposium on Information Theory in Biology*, eds H. P. Yockey, R. Platzman & H. Quastler, pp. 297–316. New York; London: Pergamon Press.

Yockey, H. P. (1974). An application of information theory to the Central Dogma and the sequence hypothesis. *Journal of Theoretical Biology* **46**, 369–406.

Yockey, H. P. (1977a). A prescription which predicts functionally equivalent residues at given sites in protein sequences. *Journal of Theoretical Biology* **67**, 337–43.

Yockey, H. P. (1977b). On the information content of cytochrome c. *Journal of Theoretical Biology* **67**, 345–76.

Yockey, H. P. (1977c). A calculation of the probability of spontaneous biogenesis by information theory. *Journal of Theoretical Biology* **67**, 377–98.

Yockey, H. P. (1978). Can the Central Dogma be derived from information theory? *Journal of Theoretical Biology* **74**, 149–52.

Yockey, H. P. (1979). Do overlapping genes violate molecular biology and the theory of evolution? *Journal of Theoretical Biology* **80**, 21–6.

Yockey, H. P. (1981). Self-organization origin of life scenarios and information theory. *Journal of Theoretical Biology* **91**, 13–31.

Yockey, H. P., Platzman, R. P. & Quastler, H., eds (1958). *Symposium on Information Theory in Biology*. New York; London: Pergamon Press.

Yourno, J. (1971). Similarity of cross-suppressible frameshifts in *Salmonella typhimurium*. *Journal of Molecular Biology* **62**, 223–31.

Yourno, J. & Heath, S. (1969). Nature of the *his D3018* frameshift mutation in *Salmonella typhimurium*. *Journal of Bacteriology* **100**, 460–8.

Yourno, J. & Kohno, T. (1972). Externally suppressible proline quadruplet CCC. *Science* **175**, 650–2.

Yourno, J. & Tanemura, S. (1970). Restriction of in-phase translation by an unlinked suppressor of a frameshift mutation in *Salmonella typhimurium*. *Nature* **225**, 422–6.

Yuasa, S., Flory, D., Basile & Oró, J. (1984). Abiotic synthesis of purines and other heterocyclic compounds by the action of electrical discharges. *Journal of Molecular Biology* **21**, 76–80.

Yung, Y. L. & Pinto, J. P. (1978). Formation of channels on Mars. *Nature* **273**, 730–2.

Zahnle, K. J., Kasting, J. F. & Pollack, J. B. (1988). Evolution of a steam atmosphere during Earth's accretion. *Icarus* **74**, 62–97.

Zaug, A. & Cech, T. R. (1986). The intervening sequence RNA of *Tetrahymena* is an enzyme. *Science* **231**, 470–5.

Zhao, M. & Bada, J. L. (1989). Extraterrestrial amino acids in the Cretaceous/Tertiary boundary sediments at Stevens Klint, Denmark. *Nature* **339**, 463–5.

Zinoni, F., Birkman, A., Leinfelder, W. & Böck, A. (1987). Co-translational insertion of selenocysteine into formate dehydrogenase from *Escherichia coli* directed by a UGA codon. *Proceedings of the National Academy of Sciences USA* **84**, 3156–60.

Zinoni, F., Birkman, A., Stadtman, T. & Böck, A. (1986). Nucleotide sequence and expression of the selenocysteine-containing polypeptide of formate dehydrogenase (formate-hydrogen-lyase-linked) from *Escherichia coli*. *Proceedings of the National Academy of Sciences USA* **83**, 4650–4.

Zuckerkandl, E. (1987). On the molecular evolutionary clock. *Journal of Molecular Evolution* **26**, 34–46.

Zuckerkandl, E. & Pauling, L. (1965). Evolutionary divergence and convergence in proteins. In *Evolving Genes and Proteins*, eds V. Bryson & H. J. Vogel, pp. 97–165. New York: Academic Press.

Zurek, W. H. (1984). Reversibility and stability of information processing systems. *Physical Review Letters* **53**, 391–4.

Zurek, W. H. (1989). Thermodynamic cost of computation, algorithmic complexity and the information metric. *Nature* **341**, 119–24.

Author index

Subject index